Communications and Control Engineering

For further volumes:
www.springer.com/series/61

Juan I. Yuz • Graham C. Goodwin

Sampled-Data Models for Linear and Nonlinear Systems

Juan I. Yuz
Departamento de Electrónica
Universidad Técnica Federico Santa María
Valparaíso, Chile

Graham C. Goodwin
School of Electrical Engineering &
Computer Science
University of Newcastle
Callaghan, New South Wales, Australia

ISSN 0178-5354 Communications and Control Engineering
ISBN 978-1-4471-5561-4 ISBN 978-1-4471-5562-1 (eBook)
DOI 10.1007/978-1-4471-5562-1
Springer London Heidelberg New York Dordrecht

Printed on acid-free paper

Springer is part of Springer Science+Business Media (www.springer.com)

To Paz and Rosslyn

Preface

Most real-world systems evolve in continuous time. However, digital implementation is almost universally used in practice. Hence, a crucial ingredient in practical estimation and control is an understanding of the impact of sampling on continuous-time models and systems. In this context, the aim of this book is to reduce the gap between continuous-time and sampled-data systems theory. The subject of sampling is huge—no one book can cover all aspects. Thus, the book emphasises exact and approximate models for sampled-data systems. Questions such as the following will be addressed:

- *What can one say when the sampling rate is high relative to the dynamics of interest?*
- *Do natural convergence results apply as the sampling rate increases?*
- *Do there remain any special features of sampled systems which are not associated with underlying continuous systems?*

The authors' motivation for writing the book was threefold:

(i) Whilst most systems evolve in continuous time, all modern control and signal processing equipment is computer based. Hence, sampling arises as an inescapable aspect of all modern control and signal processing applications.
(ii) Sampling is, at first glance, a straightforward issue. However, on closer examination, if sampling is not treated properly, misleading or erroneous results can occur.
(iii) The authors have found that many aspects of sampling are not completely understood by engineers and scientists, even though these issues are central to many of the applications with which they deal.

The goal of this book is to provide a guide for students, practising engineers, and scientists who deal with sampled-data models. The book is intended to act as a catalyst for further applications in the area of nonlinear estimation and control.

Four classes of systems are treated:

(i) linear deterministic systems,
(ii) nonlinear deterministic systems,

(iii) linear stochastic systems, and
(iv) nonlinear stochastic systems.

Several applications are also presented. These applications embellish the core ideas by showing how they impact several important problems in signals and systems.

The book was written in Valparaíso (Chile) and Newcastle (Australia) during enjoyable collaborative visits by the authors. The book assembles contemporary work of the authors and others on sampled-data models. The authors hope that, by setting these ideas down in one place, we can instill confidence in readers dealing with sampled-data issues in real-world applications of signal processing and control.

The authors express their gratitude to Jayne Disney, who assisted with the typing of the manuscript, and to Diego Carrasco, who proofread the manuscript and contributed technical ideas in many places. Finally, the authors wish to thank their respective wives, Paz and Rosslyn, for their support and encouragement in the undertaking of this writing project over a seven-year period.

Valparaíso, Chile
Newcastle, Australia
September 19, 2013

Juan I. Yuz
Graham C. Goodwin

Contents

Part II Stochastic Systems

Symbols and Acronyms

$\langle \circ, \circ \rangle$	Inner product.
$\sim$	*Distributed as* (for random variables).
$*$	Complex conjugation.
T	Matrix (or vector) transpose.
γ	Complex variable associated to the δ-operator.
Δ	Sampling period.
δ	Delta operator (forward divided difference).
$\delta(t)$	Dirac delta or continuous-time impulse function.
$\delta_K[k]$	Kronecker delta or discrete-time impulse function.
$\mu(t)$	Unitary step function or Heaviside function.
$\mu[k]$	Discrete-time unitary step function.
$\rho = \frac{d}{dt}$	Time-derivative operator.
ω	Angular frequency, in [rad/s].
ω_N	Nyquist frequency, $\omega_N = \frac{\omega_s}{2}$.
ω_s	Sampling frequency, $\omega_s = \frac{2\pi}{\Delta}$.
A, B, C, D	State-space matrices in continuous time.
$A_\delta, B_\delta, C_\delta, D_\delta$	State-space matrices in discrete time using the δ-operator (i.e., incremental models).
A_q, B_q, C_q, D_q	State-space matrices in discrete time using the shift operator q.
adj	Adjoint of a matrix.
ASZ	Asymptotic sampling zeros.
$\mathbb{C}$	Set of complex numbers.
$\mathcal{C}^n$	Space of functions whose first n derivatives are continuous.
CAR	Continuous-time auto regressive.
CSZ	Corrected sampling zeros.
CT	Continuous time.
CTWN	Continuous-time white noise.
det	Determinant of a matrix.
DFT	Discrete Fourier transform.
DT	Discrete time.
DTFT	Discrete-time Fourier transform.

DTWN	Discrete-time white noise.
$E\{\cdot\}$	Expected value.
ESD	Exact sampled-data (model).
$f(t)$	Continuous-time signal ($t \in \mathbb{R}$).
f_k or $f[k]$	Discrete-time signal or sequence ($k \in \mathbb{N}$).
$\mathcal{F}\{\cdot\}$	(Continuous-time) Fourier transform.
$\mathcal{F}^{-1}\{\cdot\}$	(Continuous-time) inverse Fourier transform.
$\mathcal{F}_d\{\cdot\}$	Discrete-time Fourier transform.
$\mathcal{F}_d^{-1}\{\cdot\}$	Discrete-time inverse Fourier transform.
FDML	Frequency-domain maximum likelihood.
FOH	First order hold.
$G(\rho)$	Deterministic part of a continuous-time system.
$G(s)$	Continuous-time transfer function (s-domain of the Laplace transform).
$G_\delta(\gamma)$	Discrete-time transfer function (γ-domain of the δ operator).
$G_q(z)$	Discrete-time transfer function (z-domain of the shift operator q).
GHF	Generalized hold function.
GSF	Generalized sampling filter.
$H(\rho)$	Stochastic part of a continuous-time system.
$h_g(t)$	Impulse response of a generalized hold or sampling function.
$\Im$	Imaginary part of a complex number.
I_n	Identity matrix of dimension n.
IV	Instrumental variables.
$\mathcal{L}\{\cdot\}$	Laplace transform.
$\mathcal{L}^{-1}\{\cdot\}$	Inverse Laplace transform.
ℓ_2	Space of square summable sequences.
$\mathcal{L}_2$	Space of square integrable functions.
LQ	Linear-quadratic (optimal control problem).
LS	Least squares.
MIFZ(D)	Model incorporating fixed zero (dynamics).
MIMO	Multiple-input multiple-output (system).
MIPZ(D)	Model incorporating parameterised zero (dynamics).
ML	Maximum likelihood.
MPC	Model predictive control.
$\mathbb{N}$	Set of natural numbers (positive integers).
$N(\mu, \sigma^2)$	Normal (or Gaussian) distribution, with mean μ and variance σ^2.
NMP	Non-minimum phase.
$\mathcal{O}(\Delta^n)$	Function *of order* Δ^n.
PEM	Prediction error method(s).
PSD	Power spectral density.
QP	Quadratic programming.
q	Forward shift operator.
$\Re$	Real part of a complex number.

$\mathbb{R}$	Set of real numbers.
s	Complex variable corresponding to the Laplace transform.
SD	Sampled data.
SDE	Stochastic differential equation.
SDR (or **SDRM**)	Simple derivative replacement (model).
SISO	Single-input single-output (system).
TTS	Truncated Taylor series.
$\mathbb{Z}$	Set of integer numbers.
$\mathcal{Z}\{\cdot\}$	$\mathcal{Z}$-transform.
$\mathcal{Z}^{-1}\{\cdot\}$	Inverse $\mathcal{Z}$-transform.
z	Complex variable corresponding to the $\mathcal{Z}$-transform of the shift operator q.

Chapter 1
Introduction

Models for continuous-time dynamical systems typically arise from the application of physical laws such as conservation of mass, momentum, and energy. These models evolve in continuous time and take the form of linear or nonlinear *differential* equations. In practice, however, these kinds of models do not tell the complete story when the system is connected to digital devices. For example, when digital controllers have to act on a real system, this action can be applied (or updated) only at specific time instants. Similarly, if data is collected from a given system, the data is typically only recorded (and stored) at specific sampling instants. As a consequence, *sampling* arises as a cornerstone problem in all aspects of modern estimation and control theory.

In this context, the current book develops *sampled-data models* for linear and nonlinear systems. The focus is on describing, in discrete time, the relationship between the input signal and the samples of the continuous-time system output. In particular, issues such as the *accuracy* of sampled-data models, the *artefacts* produced by particular sampling schemes, and the relationship between the sampled-data model and the underlying continuous-time system are studied.

The sampling process for a continuous-time system is represented schematically in Fig. 1.1. There are four basic elements shown in the figure. Each of these elements plays a role in determining the appropriate discrete-time input-output description:

- *The hold*: The hold is used to convert a discrete-time sequence $\{u_k\}$ into a continuous-time input. A very commonly used hold in practical systems is the zero-order hold, where

$$u(t) = u_k \quad \text{for } k\Delta \le t < (k+1)\Delta \tag{1.1}$$

  where Δ is the sampling time.
- *The physical system* The system will typically be described by a set of linear or nonlinear differential equations.
- *The anti-aliasing filter (AAF)*: The anti-aliasing filter prepares the continuous-time signal prior to taking samples.

J.I. Yuz, G.C. Goodwin, *Sampled-Data Models for Linear and Nonlinear Systems*, Communications and Control Engineering, DOI 10.1007/978-1-4471-5562-1_1,

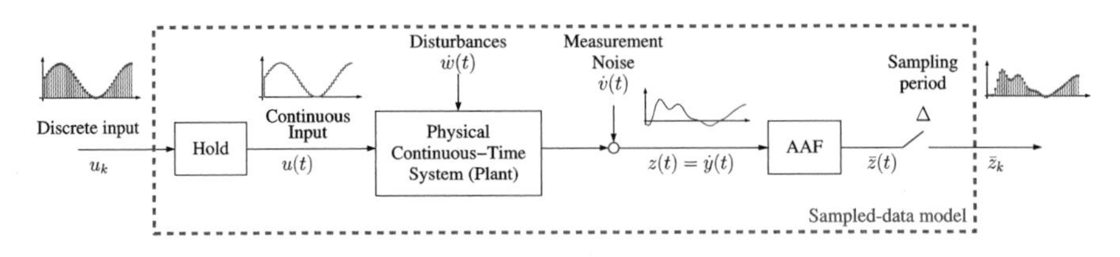

Fig. 1.1 Scheme of the sampling process of a continuous-time system

- *The sampler*: The sampler creates a discrete-time sequence $\{\bar{z}_k\}$ by instantaneous sampling, i.e.,

$$\bar{z}_k = \bar{z}(k\Delta) \tag{1.2}$$

For linear systems, it is possible to obtain exact sampled-data models from the sampling scheme shown in Fig. 1.1. In particular, given a deterministic continuous-time system, it is possible to obtain a discrete-time model which replicates the sequence of output samples. In the stochastic case, where the input of the system is assumed to be a *continuous-time white noise* process, a sampled-data model can be obtained such that its output sequence has the same second order properties as the continuous-time output at the sampling instants. Sampled models for nonlinear systems are harder to compute. For nonlinear deterministic systems, sampled-data models can be obtained such that the system output is reproduced up to some level of accuracy. In the nonlinear stochastic case, approximate sampled-data models can be obtained which approximate the stochastic properties of the output samples. Both linear and nonlinear *deterministic* systems are treated in Part I of the book, and linear and nonlinear *stochastic* systems are treated in Part II. Embellishments and extensions are presented in Part III.

Sampled-data models for continuous-time systems are used in many contexts, e.g., control, simulation, and estimation of system parameters (system identification). Most of the existing literature regarding discrete-time (and, thus, sampled-data) systems has traditionally expressed these models in terms of the shift operator q and the associated $\mathcal{Z}$-transform. However, when using this kind of model, it is not easy to relate the results to the continuous-time case. This is especially true when the sampling period is small. The inter-relationship between sampled-data models and their underlying continuous-time counterparts is more easily understood in the unified framework facilitated by the use of *incremental* models. In particular, sampled-data models *rewritten* in incremental form explicitly include the sampling period Δ in such a way that, when the sampling period is decreased, the underlying continuous-time system representation is recovered. In the same fashion, most discrete-time and continuous-time results in control, estimation, and system identification can be understood in a common framework when sampled-data models are expressed in incremental form.

It is also important to recall that, when using discrete-time models to represent continuous-time systems, there is *loss of information*. In the time domain, the intersample behaviour of signals is unknown, whereas in the frequency domain, high frequency signal components will fold back to low frequencies, making them impossible to distinguish. The loss of information is central to the difference between sampled-data and continuous-time systems. In Chap. 2, sampling and reconstruction of band-limited signals will be discussed. In Chap. 3, sampled-data models for linear deterministic systems having zero order hold inputs will be developed.

For any non-zero sampling period, there will always be a *difference* between the sampled-data model and the underlying continuous-time description. For example, discrete-time models have, in general, more zeros than the original continuous-time system. These extra zeros, called *sampling zeros*, are a result of the frequency folding effect due to the sampling process. This aspect of sampling will be discussed for deterministic systems in Chap. 5, and for linear stochastic systems in Chap. 14. The presence of sampling zeros (and their asymptotic behaviour) is determined by the continuous-time system relative degree. However, relative degree may be affected by high frequency modelling errors, even beyond the sampling frequency. Thus, robustness issues are discussed in Chaps. 7 and 15.

The book repeatedly highlights the issues and assumptions related to the use of sampled-data models. Some examples of these issues are as follows:

- Sampled-data characteristics depend not only on the continuous-time system but also on the sampling process itself. Indeed, for linear systems, the discrete-time poles depend on the continuous-time poles and the sampling period, whereas the zeros depend on the choice of the hold and the sampling devices.
- The effects of sampling *artefacts*, such as sampling zeros, play an important role in describing accurate sampled-data models. This applies both to exact sampled models for linear systems and to approximate sampled models for nonlinear systems.
- Any sampled-data description is based on some kind of *model* of the true continuous-time system. Modelling errors will usually be important at high frequencies due to the presence of unmodelled poles, zeros, or unmodelled time delays in the continuous-time system. This means that continuous-time models usually must be considered within a *bandwidth of validity* for the system.
- For stochastic models, the unknown input is assumed to be a (filtered) continuous-time white noise process. This is a mathematical abstraction that usually does not correspond to physical reality. However, it can be approximated to any desired degree of accuracy by conventional stochastic processes with broad-band spectra. This means that stochastic systems must be treated carefully. For example, the non-ideal nature of the noise can be thought of as a form of high frequency modelling error. How this aspect affects the use of sampled-data models will be discussed in Chap. 19.

The focus throughout the book is to provide answers to questions of the following type:

- When using sampled-data models, what characteristics are inherent to the underlying continuous-time system and what characteristics are a consequence of the sampling process itself?

- Is it reasonable to assume that, as the sampling period is decreased, the sampled-data model becomes indistinguishable from the underlying continuous-time system? How does this convergence occur?
- What issues are important when using a discrete-time model to represent a system?
- How do the results on sampling for linear systems apply to nonlinear systems?

Answers will be provided to these questions as the book evolves. As a preliminary step, in the next chapter, background material on sampling of signals and Fourier analysis is provided.

Part I
Deterministic Systems

Chapter 2
Background on Sampling of Signals

In this chapter, a background on sampling and Fourier analysis of signals is provided. The chapter provides a brief review of the key concepts to establish notation. Some readers may be familiar with this background material. In this case, they can proceed immediately to Chap. 3. Other readers may wish to quickly read this chapter before proceeding.

2.1 Fourier Analysis

The Fourier transform pair for a continuous-time signal $y(t)$ is

Continuous-time Fourier transform

$$\mathcal{F}\{f(t)\} = F(j\omega) = \int_{-\infty}^{\infty} f(t)e^{-j\omega t}\, dt \tag{2.1}$$

$$\mathcal{F}^{-1}\{F(j\omega)\} = f(t) = \frac{1}{2\pi}\int_{-\infty}^{\infty} F(j\omega)e^{j\omega t}\, d\omega \tag{2.2}$$

Fourier transforms allow signals to be studied in the *frequency* domain. The Fourier transform will turn out to be a very useful tool when connecting continuous-time signals with the associated sequence of samples.

Detailed issues relating to the existence, uniqueness, and other properties of Fourier transforms will not be discussed here. The reader is referred to the literature cited at the end of the chapter. Table 2.1 gives a summary of commonly used Fourier transform pairs, and Table 2.2 presents the most important properties. Several additional properties of the Fourier transform are summarised below.

J.I. Yuz, G.C. Goodwin, *Sampled-Data Models for Linear and Nonlinear Systems*, Communications and Control Engineering, DOI 10.1007/978-1-4471-5562-1_2,

Table 2.1 Fourier transform table

$f(t)\ \forall t \in \mathbb{R}$	$\mathcal{F}\{f(t)\}$
1	$2\pi\delta(\omega)$
$\delta(t)$	1
$\mu(t)$	$\pi\delta(\omega) + \frac{1}{j\omega}$
$\mu(t) - \mu(t - t_o)$	$\frac{1-e^{-j\omega t_o}}{j\omega}$
$e^{\alpha t}\mu(t),\ \Re\{\alpha\} < 0$	$\frac{1}{j\omega-\alpha}$
$te^{\alpha t}\mu(t),\ \Re\{\alpha\} < 0$	$\frac{1}{(j\omega-\alpha)^2}$
$e^{-\alpha\|t\|},\ \alpha \in \mathbb{R}^+$	$\frac{2\alpha}{\omega^2+\alpha^2}$
$\cos(\omega_o t)$	$\pi(\delta(\omega - \omega_o) + \delta(\omega - \omega_o))$
$\sin(\omega_o t)$	$j\pi(\delta(\omega + \omega_o) - \delta(\omega - \omega_o))$
$\cos(\omega_o t)\mu(t)$	$\pi(\delta(\omega - \omega_o) + \delta(\omega - \omega_o)) + \frac{j\omega}{-\omega^2+\omega_o^2}$
$\sin(\omega_o t)\mu(t)$	$j\pi(\delta(\omega + \omega_o) - \delta(\omega - \omega_o)) + \frac{\omega_o}{-\omega^2+\omega_o^2}$
$e^{-\alpha t}\cos(\omega_o t)\mu(t),\ \alpha \in \mathbb{R}^+$	$\frac{j\omega+\alpha}{(j\omega+\alpha)^2+\omega_o^2}$
$e^{-\alpha t}\sin(\omega_o t)\mu(t),\ \alpha \in \mathbb{R}^+$	$\frac{\omega_o}{(j\omega+\alpha)^2+\omega_o^2}$

Table 2.2 Basic Fourier transform properties

$f(t)$	$\mathcal{F}\{f(t)\}$	Description
$\sum_{i=1}^{l} a_i f_i(t)$	$\sum_{i=1}^{l} a_i F_i(j\omega)$	Linearity
$\frac{dy(t)}{dt}$	$j\omega Y(j\omega)$	Derivative law
$\frac{d^k y(t)}{dt^k}$	$(j\omega)^k Y(j\omega)$	High order derivative
$\int_{-\infty}^{t} y(\tau)\,d\tau$	$\frac{1}{j\omega}Y(j\omega) + \pi Y(0)\delta(\omega)$	Integral law
$y(t-\tau)$	$e^{-j\omega\tau}Y(j\omega)$	Delay
$y(at)$	$\frac{1}{\|a\|}Y(j\frac{\omega}{a})$	Time scaling
$y(-t)$	$Y(-j\omega)$	Time reversal
$\int_{-\infty}^{\infty} f_1(\tau) f_2(t-\tau)\,d\tau$	$F_1(j\omega)F_2(j\omega)$	Convolution
$y(t)\cos(\omega_o t)$	$\frac{1}{2}\{Y(j\omega - j\omega_o) + Y(j\omega + j\omega_o)\}$	Modulation (cosine)
$y(t)\sin(\omega_o t)$	$\frac{1}{j2}\{Y(j\omega - j\omega_o) - Y(j\omega + j\omega_o)\}$	Modulation (sine)
$F(t)$	$2\pi f(-j\omega)$	Symmetry
$f_1(t) f_2(t)$	$\frac{1}{2\pi}\int_{-\infty}^{\infty} F_1(j\zeta)F_2(j\omega - j\zeta)\,d\zeta$	Time-domain product
$e^{at} f_1(t)$	$F_1(j\omega - a)$	Frequency shift

Lemma 2.1 *The Fourier transform of a constant function* $f(t) = 1$ *is given by*

$$\mathcal{F}\{1\} = 2\pi\delta(\omega) \tag{2.3}$$

where $\delta(\omega)$ *is the Dirac delta function.*

Proof The definition of the inverse Fourier transform is used in Eq. (2.2) to yield

$$\mathcal{F}^{-1}\{2\pi\delta(\omega)\} = \frac{1}{2\pi}\int_{-\infty}^{\infty} 2\pi\delta(\omega)\,d\omega = 1 \tag{2.4}$$

□

Corollary 2.2 *The Fourier transform of a complex exponential* $f(t) = e^{j\omega_o t}$ *is given by*

$$\mathcal{F}\{e^{j\omega_o t}\} = 2\pi\delta(\omega - \omega_o) \tag{2.5}$$

Proof This property follows directly from Lemma 2.1 and the frequency shifting property in Table 2.2. □

The discrete-time Fourier transform (DTFT) and its associated inverse transform are defined as

Discrete-time Fourier transform

$$\mathcal{F}_d\{f_k\} = F_d(e^{j\omega\Delta}) = \Delta\sum_{k=-\infty}^{\infty} f_k e^{-j\omega k\Delta} \tag{2.6}$$

$$\mathcal{F}_d^{-1}\{F_d(e^{j\omega\Delta})\} = f_k = \frac{1}{2\pi}\int_{-\pi/\Delta}^{\pi/\Delta} F_d(e^{j\omega\Delta})e^{j\omega k\Delta}\,d\omega \tag{2.7}$$

Note that the sample period Δ appears explicitly in the above expressions. The usual DTFT sets $\Delta = 1$. In this case, the transform is the same if the sequence $\{y_k\}$ is obtained from samples of a continuous-time signal or is an inherently discrete-time signal. The period Δ has been explicitly included in the definition, as this facilitates the understanding of the connection between discrete-time and continuous-time results, for example, when y_k arises from sampling a continuous signal $y(t)$, i.e., when $y_k = y(k\Delta)$. Indeed, it is obvious that (2.6) is an approximation of the integral in (2.1).

2.2 Sampling of Continuous-Time Signals and Continuous Transforms

The connections between continuous and discrete transforms are next explored in more detail. The following result is first established:

Lemma 2.3 *Defining $s_\Delta(t)$ to be a train of impulses, i.e.,*

$$s_\Delta(t) = \Delta \sum_{k=-\infty}^{\infty} \delta(t - k\Delta) \tag{2.8}$$

the associated Fourier transform $\mathcal{F}\{s_\Delta(t)\} = S_\Delta(j\omega)$ is given by

$$S_\Delta(j\omega) = 2\pi \sum_{\ell=-\infty}^{\infty} \delta(\omega - \omega_\ell) \tag{2.9}$$

where $\omega_\ell = \frac{2\pi}{\Delta}\ell$.

Proof Notice that $s_\Delta(t)$ can also be expressed as

$$s_\Delta(t) = \sum_{\ell=-\infty}^{\infty} e^{-j\omega_\ell t} \tag{2.10}$$

where $\omega_\ell = \frac{2\pi}{\Delta}\ell$. This can be seen, for example, by noting that $s_\Delta(t)$ is a periodic signal with period Δ and hence has a Fourier series

$$s_\Delta(t) = \sum_{\ell=-\infty}^{\infty} C_\ell \, e^{-j\omega_\ell t} \tag{2.11}$$

where

$$C_\ell = \frac{1}{\Delta}\int_{-\frac{\Delta}{2}}^{\frac{\Delta}{2}} s_\Delta(t)\, e^{j\omega_\ell t}\, dt = \frac{1}{\Delta}\int_{-\frac{\Delta}{2}}^{\frac{\Delta}{2}} \Delta\delta(t)\, e^{j\omega_\ell t}\, dt = 1 \tag{2.12}$$

for all $\ell \in \mathbb{Z}$.

The associated Fourier transform is then obtained by applying the linearity property and Corollary 2.2 to (2.11), yielding

$$S_\Delta(j\omega) = \mathcal{F}\{s_\Delta(t)\} = 2\pi \sum_{\ell=-\infty}^{\infty} \delta(\omega - \omega_\ell) \tag{2.13}$$

□

Using Lemma 2.3, the (continuous-time) Fourier transform of a signal $\{y(t)\}$, defined for $t \in (-\infty, \infty)$, can be related to the discrete-time Fourier transform $Y_d(e^{j\omega\Delta})$ of the sequence of samples, $\{y_k\}$, where

$$y_k = y(k\Delta); \quad k \in \mathbb{Z} \tag{2.14}$$

Notice that $\{y_k\}$ defined in (2.14) is a sequence of numbers and, therefore, it does not have a *continuous* Fourier transform. However, it does have a DTFT as defined in (2.6)–(2.7). In order to associate a continuous Fourier transform with the sequence $\{y_k\}$, energy is added to the samples by defining the following *instrumental* signal, where each sample is turned into a scaled impulse:

$$y_\Delta(t) = \Delta \sum_{k=-\infty}^{\infty} y_k \delta(t - k\Delta) \tag{2.15}$$

Notice that

$$y_\Delta(t) = y(t) s_\Delta(t) \tag{2.16}$$

where $s_\Delta(t)$ is defined in (2.8). The signal $y_\Delta(t)$ in (2.15) is not usually considered in pure discrete-time analysis. However, it is useful when connecting continuous- and discrete-time Fourier transforms. The following result can now be established:

Lemma 2.4 *Consider a sequence $\{y_k\}$ arising from sampling a continuous-time signal $\{y_k\}$. Also consider $y_\Delta(t)$ as defined in* (2.15). *Then*

$$\mathcal{F}\{y_\Delta(t)\} = \mathcal{F}_d\{y_k\} \tag{2.17}$$

Proof Using the definition of the continuous Fourier transform as in (2.1), it follows that

$$\begin{aligned}
\mathcal{F}\{y_\Delta(t)\} &= \int_{-\infty}^{\infty} \Delta \sum_{k=-\infty}^{\infty} y_k \delta(t - k\Delta) e^{-j\omega t}\, dt \\
&= \Delta \sum_{k=-\infty}^{\infty} y_k \int_{-\infty}^{\infty} \delta(t - k\Delta) e^{-j\omega t}\, dt = \Delta \sum_{k=-\infty}^{\infty} y_k e^{-j\omega k\Delta}
\end{aligned} \tag{2.18}$$

which from (2.6) is $\mathcal{F}_d\{y_k\}$. □

In words, the above result establishes that the discrete-time Fourier transform of a sequence is equal to the continuous-time Fourier transform of the associated instrumental signal.

Note that the scaling by the sampling period Δ is important in obtaining this elegant result. If Δ is not specified, then one can arbitrarily set $\Delta = 1$, but this choice will then compact the (discrete) frequency range to $(-\pi, \pi)$.

Lemma 2.4 establishes an equivalence between frequency-domain analysis based on the DTFT of the discrete-time sequence of samples $\{y_k\}$ and frequency-domain analysis based on the Fourier transform of the continuous-time (instrumental) signal $y_\Delta(t)$. The result leads to the following key relationship between the Fourier transform of $y_\Delta(t)$ and the Fourier transform of the original signal $y(t)$:

Lemma 2.5 *Consider a signal $y(t)$, the associated sequence of samples $\{y_k\}$ defined in* (2.14), *and the instrumental signal $y_\Delta(t)$ defined in* (2.15). *Then*

$$Y_\Delta(j\omega) = \sum_{\ell=-\infty}^{\infty} Y\left(j\omega - j\frac{2\pi}{\Delta}\ell\right) \tag{2.19}$$

where $Y(j\omega) = \mathcal{F}\{y(t)\}$ and $Y_\Delta(j\omega) = \mathcal{F}\{y_\Delta(t)\} = \mathcal{F}_d\{y_k\}$.

Proof Using (2.16), the Fourier transform of $y_\Delta(t)$ can be obtained using the time-domain product property in Table 2.2. This yields

$$\begin{aligned}
\mathcal{F}\{y(t)s_\Delta(t)\} &= \frac{1}{2\pi} Y(j\omega) * S_\Delta(j\omega) \\
&= \frac{1}{2\pi}\int_{-\infty}^{\infty} Y(j\zeta)S_\Delta(j\omega - j\zeta)\,d\zeta \\
&= \frac{1}{2\pi}\int_{-\infty}^{\infty} Y(j\zeta)2\pi \sum_{\ell=-\infty}^{\infty} \delta\left(\omega - \zeta - \ell\frac{2\pi}{\Delta}\right) d\zeta \\
&= \sum_{\ell=-\infty}^{\infty} \int_{-\infty}^{\infty} Y(j\zeta)\delta\left(\omega - \zeta - \ell\frac{2\pi}{\Delta}\right) d\zeta \\
&= \sum_{\ell=-\infty}^{\infty} Y\left(j\omega + j\ell\frac{2\pi}{\Delta}\right)
\end{aligned} \tag{2.20}$$

which establishes (2.19). □

Equation (2.19) describes the *aliasing* effect, which is a cornerstone result in the analysis of sampled signals. Here it is presented in the context of sampling of

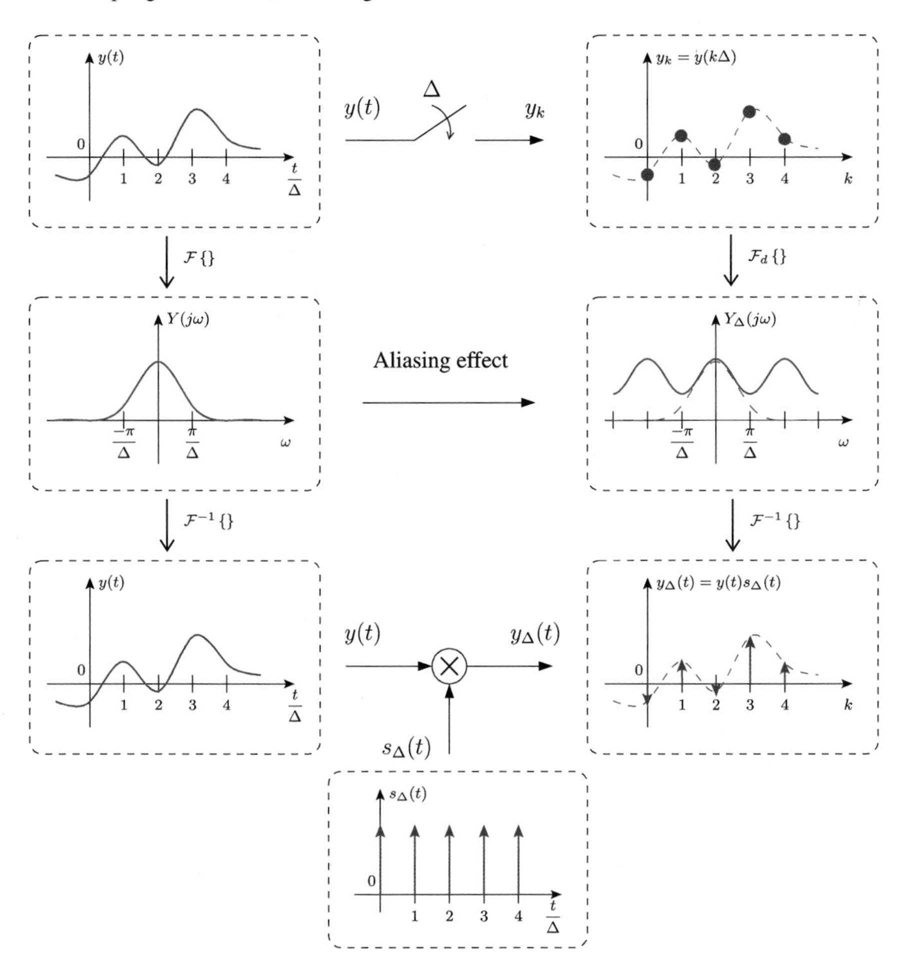

Fig. 2.1 Sampling of a continuous-time signal

deterministic signals. However, it also holds for stochastic signals, as will be shown in Chap. 11.

The aliasing effect is represented in Fig. 2.1. The figure illustrates the fact that the sampled spectrum $Y_\Delta(j\omega)$ can be either obtained as the DTFT of the sequence of samples $\{y_k\}$ or as the Fourier transform of the auxiliary signal $y_\Delta(t)$.

The result Lemma 2.5 is often stated as:

Sampling in the time domain induces folding in the frequency domain.

By virtue of the symmetry inherent in the definition of the Fourier transform and the inverse Fourier transform, the converse is also true. In particular, consider a continuous-time signal, $y(t)$, having Fourier transform $Y(j\omega)$. If the associated Fourier transform, $Y(j\omega)$, is sampled to produce a sequence of equidistant frequency-domain samples $\{Y_\ell = Y(j\ell\frac{2\pi}{\Delta})\}$, at frequency-domain spacing $\omega_s = \frac{2\pi}{\Delta}$, then an associated instrumental form $\{Y_\ell\}$ can be defined via

$$Y_\Delta(j\omega) = 2\pi \sum_{\ell=-\infty}^{\infty} Y_\ell \, \delta\left(\omega - \ell\frac{2\pi}{\Delta}\right) \tag{2.21}$$

The following result is obtained.

Lemma 2.6 *Consider a Fourier transform $Y(j\omega)$, the sequence of (frequency-domain) samples $\{Y_\ell = Y(j\ell\frac{2\pi}{\Delta})\}$, and the instrumental transform defined in (2.21). Then*

$$\mathcal{F}^{-1}\{Y_\Delta(j\omega)\} = \Delta \sum_{\ell=-\infty}^{\infty} y(t+\ell\Delta) \tag{2.22}$$

where $y(t) = \mathcal{F}^{-1}\{Y(j\omega)\}$.

Proof Notice that

$$Y_\Delta(j\omega) = Y(j\omega)S_\Delta(j\omega) \tag{2.23}$$

where $S_\Delta(j\omega)$ is defined in (2.9). Hence, from the convolution property in Table 2.2, it follows that

$$\begin{aligned} \mathcal{F}^{-1}\{Y_\Delta(j\omega)\} &= \int_{-\infty}^{\infty} y(t)s_\Delta(t-\tau)\,d\tau \\ &= \Delta \int_{-\infty}^{\infty} y(t) \sum_{\ell=-\infty}^{\infty} \delta(t-\tau-\ell\Delta)\,d\tau = \Delta \sum_{\ell=-\infty}^{\infty} y(t+\ell\Delta) \end{aligned} \tag{2.24}$$

□

The result Lemma 2.6 can be stated as

Sampling in the frequency domain induces folding in the time domain.

2.3 Signal Reconstruction from Samples

In this section, the conditions under which a continuous-time signal can be perfectly reconstructed from the sequence of associated samples are discussed. A special case of interest is where the continuous-time signal has a Fourier transform which is strictly limited to the range $\omega \in (\frac{-\pi}{\Delta}, \frac{\pi}{\Delta})$. Such a signal is said to be a *band-limited signal*.

For this kind of signal, it follows from Lemma 2.5 that the Fourier transforms of $y_\Delta(t)$ and $y(t)$ are identical on the frequency range $(\frac{-\pi}{\Delta}, \frac{\pi}{\Delta})$. It is then a straightforward matter to recover the original signal $y(t)$ from the sequence of samples $\{y_k = y(k\Delta)\}$. Note that the Fourier transform of $\{y_k\}$ is periodic (in the frequency domain). Hence, to recover the original spectrum, it suffices to extract the part of the Fourier transform of $y_\Delta(t)$ lying in the range $(\frac{-\pi}{\Delta}, \frac{\pi}{\Delta})$.

To convert this idea into a reconstruction formula, the following result is first established:

Lemma 2.7 *Consider an ideal low-pass filter defined as*

$$H_{LP}(j\omega) = \begin{cases} 1, & \omega \in (-\frac{\pi}{\Delta}, \frac{\pi}{\Delta}) \\ 0, & \omega \notin (-\frac{\pi}{\Delta}, \frac{\pi}{\Delta}) \end{cases} \tag{2.25}$$

Then, the (non-causal) impulse response of this filter is given by

$$h_{LP}(t) = \frac{\sin(\frac{\pi}{\Delta}t)}{\pi t} \tag{2.26}$$

Proof The result follows by applying the inverse Fourier transform

$$\begin{aligned} h_{LP}(t) &= \mathcal{F}^{-1}\{H_{LP}(j\omega)\} = \frac{1}{2\pi}\int_{-\infty}^{\infty} H_{LP}(j\omega)e^{j\omega t}\,d\omega \\ &= \frac{1}{2\pi}\int_{-\frac{\pi}{\Delta}}^{\frac{\pi}{\Delta}} e^{j\omega t}\,d\omega = \frac{1}{2\pi}\left[\frac{e^{j\frac{\pi}{\Delta}t} - e^{-j\frac{\pi}{\Delta}t}}{jt}\right] = \frac{\sin(\frac{\pi}{\Delta}t)}{\pi t} \end{aligned} \tag{2.27}$$

□

The ideal low-pass filter impulse response (2.26) can be conveniently expressed using the sinc function as

$$h_{LP}(t) = \frac{1}{\Delta}\operatorname{sinc}\left(\frac{\pi}{\Delta}t\right) \tag{2.28}$$

where $\operatorname{sinc}(x) = \sin(x)/x$. The following result can now be established.

Lemma 2.8 (Nyquist-Shannon Reconstruction Theorem) *Let $\{y(t)\}$ be a continuous-time signal which is strictly band-limited to $(\frac{-\pi}{\Delta}, \frac{\pi}{\Delta})$. Let $\{y_k = y(k\Delta)\}$ be the sequence of samples of $y(t)$ with sample period Δ. Then, the signal can be reconstructed from its samples as follows:*

$$y(t) = \sum_{k=-\infty}^{\infty} y_k \frac{\sin(\omega_N(t-k\Delta))}{\omega_N(t-k\Delta)} \tag{2.29}$$

where $\omega_N = \frac{\pi}{\Delta}$.

Proof Since $y(t)$ is a band-limited signal, its spectrum is preserved if it is passed through the ideal low-pass filter (2.25). Then

$$Y(j\omega) = H_{LP}(j\omega)Y_d\left(e^{j\omega\Delta}\right) = H_{LP}(j\omega)Y_\Delta(j\omega) \tag{2.30}$$

where $H_{LP}(j\omega)$ is the filter frequency response defined in (2.25). Applying the inverse Fourier transform and using the property that a product in the frequency domain translates to convolution in the time domain leads to:

$$\begin{aligned} y(t) = h_{LP}(t) * y_\Delta(t) &= \int_{-\infty}^{\infty} h_{LP}(t-\sigma) y_\Delta(\sigma)\, d\sigma \\ &= \int_{-\infty}^{\infty} \frac{\sin(\frac{\pi}{\Delta}(t-\sigma))}{\pi(t-\sigma)} \Delta \sum_{k=-\infty}^{\infty} y_k \delta(\sigma - k\Delta)\, d\sigma \\ &= \sum_{k=-\infty}^{\infty} y_k \int_{-\infty}^{\infty} \frac{\sin(\frac{\pi}{\Delta}(t-\sigma))}{\frac{\pi}{\Delta}(t-\sigma)} \delta(\sigma - k\Delta) d\sigma \\ &= \sum_{k=-\infty}^{\infty} y_k \frac{\sin(\frac{\pi}{\Delta}(t-k\Delta))}{\frac{\pi}{\Delta}(t-k\Delta)} \end{aligned} \tag{2.31}$$

□

The frequency $\omega_N = \frac{\pi}{\Delta}$ which appears in Lemma 2.3 is called the Nyquist frequency and corresponds to one half of the sampling frequency $\omega_s = \frac{2\pi}{\Delta}$. Figure 2.2 shows the instantaneous sampling of a band-limited signal. Figure 2.3 illustrates the reconstruction of a signal as the sum of *sinc* functions according to Lemma 2.8.

Lemma 2.8 establishes the *ideal* conditions under which a continuous-time signal can be perfectly reconstructed from its sequence of discrete samples. Note that the result requires the signal to be band-limited and the sampling frequency to be at least twice the highest frequency component of the signal. Additionally, the formula given in (2.29) implies a *non-causal* reconstruction based on *all* the samples of

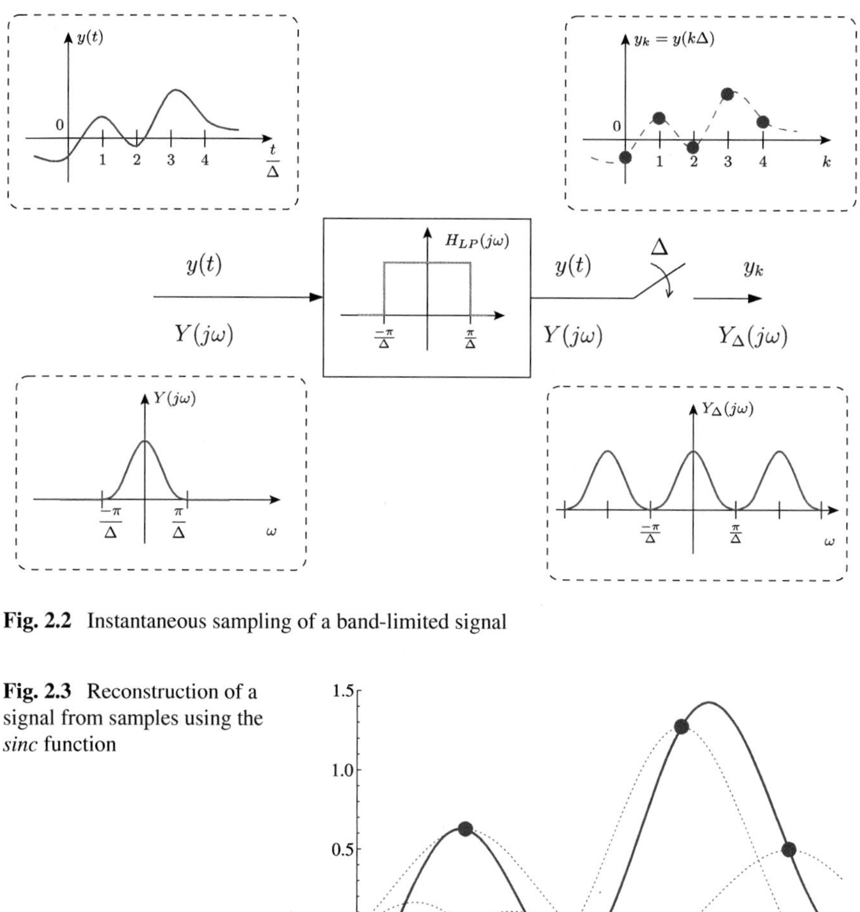

Fig. 2.2 Instantaneous sampling of a band-limited signal

Fig. 2.3 Reconstruction of a signal from samples using the *sinc* function

the signal, since k goes from $-\infty$ to ∞. This kind of reconstruction cannot be performed in practice, but it sets the minimal requirements to perfectly reconstruct the signal. Practical reconstruction strategies use, for example, simple interpolation or extrapolation techniques to approximate $y(t)$ between the samples $\{y_k = y(k\Delta)\}$.

2.4 Anti-aliasing Filters

The material in Sects. 2.2 and 2.3 provides strong motivation to use an anti-aliasing filter prior to taking samples. In particular, say that a continuous signal $y(t)$ is the sum of two other signals, namely $s(t)$ and $n(t)$. Let $s(t)$ be the signal of interest and assume that it is band-limited to the range $[-\frac{\pi}{\Delta}, \frac{\pi}{\Delta}]$. Lemma 2.8 implies that $s(t)$ can be perfectly reconstructed from samples taken at period Δ. On the other hand, the signal $n(t)$ is assumed to be a contamination signal. This signal is assumed to have relatively small spectral content in the range $[-\frac{\pi}{\Delta}, \frac{\pi}{\Delta}]$ but significant spectral components outside of that range. Then considering Lemma 2.5, it follows that the spectral content of the samples $y(t)$, i.e., $Y_\Delta(j\omega)$, will be corrupted in a major way by the *out-of-band* components in $n(t)$. In this case, the samples of $y(t)$ will be a poor representation of the samples of $s(t)$. This leads to the conclusion that it is desirable, in practice, to band-pass filter $y(t)$ prior to taking samples so that the corruption of the samples by out-of-band components in $n(t)$ is mitigated. Such a filter is commonly called an *anti-aliasing filter* (AAF).

The above argument suggests that an ideal band-pass filter should be used. In practice, this is often replaced by a more realistic causal low pass-filter. Note that in Part II of the book, an alternative stochastic interpretation will be given to the above argument.

2.5 Summary

The key points covered in this chapter are:

- The definition of the continuous Fourier transform pair (2.1)–(2.2).
- The definition of the discrete-time Fourier transform pair (2.6)–(2.7).
- The definition of the instrumental signal $y_\Delta(t)$ associated with a sequence $\{y_k\}$, i.e.,

$$y_\Delta(t) = \Delta \sum_{k=-\infty}^{\infty} y_k \delta(t - k\Delta) \tag{2.32}$$

- The fact that the continuous Fourier transform of $y_\Delta(t)$ is *equal* to the discrete-time Fourier transform of $\{y_k\}$.
- The folding or *aliasing* formula for signals, i.e.,

$$Y_\Delta(j\omega) = \sum_{\ell=-\infty}^{\infty} Y\left(j\omega - j\frac{2\pi}{\Delta}\ell\right) \tag{2.33}$$

  where $Y(j\omega) = \mathcal{F}\{y(t)\}$ and $Y_\Delta(j\omega) = \mathcal{F}\{y_\Delta(t)\}$.
- The reconstruction formula for band-limited signals

$$y(t) = \sum_{k=-\infty}^{\infty} y_k \frac{\sin(\omega_N(t - k\Delta))}{\omega_N(t - k\Delta)} \tag{2.34}$$

  where $\omega_N = \frac{\pi}{\Delta}$ is the Nyquist frequency.
- The desirability of including an anti-aliasing filter prior to sampling so that the samples are not dominated by out-of-band contamination.

Further Reading

Further details of Fourier analysis can be found in many textbooks, including:

Gasquet C, Witomski P (1998) Fourier analysis and applications: filtering, numerical computations, wavelets. Springer, Berlin

Lathi BP (2004) Linear systems and signals, 2nd edn. Oxford University Press, Oxford

Oppenheim AV, Schafer RW (1999) Discrete-time signal processing, 2nd edn. Prentice Hall International, New York

Stein E, Shakarchi R (2003) Fourier analysis: an introduction. Princeton University Press, Princeton

Churchill RV, Brown JW (2008) Complex variables and applications, 8th edn. McGraw-Hill, Englewood Cliffs

Chapter 3
Sampled-Data Models for Linear Deterministic Systems

Chapter 2 of the book was concerned with continuous-time signals and sequences of samples. The current chapter considers the main topic of interest in this book: how to obtain a sampled-data model for a given continuous-time system.

Sampled models for linear systems with deterministic inputs are first considered. In this case, it is possible to obtain an exact sampled-data model whose output sequence is equal to the samples of the continuous-time system output. Unless stated explicitly otherwise, it is assumed that the anti-aliasing filter, if present, is included as part of the continuous-time system model.

3.1 Continuous-Time Model

A general linear time-invariant *continuous-time* system can be described in state-space form as follows:

(Linear deterministic system)

$$\frac{dx(t)}{dt} = Ax(t) + Bu(t) \tag{3.1}$$

$$\bar{y}(t) = \dot{y}(t) = Cx(t) \tag{3.2}$$

Remark 3.1 The reader may be concerned about the notation $\bar{y}(t) = \dot{y}(t)$. This notation is simply a definition at this stage. However, it is introduced here to set the scene for future use with continuous-time stochastic models covered in Part II. It is also insightful to express the model (3.1)–(3.2) in *incremental form* as:

J.I. Yuz, G.C. Goodwin, *Sampled-Data Models for Linear and Nonlinear Systems*, Communications and Control Engineering, DOI 10.1007/978-1-4471-5562-1_3,

(Incremental form)

$$dx = Ax(t)\,dt + Bu(t)\,dt \tag{3.3}$$

$$dy = Cx(t)\,dt \tag{3.4}$$

where dx, dy, and dt denote infinitesimal increments in the state, output, and time, respectively.

The associated input-output transfer function can be described as in the following lemma.

Lemma 3.2 *The Laplace transform of the system output $\bar{y}(t)$ in* (3.2), *with zero initial conditions, is given by*

$$\bar{Y}(s) = G(s)U(s) \tag{3.5}$$

where $U(s)$ and $\bar{Y}(s)$ are the Laplace transform of the system input $u(t)$ and output $\bar{y}(t)$, respectively, and $G(s)$ is the system transfer function:

$$G(s) = C(sI_n - A)^{-1}B \tag{3.6}$$

where $I_n \in \mathbb{R}^{n \times n}$ is the identity matrix.

Proof Applying the Laplace transform to Eq. (3.1) yields

$$sX(s) - x(0) = A\,X(s) + B\,U(s) \tag{3.7}$$

Assuming zero initial conditions,

$$X(s) = (sI_n - A)^{-1}B\,U(s) \tag{3.8}$$

Substituting into (3.2) shows that the Laplace transform of the output is given by

$$\bar{Y}(s) = C\,X(s) = C(sI_n - A)^{-1}B\,U(s) \tag{3.9}$$

□

In the case of single-input single-output (SISO) systems, the transfer function (3.6) can be represented as a quotient of polynomials:

$$G(s) = \frac{F(s)}{E(s)} \tag{3.10}$$

where the numerator and denominator polynomials can be expanded as:

$$F(s) = f_m s^m + f_{m-1} s^{m-1} + \cdots + f_o = \sum_{\ell=0}^{m} f_\ell s^\ell \tag{3.11}$$

$$E(s) = s^n + e_{n-1} s^{n-1} + \cdots + e_o = \sum_{\ell=0}^{n} e_\ell s^\ell \tag{3.12}$$

Alternatively, the transfer function (3.10) can be factorised as follows:

$$G(s) = \frac{K(s-\sigma_1)\cdots(s-\sigma_m)}{(s-p_1)\cdots(s-p_n)} = \frac{K\prod_{\ell=1}^{m}(s-\sigma_\ell)}{\prod_{\ell=1}^{n}(s-p_\ell)} \tag{3.13}$$

The m roots $\{\sigma_\ell\}$ of $F(s)$ are called the zeros of the system. The n roots $\{p_\ell\}$ of $E(s)$ are called the poles of the system. The relative degree of the system is $r = n - m > 0$.

If, in (3.1)–(3.2), the pair (A, B) is completely controllable and (C, A) is completely observable, then the state-space model is said to be a minimal realization. This implies that there are no zero/pole cancellations in (3.10), and, thus, the numerator and denominator polynomials are given by:

$$F(s) = C\,\mathrm{adj}(sI_n - A)B \tag{3.14}$$

$$E(s) = \det(sI_n - A) \tag{3.15}$$

Lemma 3.3 *The numerator* (3.14) *can also be expressed as*

$$F(s) = \det\begin{bmatrix} sI_n - A & -B \\ C & 0 \end{bmatrix} \tag{3.16}$$

Proof The matrix inversion lemma can be used to obtain the determinant of a block matrix:

$$\det\begin{bmatrix} P & Q \\ R & S \end{bmatrix} = \det(P)\det\bigl(S - RP^{-1}Q\bigr) \tag{3.17}$$

where P and S are square matrices. The result follows on choosing $P = sI_n - A$, $Q = -B$, $R = C$, and $S = 0$, and then substituting into (3.17) to obtain

$$\det \begin{bmatrix} sI_n - A & -B \\ C & 0 \end{bmatrix} = \det(sI_n - A)\det\bigl(C(sI_n - A)^{-1}B\bigr)$$

$$= \det(sI_n - A)C\frac{\operatorname{adj}(sI_n - A)}{\det(sI_n - A)}B \tag{3.18}$$

which corresponds to (3.14). □

3.2 The Hold Device

Recall from Fig. 1.1 on p. 2 that a hold device is needed to generate a continuous-time input $u(t)$ to the system from the sequence of values u_k, defined at specific time instants $t_k = k\Delta$. The most commonly used hold devices are the following:

Zero-order hold (ZOH). This device simply holds the output constant between sampling instants, i.e.,

$$u(t) = u_k; \quad k\Delta \le t < (k+1)\Delta \tag{3.19}$$

First-order hold (FOH). This device is based on a linear *extrapolation* using the current and the previous elements of the input sequence, i.e.,

$$u(t) = u_k + \frac{u_k - u_{k-1}}{\Delta}(t - k\Delta); \quad k\Delta \le t < (k+1)\Delta \tag{3.20}$$

There are other and more general options for the hold device, for example, fractional order and generalised holds. (These alternative holds will be discussed in Chap. 6.) Any of these can be uniquely characterised by their *impulse response* $h(t)$, defined as the continuous-time output (of the hold device) obtained when u_k is the Kronecker delta function. By way of illustration, Fig. 3.1 shows the *impulse responses* corresponding to the ZOH, FOH, and a more general hold function.

3.3 The Sampled-Data Model

There are different ways to obtain the sampled-data model corresponding to the sampling scheme in Fig. 1.1 on p. 2. The model can be derived from the state-space model (3.1)–(3.2), or directly from the transfer function (3.10).

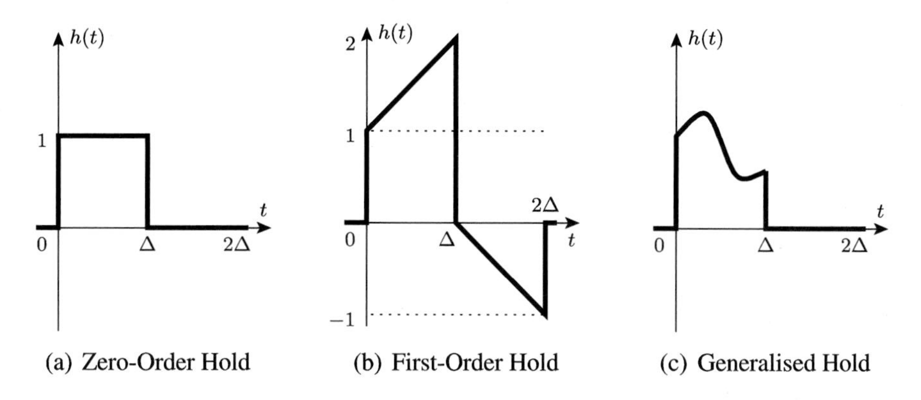

Fig. 3.1 Impulse response of some hold devices

The following result describes a state-space representation of the sampled-data model when the continuous-time system input is generated using a ZOH.

Lemma 3.4 *If the input of the continuous-time system* (3.1)–(3.2) *is generated from the input sequence* u_k *using a ZOH, then a state-space representation of the resulting sampled-data model is given by*:

$$qx_k = x_{k+1} = A_q x_k + B_q u_k \tag{3.21}$$

$$\bar{y}_k = C x_k \tag{3.22}$$

where the sampled output is $\bar{y}_k = \bar{y}(k\Delta)$, *and the sampled-data model matrices are*:

$$A_q = e^{A\Delta}, \qquad B_q = \int_0^{\Delta} e^{A\eta} B \, d\eta \tag{3.23}$$

Proof The state evolution starting at $t = t_k = k\Delta$ satisfies

$$x(k\Delta + \tau) = e^{A\tau} x(k\Delta) + \int_{k\Delta}^{k\Delta+\tau} e^{A(k\Delta+\tau-\eta)} B u(\eta) \, d\eta \tag{3.24}$$

Replacing $\tau = \Delta$, and noticing that the ZOH input satisfies $u(\eta) = u_k$, when $k\Delta \leq \eta < k\Delta + \Delta$, we can arrive at (3.21). Equation (3.22) is obtained directly from the instantaneous relation (3.2). □

Note that the discrete-time transfer function representation of the sampled-data system (3.21)–(3.22) is

$$G_q(z) = C(zI_n - A_q)^{-1} B_q \tag{3.25}$$

It is important to notice that the sampled-data model $G_q(z)$ in (3.25) is *exact*. In fact, from (3.2) and (3.22), it follows that the output sequence $\bar{y}_k$ is equal to the continuous-time system output $\bar{y}(t)$ at the sampling instants $t = k\Delta$, i.e.,

$$\bar{y}_k = \bar{y}(k\Delta) \tag{3.26}$$

As an alternative, several equivalent expressions can be obtained for the sampled transfer function $G_q(z)$ in (3.25). In fact, the sampled model given by Lemma 3.4 is equivalent to the pulse transfer function obtained directly from the continuous-time transfer function (3.10) as presented in the following result.

Lemma 3.5 *The sampled-data transfer function* (3.25) *can be obtained using the following expression:*

$$G_q(z) = (1 - z^{-1})\mathcal{Z}\left\{\mathcal{L}^{-1}\left\{\frac{G(s)}{s}\right\}_{t=k\Delta}\right\} \tag{3.27}$$

$$= (1 - z^{-1})\frac{1}{2\pi j}\int_{\gamma - j\infty}^{\gamma + j\infty} \frac{e^{s\Delta}}{z - e^{s\Delta}} \frac{G(s)}{s}\, ds \tag{3.28}$$

where Δ is the sampling period and $\gamma \in \mathbb{R}$ is such that all poles of $G(s)/s$ have real part less than γ. Moreover, evaluating the complex integral in (3.28), *the following two equivalent expressions can be obtained:*

$$G_q(z) = (1 - z^{-1}) \sum_{\ell=-\infty}^{\infty} \frac{G((\log z + 2\pi j\ell)/\Delta)}{\log z + 2\pi j\ell} \tag{3.29}$$

$$= (1 - z^{-1}) \sum_{\ell=0}^{n} \operatorname{Res}_{s=p_\ell}\left\{\frac{G(s)}{s} \frac{e^{s\Delta}}{z - e^{s\Delta}}\right\} \tag{3.30}$$

where the residues are computed at $p_o = 0$ and at the system poles $\{p_\ell\}_{\ell=1,\dots,n}$.

Proof Let u_k be a step sequence. Then $u(t)$, generated by the ZOH (3.19), is a continuous-time step signal. Thus, the discrete-time transfer function can be obtained from the inverse Laplace transform of the continuous-time step response, computing the $\mathcal{Z}$-transform of the sequence of output samples, and finally dividing it by the $\mathcal{Z}$-transform of a discrete-time step:

$$G_q(z) = \frac{\mathcal{Z}\{y(k\Delta)\}}{\mathcal{Z}\{u_k\}} = \frac{\mathcal{Z}\{\mathcal{L}^{-1}\{\frac{G(s)}{s}\}_{t=k\Delta}\}}{\frac{z}{z-1}} \tag{3.31}$$

which corresponds to (3.27). From the inverse Laplace transform definition, it follows that

$$\begin{aligned} G_q(z) &= \left(1 - z^{-1}\right) \sum_{k=0}^{\infty} z^{-k} \frac{1}{2\pi j} \int_{\gamma - j\infty}^{\gamma + j\infty} \frac{G(s)}{s} e^{sk\Delta} \, ds \\ &= \left(1 - z^{-1}\right) \frac{1}{2\pi j} \int_{\gamma - j\infty}^{\gamma + j\infty} \left[\sum_{k=1}^{\infty} z^{-k} e^{sk\Delta} \right] \frac{G(s)}{s} \, ds \end{aligned} \tag{3.32}$$

Note that the sum in brackets starts from $k = 1$ because the continuous-time step response given by the inverse Laplace is equal to zero at time $t = 0$ (i.e., for $k = 0$), since $G(s)$ has more poles than zeros (i.e., it is a strictly proper transfer function). Thus, in (3.32), the infinite sum in brackets can be computed for $|z^{-1} e^{s\Delta}| < 1$. Note that, since $\text{Re}\{s\} \geq \gamma$, it follows that, for $|z| \geq e^{\gamma\Delta}$, the pulse transfer function is given by

$$G_q(z) = \left(1 - z^{-1}\right) \frac{1}{2\pi j} \int_{\gamma - j\infty}^{\gamma + j\infty} \left[\frac{z^{-1} e^{s\Delta}}{1 - z^{-1} e^{s\Delta}} \right] \frac{G(s)}{s} \, ds \tag{3.33}$$

which corresponds to (3.28).

Note that the integrand

$$I(s) = \frac{e^{s\Delta}}{(z - e^{s\Delta})} \frac{G(s)}{s} \xrightarrow{|s| \to \infty} 0 \tag{3.34}$$

since $G(s)$ is strictly proper. This implies that the integral path from $s = \gamma - j\infty$ to $s = \gamma + j\infty$ can be closed to the left or to the right of the complex plane with $|s| \to \infty$, as shown in Fig. 3.2, without modifying the result of the complex integral in (3.28). If the integration path is closed to the left of the complex plane using Γ_1 (the left curve in Fig. 3.2), this curve encloses the poles of $G(s)$ and an extra pole at the origin $s = p_o = 0$; i.e., the residues can be evaluated as follows:

$$\int_{\gamma - j\infty}^{\gamma + j\infty} I(s) \, ds = \oint_{\Gamma_1} I(s) \, ds = 2\pi j \sum_{\ell=0}^{n} \text{Res}_{s = p_\ell} I(s) \tag{3.35}$$

Substituting into (3.33) leads to (3.30).

On the other hand, if the integration path is closed to the right of the complex plane using Γ_2 (the right curve in Fig. 3.2), then the residues of $I(s)$ are needed at

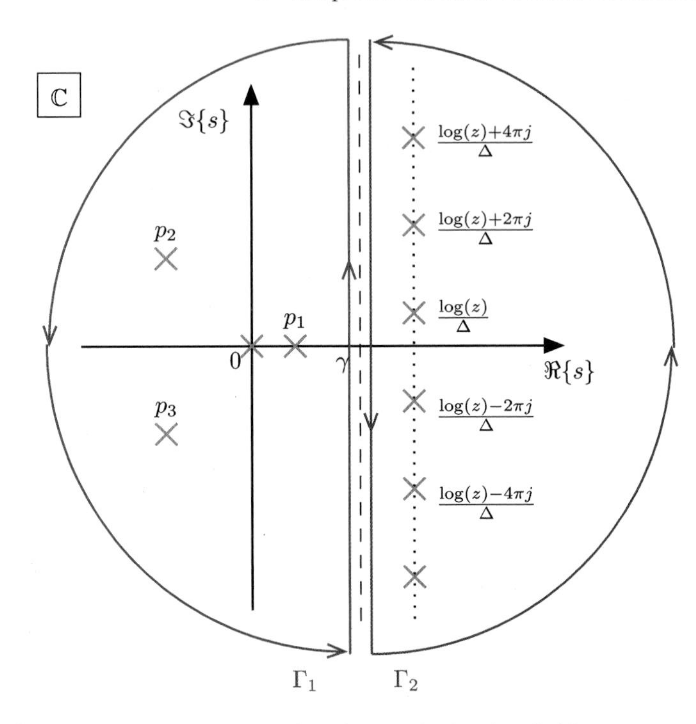

Fig. 3.2 Integral paths used to compute the pulse transfer function $G_q(z)$

the roots of

$$z - e^{s\Delta} = 0 \quad \Rightarrow \quad s = s_\ell = \frac{\log(z) + 2\pi j \ell}{\Delta} \tag{3.36}$$

where log is the natural logarithm and $\ell \in \mathbb{Z}$. These roots are schematically shown in Fig. 3.2 for an arbitrary value of z inside the region of convergence $\mathrm{Re}\{s\} = \frac{\log|z|}{\Delta} > \gamma$. It follows that

$$\int_{\gamma-j\infty}^{\gamma+j\infty} I(s)\, ds = -\oint_{\Gamma_2} I(s)\, ds = -2\pi j \operatorname{Res}_{s=s_\ell} I(s) \tag{3.37}$$

where

$$\operatorname{Res}_{s=s_\ell} I(s) = \lim_{s\to s_\ell} \left\{ I(s)(s - s_\ell) \right\} = \frac{e^{s_\ell \Delta} G(s_\ell)}{s_\ell} \lim_{s\to s_\ell} \left[\frac{s - s_\ell}{z - e^{s\Delta}} \right] \tag{3.38}$$

Finally, using l'Hôpital's rule for limits, we obtain

$$\operatorname{Res}_{s=s_\ell} I(s) = \frac{e^{s_\ell \Delta} G(s_\ell)}{s_\ell} \lim_{s\to s_\ell} \left[\frac{1}{-\Delta e^{s\Delta}} \right] = -\frac{G(s_\ell)}{\Delta s_\ell} \tag{3.39}$$

Substituting (3.39) into (3.37) and (3.33), replacing s_ℓ defined in (3.36), leads to (3.29). This completes the proof. □

The representations obtained in (3.30) and (3.29) provide different types of insight regarding the sampled-data model:

- Equation (3.30) shows a clear mapping of the system poles from the continuous-time to the discrete-time domain; namely, every continuous-time pole at $s = p_i$ is mapped to $z = e^{p_i \Delta}$.
- On the other hand, expression (3.29), when considered in the frequency domain by replacing $z = e^{j\omega\Delta}$, illustrates the well-known *aliasing* effect: the frequency response of the sampled-data system is obtained by folding the continuous-time frequency response, i.e.,

$$G_q\left(e^{j\omega\Delta}\right) = \frac{1}{\Delta}\sum_{\ell=-\infty}^{\infty} H_{\mathrm{ZOH}}\left(j\omega + j\frac{2\pi}{\Delta}\ell\right) G\left(j\omega + j\frac{2\pi}{\Delta}\ell\right) \tag{3.40}$$

where $H_{\mathrm{ZOH}}(s)$ is the Laplace transform of the ZOH impulse response in Fig. 3.1(a), i.e.,

$$H_{\mathrm{ZOH}}(s) = \frac{1 - e^{-s\Delta}}{s} \tag{3.41}$$

The above analysis shows that the sampled-data model for a given continuous-time system depends on the choice of the hold device. By way of illustration, the following lemma establishes the sampled-data model obtained when, instead of the ZOH considered above, the continuous-time input is generated by an FOH.

Lemma 3.6 *If the continuous-time plant input is generated using an FOH as in* (3.20), *the corresponding sampled-data model can be represented in the following state-space form*:

$$\begin{bmatrix} x_{k+1} \\ u_k \end{bmatrix} = \begin{bmatrix} A_q & B_q^1 \\ 0 & 0 \end{bmatrix} \begin{bmatrix} x_k \\ u_{k-1} \end{bmatrix} + \begin{bmatrix} B_q^2 \\ 1 \end{bmatrix} u_k \tag{3.42}$$

$$y_k = \begin{bmatrix} C & 0 \end{bmatrix} \begin{bmatrix} x_k \\ u_{k-1} \end{bmatrix} \tag{3.43}$$

where:

$$A_q = e^{A\Delta} \tag{3.44}$$

$$B_q^1 = \int_0^\Delta \left(\frac{\eta}{\Delta} - 1\right) e^{A\eta} B \, d\eta \tag{3.45}$$

$$B_q^2 = \int_0^\Delta \left(2 - \frac{\eta}{\Delta}\right) e^{A\eta} B \, d\eta \tag{3.46}$$

The discrete-time transfer function can be obtained from the state-space representation as

$$G_q(z) = \begin{bmatrix} C & 0 \end{bmatrix} \left(zI_{n+1} - \begin{bmatrix} A_q & B_q^1 \\ 0 & 0 \end{bmatrix} \right)^{-1} \begin{bmatrix} B_q^2 \\ 1 \end{bmatrix} \tag{3.47}$$

Equivalently, the discrete-time transfer function can be obtained from the continuous-time transfer function $G(s)$, i.e.,

$$G_q(z) = \left(1 - z^{-1}\right)^2 \mathcal{Z}\left\{\mathcal{L}^{-1}\left\{\frac{(1 + \Delta s)}{\Delta s^2} G(s)\right\}\right\} \tag{3.48}$$

Proof Substituting the FOH input defined in (3.20) into the state evolution equation (3.24), it follows that

$$x(k\Delta + \tau) = e^{A\tau} x(k\Delta) + \int_{k\Delta}^{k\Delta+\tau} e^{A(k\Delta+\tau-\bar{\eta})} B\left[u_k + \frac{u_k - u_{k-1}}{\Delta}(\bar{\eta} - k\Delta)\right] d\bar{\eta} \tag{3.49}$$

Changing variables inside the integral (using $\eta = k\Delta + \tau - \bar{\eta}$) and setting $\tau = \Delta$, it follows that

$$\begin{aligned} x(k\Delta + \Delta) &= e^{A\Delta} x(k\Delta) - \int_{\Delta}^{0} e^{A\eta} B\left[u_k + \frac{u_k - u_{k-1}}{\Delta}(\Delta - \eta)\right] d\eta \\ &= e^{A\Delta} x(k\Delta) + \int_{0}^{\Delta} e^{A\eta} B\left[\left(\frac{\eta}{\Delta} - 1\right) u_{k-1} + \left(2 - \frac{\eta}{\Delta}\right) u_k\right] d\eta \\ &= A_q x(k\Delta) + B_q^1 u_{k-1} + B_q^2 u_k \end{aligned} \tag{3.50}$$

which corresponds to the first line of Eq. (3.42). The second line follows by defining an auxiliary state variable for the *previous input* $x_k^a = u_{k-1}$. Hence, $x_{k+1}^a = u_k$. The output equation (3.43) is equivalent to the continuous output equation (3.2).

To show (3.48), consider the following input sequence:

$$u_k = \begin{cases} 0; & k < 0 \\ (k+1)\Delta; & k \geq 0 \end{cases} \tag{3.51}$$

From the FOH definition (3.20), the associated continuous-time input is given by

$$u(t) = \begin{cases} 0; & t < 0 \\ t + \Delta; & t \geq 0 \end{cases} \tag{3.52}$$

The pulse transfer function is then obtained from

$$G_q(z) = \frac{\mathcal{Z}\{\mathcal{L}^{-1}\{G(s)U(s)\}_{t=k\Delta}\}}{\mathcal{Z}\{u_k\}} \tag{3.53}$$

where, from (3.51),

$$\mathcal{Z}\{u_k\} = \frac{\Delta z^2}{(z-1)^2} \tag{3.54}$$

and, from (3.52),

$$U(s) = \frac{1}{s^2} + \frac{\Delta}{s} \tag{3.55}$$

Substituting (3.54) and (3.55) into (3.53) leads to (3.48). This completes the proof. □

Sampled-data models obtained using more general hold devices, such as the generalised hold with impulse response as shown in Fig. 3.1(c), will be discussed in Chap. 6.

3.4 Poles and Zeros

The relationship between the poles and zeros of the sampled-data model (3.25) and the poles and zeros of the continuous-time system (3.6) will next be considered.

The discrete-time model (3.25) can be rewritten as a quotient of polynomials as

$$G_q(z) = \frac{F_q(z)}{E_q(z)} \tag{3.56}$$

where:

$$F_q(z) = C\,\mathrm{adj}(zI_n - A_q)B_q = \det\begin{bmatrix} zI_n - A_q & -B_q \\ C & 0 \end{bmatrix} \tag{3.57}$$

$$E_q(z) = \det(zI_n - A_q) \tag{3.58}$$

where the second equality in (3.57) is obtained as in Lemma 3.3 on p. 23.

The relationship between the poles of the pulse transfer function (3.56) and the poles of the underlying continuous-time system can be established from Eq. (3.23). Indeed, if λ_i is an eigenvalue of A (i.e., a pole of $G(s)$), then $e^{\lambda_i \Delta}$ is an eigenvalue of $A_q = e^{A\Delta}$, and, thus, a pole of $G_q(z)$. The same result is obtained immediately from the expression in (3.30) of Lemma 3.5.

On the other hand, the relationship between the zeros in discrete and in continuous time is much more involved, as can be seen from the numerator polynomial $F_q(z)$. Actually, the discrete-time transfer function (3.56) will generically have relative degree 1, independent of the relative degree of the continuous-time system. Thus, extra zeros appear in the sampled-data model with no continuous-time counterpart. These so-called *sampling zeros* can be asymptotically characterised as shown later in Chap. 5.

Similar relationships between discrete- and continuous-time poles and zeros can be established when using a non-ZOH input. For example, consider the sampled-data model obtained in Lemma 3.6 for the FOH case; then the discrete-time poles are given by the eigenvalues of $e^{A\Delta}$ (as in the ZOH case) plus one pole at the origin $z = 0$. On the other hand, the discrete-time zeros will generally be different from the ones obtained when using a ZOH.

Example 3.7 Consider a continuous-time system with transfer function

$$G(s) = \frac{6(s+5)}{(s+2)(s+3)(s+4)} \tag{3.59}$$

The exact sampled-data model, when the input is generated using a ZOH, for a sampling period $\Delta = 0.01$, is given by

$$G_q(z) = \frac{2.96 \times 10^{-4}(z+0.99)(z-0.95)}{(z-0.98)(z-0.97)(z-0.96)} \tag{3.60}$$

3.5 Relative Degree

In the previous section, it was shown that, independent of the continuous-time relative degree and provided that there are no pure time delays in the original model, the relative degree of the exact sampled-data model for a linear system is generally equal to 1. This apparent mismatch between the sampled-data model and the underlying continuous system is based on the following two results.

Lemma 3.8 *The relative degree $r = n - m$ of a continuous-time transfer function*

$$G(s) = \frac{b_m s^m + \cdots + b_o}{s^n + a_{n-1}s^{n-1} + \cdots + a_o} \quad (b_m \neq 0) \tag{3.61}$$

describes the smoothness *of the system output, in the sense that it corresponds to the number of times that the output has to be differentiated in order to make the input appear explicitly.*

Proof Assuming that the system output and its derivatives are all zero just before the step input is applied (i.e., at $t = 0_-$), the step response of the system is given by

$$\bar{Y}(s) = \frac{G(s)}{s} = \frac{1}{s}\frac{b_m s^m + \cdots + b_o}{(s^n + a_{n-1}s^{n-1} + \cdots + a_o)} \tag{3.62}$$

According to the initial value theorem, just after the step is applied (i.e., at $t = 0_+$), the ℓ-th derivative of the output signal can be evaluated by

$$\left.\frac{d^\ell \bar{y}}{dt^\ell}\right|_{t=0_+} = \lim_{s\to\infty} s\big(s^\ell \bar{Y}(s)\big) = \lim_{s\to\infty} s^\ell G(s)$$

$$= \lim_{s\to\infty} \frac{s^\ell (b_m s^m + \cdots + b_o)}{s^n + a_{n-1}s^{n-1} + \cdots + a_o} = \begin{cases} 0; & \ell < r \\ b_m \neq 0; & \ell = r \end{cases} \tag{3.63}$$

where $r = n - m$ is the system relative degree. □

Corollary 3.9 *Consider a state-space representation of the transfer function* (3.61), *i.e.*,

$$G(s) = C(sI_n - A)^{-1}B \tag{3.64}$$

Then the system relative degree is r, if and only if

$$CA^{\ell-1}B = \begin{cases} 0; & \ell < r \\ b_m \neq 0; & \ell = r \end{cases} \tag{3.65}$$

Proof The transfer function $G(s)$ can be expanded in a power series around $s = \infty$, i.e.,

$$G(s) = \frac{1}{s}CB + \frac{1}{s^2}CAB + \cdots + \frac{1}{s^\ell}CA^{\ell-1}B + \cdots \tag{3.66}$$

The result then follows from (3.63). □

The previous result shows that, for any system of relative degree r, when an excitation such as the ZOH input (3.19) is applied, the output and its first $r - 1$ derivatives are continuous signals. The rth derivative is the first derivative such that a *jump* appears when the step input is introduced. It will be shown later in Chap. 9 that the result in Lemma 3.8 has a nonlinear counterpart.

In the next result, it is shown that, for discrete-time models, the concept of relative degree can be given a different interpretation.

Lemma 3.10 *The relative degree of a discrete-time transfer function*

$$G_q(z) = \frac{b_{q,m}z^m + \cdots + b_{q,o}}{z^n + a_{q,n-1}z^{n-1} + \cdots + a_{q,o}} \quad (b_m \neq 0) \tag{3.67}$$

describes the input-output delay *of the system output, i.e., it corresponds to the number of sampling instants before any change in the input appears at the system output.*

Proof The transfer function (3.67) can be expanded in powers of z^{-1} as follows:

$$\begin{aligned} G_q(z) &= \frac{b_{q,m}z^m + \cdots + b_{q,o}}{z^n(1 + a_{q,n-1}z^{-1} + \cdots + a_{q,o}z^{-n})} \\ &= \left(b_{q,m}z^{m-n} + \cdots + b_{q,o}z^{-n}\right)\left(1 + \alpha z^{-1} + \cdots\right) \\ &= b_{q,m}z^{-r} + \beta z^{-r-1} + \cdots \end{aligned} \tag{3.68}$$

which clearly shows that any input sequence to the system will show an effect in the output only $r = n - m$ sampling instants later. □

Corollary 3.11 *Consider a state-space representation of the discrete transfer function* (3.67), *i.e.*,

$$G_q(z) = C_q(sI_n - A_q)^{-1}B_q \tag{3.69}$$

Then the system relative degree is r, *if and only if*

$$C_q A_q^{\ell-1} B_q = \begin{cases} 0; & \ell < r \\ b_{q,m} \neq 0; & \ell = r \end{cases} \tag{3.70}$$

Proof The proof follows the same lines as in Corollary 3.9. □

The previous corollary can be used to obtain the relative degree of a discrete-time transfer function that arises from the sampling of a continuous-time system having relative degree r. In this case, the discrete-time state-space representation is given by (3.21)–(3.23). Thus, the relative degree of the sampled-data transfer function is obtained from (3.70), i.e.,

$$\begin{aligned} C_q A_q^{\ell-1} B_q &= C e^{A\Delta(\ell-1)} \int_0^\Delta e^{A\eta}\, d\eta\, B \\ &= C\left(I_n + A\Delta(\ell-1) + A^2\frac{(\Delta(\ell-1))^2}{2} + \cdots\right) \\ &\quad \times \Delta\left(I_n + A\frac{\Delta}{2} + A^2\frac{\Delta^2}{3} + \cdots\right)B \end{aligned} \tag{3.71}$$

In particular, considering the case $\ell = 1$ leads to

$$\begin{aligned} C_q B_q &= \Delta C\left(I_n + A\frac{\Delta}{2} + A^2\frac{\Delta^2}{3!} + \cdots\right)B \\ &= \Delta C\left(A^{r-1}\frac{\Delta^{r-1}}{r!} + \cdots\right)B = \frac{\Delta^r}{r!}CA^{r-1}B + \cdots \end{aligned} \tag{3.72}$$

which will generally be different from 0. Thus, the relative degree of the sampled-data models will, in general, be equal to 1, independent of the continuous-time relative degree r.

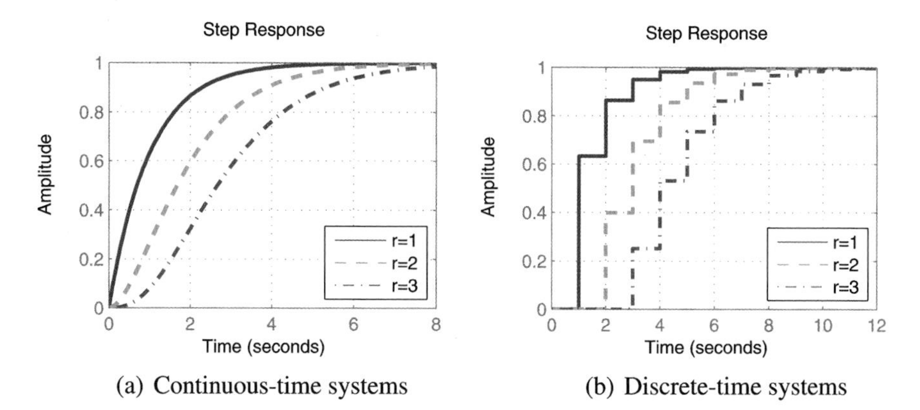

Fig. 3.3 Step responses of continuous-time and discrete-time systems having relative degree equal to 1, 2, and 3

Remark 3.12 Lemmas 3.8 and 3.10 explain the apparent mismatch between having a continuous-time system of relative degree $r \geq 1$ and an associated sampled-data model with relative degree equal to 1. In fact, independent of the continuous relative degree r, when the ZOH input is updated, the system output will respond so that the effect is usually seen at the next sampling instant (provided that there are no pure time delays). Thus, sampled-data models will generally have relative degree equal to 1. When the continuous-time relative degree is *larger*, then the continuous-time system output is *smoother*; however, there will, in general, be an effect at the next sampling instant—see also Fig. 3.3.

The previous remark clarifies the difference between relative degree in continuous-time and discrete-time domains. However, it also emphasises a very important fact when using a sampled-data model for prediction or control: the true continuous-time system generally exhibits a response one sampling instant after an input excitation is applied. This is particularly important when using approximate sampled-data models. It will be shown in Chaps. 8 and 9 that an approximate sampled model that accurately describes the true continuous time, in particular, for high frequencies, must have discrete-time relative degree equal to 1. The above discussion has used the phrase 'generally' when describing the discrete-time relative degree 1. The following example shows that, for certain sampling periods, the discrete relative degree can be greater than 1.

Example 3.13 Consider the continuous system

$$G(s) = \frac{-s+1}{(s+1)(s+2)(s+3)} \tag{3.73}$$

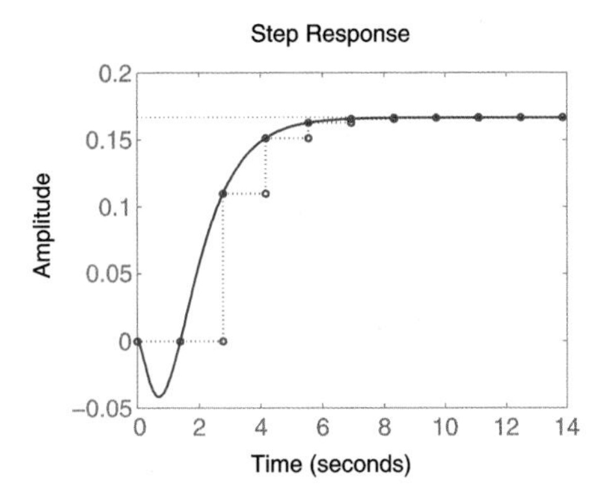

Fig. 3.4 Step responses of the continuous and sampled-data models in Example 3.13

This system has continuous relative degree 2. The associated sampled-data model is given by

$$G_q(z) = \frac{(1-e^{-\Delta})^2((4e^{-3\Delta} - e^{-4\Delta}) + (5e^{-\Delta} - 5e^{-3\Delta})z + (1-4e^{-\Delta})z^2)}{6(z-e^{-\Delta})(z-e^{-2\Delta})(z-e^{-3\Delta})} \tag{3.74}$$

This model has relative degree 1 for almost all Δ. However, since the continuous system has a real right half-plane zero, the step response will undershoot. Hence, by choosing the sampling period exactly equal to the time the response crosses zero, one obtains, at this special value of Δ, discrete relative degree 2. From (20.15) it follows that this corresponds to a sampling period such that the numerator becomes a first order polynomial, i.e.,

$$1 - 4e^{-\Delta} = 0 \quad \Longleftrightarrow \quad \Delta = \log 4 \approx 1.386 \tag{3.75}$$

Figure 3.4 shows the step responses of the continuous and sampled-data system for this particular choice of the sampling period, which confirms the above claim.

One of the recurring themes in this book will be that, when viewed from an appropriate perspective, the sampled-data and continuous-time results are consistent, in particular, as the sampling period is reduced to 0. In this framework, the following result is presented. The result is equivalent to Lemma 3.10, but is closer in spirit to the result given in Lemma 3.8 for continuous-time systems.

Lemma 3.14 *The relative degree of a discrete-time transfer function* (3.67) *describes the number of times that the system output has to be* forward-differenced *in order to make the input appear explicitly.*

Proof Let $\tilde{y}_k^\ell$ denote the sequence obtained by differencing the output ℓ times. Then,

$$\tilde{y}_k^\ell = (q-1)^\ell y_k \quad \Rightarrow \quad \tilde{Y}_q^\ell(z) = (z-1)^\ell G_q(z) U(z) \tag{3.76}$$

where, using (3.68), it follows that

$$(z-1)^\ell G_q(z) = b_{q,m} z^{\ell - r} + \cdots \tag{3.77}$$

which clearly shows that the output must be differenced r times to obtain a non-delayed response. □

In the following chapter, the issue of relative degree in sampled-data models will be revisited using the concept of incremental models. This kind of model will enable the results in Lemmas 3.8 and 3.10 (or 3.14) to be presented in a unified framework.

3.6 Summary

The key points covered in this chapter are:

- The deterministic linear continuous-time state-space model (3.1)–(3.2)
- The incremental form of the deterministic linear continuous-time state-space model (3.3)–(3.4)
- The definition of the zero order hold (ZOH) in (3.19).
- The equivalent discrete model assuming a ZOH input (Lemma 3.4):

$$qx_k = x_{k+1} = A_q x_k + B_q u_k \tag{3.78}$$

$$\bar{y}_k = Cx_k \tag{3.79}$$

where

$$A_q = e^{A\Delta}, \qquad B_q = \int_0^\Delta e^{A\eta} B \, d\eta \tag{3.80}$$

- The equivalent discrete transfer function (Lemma 3.5):

$$G_q(z) = (1 - z^{-1}) \mathcal{Z}\left\{ \mathcal{L}^{-1}\left\{ \frac{G(s)}{s} \right\}_{t=k\Delta} \right\} \tag{3.81}$$

$$= (1 - z^{-1}) \sum_{\ell=0}^{n} \mathrm{Res}_{s=p_\ell} \left\{ \frac{G(s)}{s} \frac{e^{s\Delta}}{z - e^{s\Delta}} \right\} \tag{3.82}$$

$$= (1 - z^{-1}) \sum_{\ell=-\infty}^{\infty} \frac{G((\log z + 2\pi j\ell)/\Delta)}{\log z + 2\pi j\ell} \tag{3.83}$$

- The observation that discrete models generally have relative degree 1, irrespective of the relative degree of the underlying continuous-time system.

Further Reading

Further background on linear sampled-data models can be found in

Åström KJ, Wittenmark B (1997) Computer controlled systems. Theory and design, 3rd edn. Prentice Hall, Englewood Cliffs

Feuer A, Goodwin GC (1996) Sampling in digital signal processing and control. Birkhäuser, Boston
Franklin GF, Powell JD, Workman ML (1997) Digital control of dynamic systems, 3rd. edn. Addison-Wesley, Reading
Middleton RH, Goodwin GC (1990) Digital control and estimation. A unified approach. Prentice Hall, Englewood Cliffs

Chapter 4
Incremental Sampled-Data Models

This chapter explores, in greater detail, the relationship between sampled-data models and the associated continuous-time system. In particular, the chapter will show how the sampled-data model relates to the original continuous-time model, when a fast sampling rate is used, i.e., when making the sampling period arbitrarily small. It seems intuitively clear that, if the sampling period is small, then the continuous-time model should somehow be recovered. However, there are some unexpected artefacts which are a result of the sampling process. Indeed, a first (and perhaps obvious) attempt to study the asymptotic behaviour of the discrete-time model utilizing Lemma 3.4 on p. 25 does not lead to meaningful results.

4.1 Sampled-Data Models for Fast Sampling Rates

The model given by (3.21)–(3.23) cannot be directly linked back to the underlying continuous-time system. The reason is that, if the limit is taken, as the sampling period Δ goes to 0, then all information about the underlying continuous-time system is lost. Specifically, it can be seen from (3.23) that the following result holds for *all* linear systems:

$$\lim_{\Delta \to 0} A_q = I \quad \text{and} \quad \lim_{\Delta \to 0} B_q = 0 \tag{4.1}$$

The above result reflects the well-known fact that all the poles of a sampled-data model asymptotically approach the point $z = 1$ for fast sampling rates.

The source of this difficulty is that (3.21) describes the *next value* of x_k. Also, it is intuitively clear that $x_{k+1} \to x_k$ as the sampling period $\Delta \to 0$. This result is consistent with (4.1). The difficulty is fundamental and intrinsic to shift operator model descriptions. This seems, at first sight, paradoxical. However, the paradox

J.I. Yuz, G.C. Goodwin, *Sampled-Data Models for Linear and Nonlinear Systems*, Communications and Control Engineering, DOI 10.1007/978-1-4471-5562-1_4,

is readily resolved if (3.21) is expressed in *incremental* form. In order to achieve this, x_k is subtracted from both sides of the equation. Also Δ is factored out of the right-hand side. This leads to

$$dx_k^+ = x_{k+1} - x_k = A_i\, x_k \Delta + B_i\, u_k \Delta \tag{4.2}$$

where A_i and B_i are simple functions of A_q and B_q, i.e.,

$$A_i = \frac{A_q - I}{\Delta} \quad \text{and} \quad B_i = \frac{B_q}{\Delta} \tag{4.3}$$

Remark 4.1 A key feature of the above model is that it has the same structure as the incremental continuous-time model (3.3). Also, taking the limit as the sampling period tends to zero leads to

$$A_i \xrightarrow{\Delta \to 0} A \quad \text{and} \quad B_i \xrightarrow{\Delta \to 0} B \tag{4.4}$$

where A and B are the corresponding continuous-time matrices! This is a pleasing by-product of the use of the incremental model (4.2). Indeed, it will be shown in the sequel that use of the incremental model is crucial in obtaining meaningful connections between continuous-time models and their discrete-time counterparts at fast sampling rates.

Incremental models lead to a well-defined limit as the sampling rate is increased. Also, the limiting model takes the sampled-data model back to the underlying continuous-time model in a heuristically satisfactory fashion.

4.2 The Delta Operator

In the context of incremental models, it will also be convenient, for future use, to define the discrete δ-*operator* as

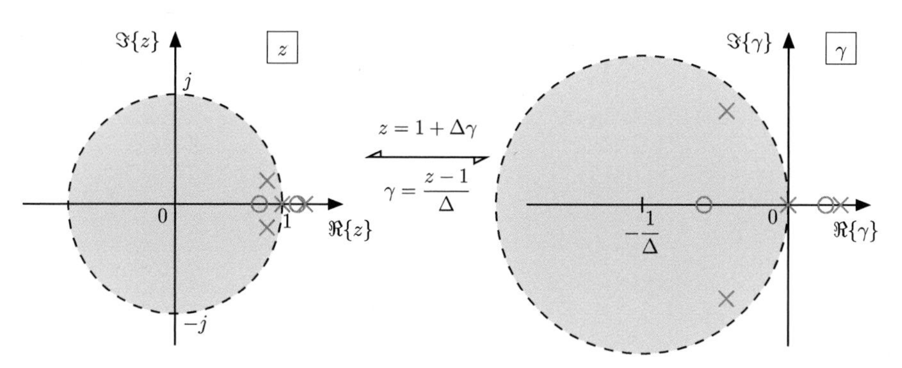

Fig. 4.1 Complex planes associated with the z and γ variables, corresponding to shift operator q and incremental operator δ, respectively

$$\delta x_k = \frac{dx_k^+}{\Delta} = \frac{x_{k+1} - x_k}{\Delta} = \frac{q-1}{\Delta} x_k \tag{4.5}$$

where q is the forward shift operator. The variable γ is also introduced as the complex variable associated with the δ-operator

$$\delta = \frac{q-1}{\Delta} \quad \Longleftrightarrow \quad \gamma = \frac{z-1}{\Delta} \tag{4.6}$$

When using the δ-operator, the state-space model of a discrete-time system is given by

$$\delta x_k = A_i x_k + B_i u_k \tag{4.7}$$

$$y_k = C x_k \tag{4.8}$$

and the associated transfer function, in the domain of the complex variable γ, is given by

$$G_d(\gamma) = C(\gamma I_n - A_i)^{-1} B_i \tag{4.9}$$

Note that this transfer function is readily obtained on changing variables in the shift operator form of the discrete-time transfer function, i.e.,

$$G_d(\gamma) = G_q(z)|_{z=1+\Delta\gamma} = G_q(1 + \Delta\gamma) \tag{4.10}$$

The effect of the mapping between the complex variables z and γ is shown in Fig. 4.1.

Equation (4.3) guarantees the convergence result in (4.4). It follows that, as the sampling period goes to zero, the continuous-time transfer function is recovered,

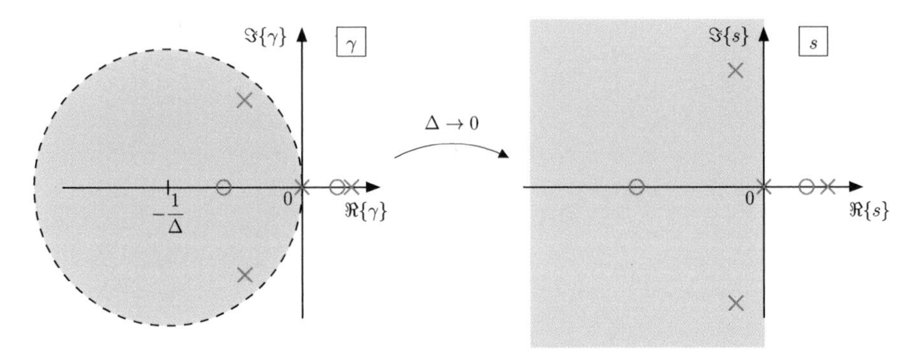

Fig. 4.2 Complex planes associated with the γ and s variables, corresponding to incremental and continuous-time models, respectively

i.e.,

$$G_d(\gamma) \xrightarrow{\Delta \to 0} G(\gamma) \tag{4.11}$$

This implies that when the sampled-data model is parameterised in incremental form, the poles and zeros of the discrete-time transfer function (in the γ-variable) tend to their continuous-time locations in the s-plane. This fact is represented in Fig. 4.2. In particular, the poles of the discrete-time transfer function are located at

$$\gamma_i = \frac{e^{p_i \Delta} - 1}{\Delta} \xrightarrow{\Delta \to 0} p_i \tag{4.12}$$

for each continuous-time pole p_i. The use of the delta operator makes the sampling period Δ explicit and is crucial in showing how discrete-time results revert back to their continuous-time counterparts when Δ goes to 0. Appropriate scaling (i.e., inclusion of Δ) is important in obtaining the result.

A key observation is that the delta operator is not related to Euler integration. Indeed, the use of the *delta* operator is simply a way of reparameterising any discrete-time model via the transformation $q = \delta\Delta + 1$ or $\delta = (q-1)/\Delta$. This reparameterisation has the advantage of highlighting the link between discrete-time and continuous-time domains and also achieving improved numerical properties. Of course, any shift-domain model can be converted to delta form and vice versa (see (4.3)). This is completely independent of the use of Euler integration.

From (4.2)–(4.5) it follows that the δ-operator is simply an alternative way of writing incremental models. Hence, in the sequel, the terms *incremental model* and *δ-domain model* will be used interchangeably.

Example 4.2 Consider the exact sampled-data model (3.60) obtained in Example 3.7. The system can be reparameterised to incremental form by using the change of variable $z = (1 + \Delta\gamma)$ where, for this particular example, the sampling period is chosen as $\Delta = 0.01$. This yields the following *exact* delta domain model:

$$G_\delta(\gamma) = G_q(1 + \Delta\gamma)$$
$$= \frac{5.882(0.005\gamma + 1)(\gamma + 4.88)}{(\gamma + 1.98)(\gamma + 2.96)(\gamma + 3.92)} \tag{4.13}$$

The reader will see that (4.13) closely resembles (3.59) save for the extra term $(0.005\gamma + 1)$ in the numerator. The origins of this extra term lie at the very heart of sampled-data models. The associated issues will be discussed in the next chapter in detail.

4.3 Relative Degree Revisited

Section 3.5 discussed the apparent mismatch between relative degree of continuous-time and discrete-time transfer functions. For continuous-time systems, the relative degree of a transfer function $G(s)$ corresponds to the smoothness of the step response or, equivalently, to the number of times the output must be differentiated to make the input appear explicitly. On the other hand, for discrete-time systems, the relative degree of a transfer function $G(z)$ corresponds to the number of delays between input and output.

Additionally, in the case of sampled-data models, the relative degree of the discrete-time transfer function is generally equal to 1, independent of the continuous-time relative degree. In fact, this corresponds to the presence of additional *sampling zeros*, which are discussed further in Chap. 5. However, from (4.11), it follows that, when using incremental models, the continuous relative degree is recovered when the sampling period goes to zero.

The connections between the above-mentioned results are clarified by revisiting Lemma 3.14 and analysing the behaviour of divided differences as the sampling period goes to zero. Indeed, preservation of the idea of continuous-time relative degree is clarified in the next result.

Lemma 4.3 *Consider a continuous-time system* (3.1)–(3.2) *having relative degree r and the associated sampled-data model* (3.21)–(3.23) (*or equivalently in incremental form* (4.7)–(4.8)). *Then, the output sequence must be differenced and scaled by the sampling period r times in order to make the input appear of order* 1 (*i.e.*, $\mathcal{O}(1)$).

Proof As in Lemma 3.14, consider the first element in the expansion of the differenced and scaled transfer function

$$
\begin{aligned}
\left(\frac{z-1}{\Delta}\right)^{\ell} G_q(z) &= \left(\frac{z-1}{\Delta}\right)^{\ell} C(zI_n - A_q)^{-1} B_q \\
&= \frac{(z-1)^{\ell}}{\Delta^{\ell}} C\left(z^{-1} + A_q z^{-2} + \cdots\right) B_q \\
&= \frac{z^{-1}}{\Delta^{\ell}} C B_q + \cdots
\end{aligned} \tag{4.14}
$$

where B_q can be expanded as follows:

$$
\begin{aligned}
\left(\frac{z-1}{\Delta}\right)^{\ell} G_q(z) &= \frac{z^{-1}}{\Delta^{\ell}} C\Delta\left(I + A\frac{\Delta}{2} + A^2\frac{\Delta^2}{3!} + \cdots\right) B + \cdots \\
&= \frac{z^{-1}}{\Delta^{\ell}} C\Delta\left(A^{r-1}\frac{\Delta^{r-1}}{r!} + \cdots\right) B + \cdots \\
&= z^{-1}\frac{\Delta^{r-\ell}}{r!} C A^{r-1} B + \cdots
\end{aligned} \tag{4.15}
$$

which shows that, in general, the differenced (and scaled) output sequence is of the order of 1 (i.e., $\in \mathcal{O}(1)$ as the sampling period goes to zero) if and only if $\ell = r$. □

Remark 4.4 Lemma 4.3 embellishes Remark 3.12 a bit further. Although it is true that a continuous-time system of relative degree r will generally result in a sampled-data system of relative degree 1, more can be said about the behaviour as the sampling period Δ goes to zero. In particular, if the differences are scaled appropriately, then if the output of the resultant sampled-data system is differenced $1, \ldots, r-1$ times, the input will appear, without delay, but the weighting of the input will be $\mathcal{O}(\Delta)$ and will go to zero as Δ goes to 0. However, if the output is differenced and scaled r times, then the weighting on the undelayed input will not go to 0 with the sampling period. This shows that the continuous-time and sampled-data results are not as far apart as might have previously seemed the case.

4.4 Summary

The key points covered in this chapter are:

- The misleading convergence properties obtained when the shift operator model in (4.1) is used.

- The alternative incremental model

$$d x_k^+ = x_{k+1} - x_k = A_i x_k \Delta + B_i u_k \Delta \tag{4.16}$$

and the associated (well-behaved) convergence properties

$$A_i \xrightarrow{\Delta \to 0} A \quad \text{and} \quad B_i \xrightarrow{\Delta \to 0} B \tag{4.17}$$

- The δ-operator

$$\delta = \frac{q - 1}{\Delta} \tag{4.18}$$

- The convergence result for transfer functions

$$G_d(\gamma) \xrightarrow{\Delta \to 0} G(\gamma) \tag{4.19}$$

- The role of relative degree in incremental models and the unifying result in Lemma 4.3: differencing and scaling r times the output sequence, corresponds, in fact, to differentiating r times the output of the continuous system, as the sampling period goes to zero.

Further Reading

Additional background on incremental models can be found in

Feuer A, Goodwin GC (1996) Sampling in digital signal processing and control. Birkhäuser, Boston

Goodwin GC, Middleton RH, Poor HV (1992) High-speed digital signal processing and control. Proc IEEE 80(2):240–259

Mansour M (1993) Stability and robust stability of discrete-time systems in the δ-transform. In: Jamshidi M et al (eds) Fundamentals of discrete-time systems: a tribute to Prof. Eliahu I. Jury. TSI Press, San Antonio

Middleton RH, Goodwin GC (1990) Digital control and estimation. A unified approach. Prentice Hall, Englewood Cliffs

Premaratne K, Jury EI (1994) Tabular method for determining root distribution of delta-operator formulated real polynomials. IEEE Trans Autom Control 39(2):352–355

Salgado ME, Middleton RH, Goodwin GC (1988) Connection between continuous and discrete Riccati equations with applications to Kalman filtering. IEE Proc Part D, Control Theory Appl 135(1):28–34

Chapter 5
Asymptotic Sampling Zeros

Abstract The location of poles of sampled models depends on the poles of the underlying continuous-time system and the sampling period. However, the relation between the zeros in discrete-time and continuous-time is much more involved. The sampled model can have extra zeros which have no continuous-time counterpart. This chapter examines the asymptotic behaviour of these sampling zeros in the discrete-time model, as the sampling period goes to zero.

It has been shown in Sect. 3.4 that the poles of a sampled-data model can be readily characterised in terms of the sampling period, Δ, together with the continuous-time system poles. However, the relation between the zeros in discrete time and continuous time is much more involved. The discrete-time model can have extra zeros which have no continuous-time counterpart. These extra zeros are called the *sampling zeros* of the discrete-time model.

This chapter examines the asymptotic behaviour of the sampling zeros in sampled-data models as the sampling period goes to zero.

5.1 Pure Integrator Model

First consider an rth order integrator. It will be shown later that this special case is the key to understanding more general results. Indeed, as the sampling rate increases, a system of relative degree r *behaves* as an rth order integrator beyond the sampling frequency. This will be a recurrent and insightful interpretation for deterministic and stochastic systems.

The connection is established in the following theorem.

Theorem 5.1 *Let $G(s)$ be a linear system having relative degree $r = n - m$ given by*

$$G(s) = \frac{b_m s^m + b_{m-1} s^{m-1} + \cdots + b_o}{a_n s^n + a_{n-1} s^{n-1} + \cdots + a_o} \tag{5.1}$$

J.I. Yuz, G.C. Goodwin, *Sampled-Data Models for Linear and Nonlinear Systems*, Communications and Control Engineering, DOI 10.1007/978-1-4471-5562-1_5,

When the sampling period tends to zero, the discrete system behaves as an integrator of order r, i.e.,

$$G_d(z) \xrightarrow{\Delta \to 0} \mathcal{ZOH}\left\{\frac{1}{s^r}\right\} = \left(1 - z^{-1}\right)\frac{\Delta^r}{\Gamma(r+1)}\sum_{k=1}^{\infty} k^{\alpha} z^{-k} \tag{5.2}$$

Proof The proof follows by applying a change of variables $\eta = s\Delta$ in (3.28), i.e.,

$$G_q(z) = \frac{(z-1)}{z}\int_{\gamma\Delta - j\infty}^{\gamma\Delta + j\infty} \frac{e^{\eta}}{z - e^{\eta}} G\left(\frac{\eta}{\Delta}\right)\frac{d\eta}{\eta} \tag{5.3}$$

where, using (5.1),

$$G\left(\frac{\eta}{\Delta}\right) = \frac{b_m}{a_n}\left(\frac{\Delta}{\eta}\right)^r \left(\frac{1 + \cdots + \frac{b_0}{b_m}(\frac{\Delta}{\eta})^m}{1 + \cdots + \frac{a_0}{a_n}(\frac{\Delta}{\eta})^n}\right) \tag{5.4}$$

Hence, letting the sampling period tend to zero, it follows that

$$\lim_{\Delta \to 0} \Delta^{-r} G_q(z) = \frac{(z-1)}{z}\int_{\gamma\Delta - j\infty}^{\gamma\Delta + j\infty} \frac{e^{\eta}}{z - e^{\eta}}\frac{b_m}{a_n}\frac{1}{\eta^r}\frac{d\eta}{\eta} \tag{5.5}$$

which corresponds to the sampled model of an rth order integrator, where $r = (n - m)$ is the system relative degree. □

Based on the above result, the discussion of sampling zeros is appropriately initiated by restricting attention to an rth order integrator. For this case, the following result holds.

Lemma 5.2 *For sampling period Δ, the pulse transfer function corresponding to the rth order integrator $G(s) = s^{-r}$ is given by*

$$G_q(z) = \frac{\Delta^r}{r!}\frac{B_r(z)}{(z-1)^r} \tag{5.6}$$

where:

$$B_r(z) = b_1^r z^{r-1} + b_2^r z^{r-2} + \cdots + b_r^r \tag{5.7}$$

$$b_k^r = \sum_{\ell=1}^{k} (-1)^{k-\ell} \ell^r \binom{r+1}{k-\ell} \tag{5.8}$$

Proof Substitute $G(s) = s^{-r}$ into the pulse transfer function expression in (3.30) on p. 26, where the $r + 1$ poles of $G(s)/s$ at $s = 0$ are mapped to $z = 1$. As a consequence, the pulse transfer function is rational, i.e.,

$$G_q(z) = \frac{\beta_{r-1}z^{r-1} + \beta_{r-2}z^{r-2} + \cdots + \beta_0}{\alpha_r z^r + \alpha_{r-1}z^{r-1} + \cdots + \alpha_0} \tag{5.9}$$

and

$$\alpha_r z^r + \alpha_{r-1}z^{r-1} + \cdots + \alpha_0 = (z-1)^r \tag{5.10}$$

Hence,

$$\alpha_k = \binom{r}{k}(-1)^{r-k}; \quad k \in \{0, \ldots, r\} \tag{5.11}$$

If the input signal is a step sequence, then the zero order hold (ZOH) input to the system is a continuous step signal $u(t) = 1$, for all $t > 0$. The output of the rth order integrator is then given by

$$y(t) = \frac{t^r}{r!}; \quad t \geq 0 \tag{5.12}$$

Thus, from (5.9), when the input is a step signal, it follows that

$$\sum_{\ell=1}^{k} \alpha_{r-k+\ell}\, y(\ell\Delta) = \sum_{\ell=0}^{k-1} \beta_{r-k+\ell}; \quad k \in \{1, \ldots, r\} \tag{5.13}$$

On changing variables $k = r - \tilde{k} + 1$, it follows that

$$\sum_{\ell=1}^{r-\tilde{k}+1} \alpha_{\tilde{k}-1+\ell}\, y(\ell\Delta) = \sum_{\ell=0}^{r-\tilde{k}} \beta_{\tilde{k}-1+\ell}; \quad \tilde{k} \in \{1, \ldots, r\} \tag{5.14}$$

Notice that

$$\begin{aligned} \beta_{\tilde{k}-1} &= \sum_{\ell=0}^{r-\tilde{k}} \beta_{\tilde{k}-1+\ell} - \sum_{\ell=0}^{r-\tilde{k}-1} \beta_{\tilde{k}+\ell} \\ &= \sum_{\ell=1}^{r-\tilde{k}+1} \alpha_{\tilde{k}-1+\ell}\, y(\ell\Delta) - \sum_{\ell=1}^{r-\tilde{k}} \alpha_{\tilde{k}+\ell}\, y(\ell\Delta) \end{aligned} \tag{5.15}$$

Substituting (5.11) and (5.12) into the last equation leads to:

$$\begin{aligned} \beta_{\tilde{k}-1} &= \alpha_r y\big((r-\tilde{k}+1)\Delta\big) - \sum_{\ell=1}^{r-\tilde{k}} [\alpha_{\tilde{k}+\ell} - \alpha_{\tilde{k}-1+\ell}]\, y(\ell\Delta) \\ &= \frac{((r-\tilde{k}+1)\Delta)^r}{r!} - \sum_{\ell=1}^{r-\tilde{k}} \left[\binom{r}{\tilde{k}+\ell} + \binom{r}{\tilde{k}-1+\ell}\right](-1)^{r-\tilde{k}-\ell} \frac{(\ell\Delta)^r}{r!} \end{aligned}$$

$$= \frac{((r-\tilde{k}+1)\Delta)^r}{r!} - \sum_{\ell=1}^{r-\tilde{k}} \binom{r+1}{\tilde{k}+\ell} (-1)^{r-\tilde{k}-\ell} \frac{(\ell\Delta)^r}{r!}$$

$$= \frac{\Delta^r}{r!} \sum_{\ell=1}^{r-\tilde{k}+1} \binom{r+1}{\tilde{k}+\ell} (-1)^{r-\tilde{k}+1-\ell} \ell^r \tag{5.16}$$

From the last equation, it follows that $\beta_{\tilde{k}-1} = \frac{\Delta^r}{r!} b_k^r$ as defined in (5.8), using $k = r - \tilde{k} + 1$. □

Remark 5.3 The polynomials defined in (5.7)–(5.8) correspond, in fact, to the Euler–Frobenius polynomials (also called *reciprocal polynomials*). These polynomials are known to satisfy several properties:

1. Their coefficients can be computed recursively, i.e., $b_1^r = b_r^r = 1$, for all $r \geq 1$, and

$$b_k^r = k b_k^{r-1} + (r - k + 1) b_{k-1}^{r-1} \tag{5.17}$$

   for $k = 2, \ldots, r-1$.
2. From the symmetry of the coefficients it follows that, if $B_r(z_0) = 0$, then $B_r(z_0^{-1}) = 0$.
3. Their roots are always negative real.
4. They satisfy an interlacing property, namely, every root of the polynomial $B_{r+1}(z)$ lies between every two adjacent roots of $B_r(z)$, for $r \geq 2$.
5. The following recursive relation holds:

$$B_{r+1}(z) = z(1-z) B_r{}'(z) + (rz + 1) B_r(z) \tag{5.18}$$

   for all $r \geq 1$, and where $B_r{}' = \frac{dB_r}{dz}$.

A more complete description of Euler–Frobenius polynomials is given in Chap. 20. The first few polynomials are:

$$B_1(z) = 1 \tag{5.19}$$

$$B_2(z) = z + 1 \tag{5.20}$$

$$B_3(z) = z^2 + 4z + 1 = (z + 2 + \sqrt{3})(z + 2 - \sqrt{3}) \tag{5.21}$$

$$B_4(z) = z^3 + 11z^2 + 11z + 1 = (z+1)(z + 5 + 2\sqrt{6})(z + 5 - 2\sqrt{6}) \tag{5.22}$$

The following result gives an alternative interpretation of the Euler–Frobenius polynomials. This result will be used later in the book.

Lemma 5.4 *The polynomials defined in Lemma 5.2 satisfy the following equation*:

$$\sum_{k=-\infty}^{\infty} \frac{1}{(\log z + j2\pi k)^r} = \frac{z B_{r-1}(z)}{(r-1)!(z-1)^r}, \quad r \geq 2 \tag{5.23}$$

Proof The result follows from the alternative representation of the pulse transfer function (3.29) on p. 26, for the case of an rth order integrator. □

The results can also be presented in the following equivalent incremental form.

Lemma 5.5 *Given a sampling period Δ, the exact incremental sampled-data model corresponding to the n-th order integrator $G(s) = s^{-r}$, $r \geq 1$, when using a ZOH input, is given by*

$$G_\delta(\gamma) = \frac{p_r(\Delta\gamma)}{\gamma^r} \tag{5.24}$$

where the polynomial $p_r(\Delta\gamma)$ is given by

$$p_r(\Delta\gamma) = \det M_r \tag{5.25}$$

and where the matrix M_r is defined by

$$M_r = \begin{bmatrix} 1 & \frac{\Delta}{2!} & \cdots & \frac{\Delta^{r-2}}{(r-1)!} & \frac{\Delta^{r-1}}{r!} \\ -\gamma & 1 & \cdots & \frac{\Delta^{r-3}}{(r-2)!} & \frac{\Delta^{r-2}}{(r-1)!} \\ \vdots & \ddots & \ddots & \vdots & \vdots \\ 0 & \cdots & -\gamma & 1 & \frac{\Delta}{2!} \\ 0 & \cdots & 0 & -\gamma & 1 \end{bmatrix} \tag{5.26}$$

Proof The rth order integrator $G(s) = s^{-r}$ can be represented in the state-space form (3.1)–(3.2) on p. 21, where the matrices take the specific forms:

$$A = \left[\begin{array}{c|ccc} 0 & & & \\ \vdots & & I_{r-1} & \\ 0 & & & \\ \hline 0 & 0 & \cdots & 0 \end{array}\right], \quad B = \begin{bmatrix} 0 \\ \vdots \\ 0 \\ 1 \end{bmatrix}, \quad C = \begin{bmatrix} 1 & 0 & \cdots & 0 \end{bmatrix} \tag{5.27}$$

The equivalent sampled-data system can readily be obtained on noting that, by the Cayley–Hamilton theorem, any matrix A satisfies its own characteristic equation, i.e., for the A matrix given in (5.27), it follows that $A^r = 0$. As a consequence, the corresponding exponential matrix is readily obtained:

$$e^{A\Delta} = I + A\Delta + \cdots + A^{r-1}\frac{\Delta^{r-1}}{(r-1)!} \tag{5.28}$$

Substituting into (4.3) leads to:

$$\begin{aligned}
A_i &= \frac{e^{A\Delta} - I}{\Delta} = \begin{bmatrix} 0 & 1 & \frac{\Delta}{2!} & \cdots & \frac{\Delta^{r-2}}{(r-1)!} \\ 0 & 0 & 1 & \cdots & \frac{\Delta^{r-3}}{(r-2)!} \\ \vdots & & \ddots & \ddots & \vdots \\ 0 & & \cdots & 0 & 1 \\ 0 & 0 & \cdots & & 0 \end{bmatrix}, \\
B_i &= \frac{1}{\Delta}\int_0^{\Delta} e^{A\eta}\, d\eta\, B = \begin{bmatrix} \frac{\Delta^{r-1}}{r!} \\ \frac{\Delta^{r-2}}{(r-1)!} \\ \vdots \\ \frac{\Delta}{2} \\ 1 \end{bmatrix}
\end{aligned} \tag{5.29}$$

Then the zeros of the sampled-data model can be obtained from Lemma 3.3 on p. 23, i.e.,

$$\begin{aligned}
\text{num}\{G_i(\gamma)\} &= \det\begin{bmatrix} \gamma I_r - A_i & -B_i \\ C & 0 \end{bmatrix} = \det\begin{bmatrix} \gamma & -1 & -\frac{\Delta}{2!} & \cdots & -\frac{\Delta^{r-1}}{r!} \\ 0 & \gamma & -1 & \cdots & -\frac{\Delta^{r-2}}{(r-1)!} \\ \vdots & & \ddots & \ddots & \vdots \\ 0 & \cdots & & \gamma & -1 \\ 1 & 0 & \cdots & & 0 \end{bmatrix} \\
&= (-1)^{r-1}\det\begin{bmatrix} -1 & -\frac{\Delta}{2!} & \cdots & -\frac{\Delta^{r-1}}{r!} \\ \gamma & -1 & \cdots & -\frac{\Delta^{r-2}}{(r-1)!} \\ & \ddots & \ddots & \vdots \\ & & \gamma & -1 \end{bmatrix}
\end{aligned} \tag{5.30}$$

where the first determinant is evaluated across the last row. The result follows by applying the change of sign to the matrix in the last determinant, i.e.,

$$(-1)^{r-1}\det\begin{bmatrix} -1 & -\frac{\Delta}{2!} & \cdots & -\frac{\Delta^{r-1}}{r!} \\ \gamma & -1 & \cdots & -\frac{\Delta^{r-2}}{(r-1)!} \\ & \ddots & \ddots & \vdots \\ & & \gamma & -1 \end{bmatrix} = \det\begin{bmatrix} 1 & \frac{\Delta}{2!} & \cdots & \frac{\Delta^{r-1}}{r!} \\ -\gamma & 1 & \cdots & \frac{\Delta^{r-2}}{(r-1)!} \\ & \ddots & \ddots & \vdots \\ & & -\gamma & 1 \end{bmatrix} \tag{5.31}$$

□

Remark 5.6 The above result, though formally equivalent to the shift-domain expressions in Lemma 5.2, describes the results in a form which will prove useful later, especially in relation to nonlinear systems.

Remark 5.7 The polynomials $p_r(\Delta\gamma)$ in Lemma 5.5, when rewritten in terms of the z-variable using (4.6), correspond to the Euler–Frobenius polynomials. In fact, the following relation holds:

$$p_r(\Delta\gamma)\big|_{\gamma=\frac{z-1}{\Delta}} = p_r(z-1) = \frac{B_r(z)}{r!} \tag{5.32}$$

The first of the Euler–Frobenius polynomials in the γ-domain (corresponding to those in (5.19)–(5.21), on p. 50) are given by:

$$p_1(\Delta\gamma) = 1 \tag{5.33}$$

$$p_2(\Delta\gamma) = 1 + \frac{\Delta}{2}\gamma \tag{5.34}$$

$$p_3(\Delta\gamma) = 1 + \Delta\gamma + \frac{\Delta^2}{6}\gamma^2 \tag{5.35}$$

Remark 5.8 Note that, in the γ-domain, the Euler–Frobenius polynomials are an explicit function of the argument $\Delta\gamma$. This means that the roots of the polynomials, in the complex plane (of the variable γ), all go to infinity as Δ goes to zero. This reveals a close connection to the continuous-time case. Indeed, as $\Delta \to 0$, the discrete-time sampling zeros converge to $-\infty$. Hence, in the limit, the sampling zeros simply contribute to the continuous-time relative degree.

A second result is now presented, which establishes a recursive relation between the Euler–Frobenius polynomials in the γ-domain. Note that this recursion, together with (5.18) for the z-domain formulation, is useful when computing the polynomial coefficients.

Lemma 5.9 *The polynomials $p_r(\Delta\gamma)$ defined by* (5.25)–(5.26) *satisfy the recursion*:

$$p_0(\Delta\gamma) \triangleq 1 \tag{5.36}$$

$$p_r(\Delta\gamma) = \sum_{\ell=1}^{r} \frac{(\Delta\gamma)^{\ell-1}}{\ell!} p_{r-\ell}(\Delta\gamma); \quad r \geq 1 \tag{5.37}$$

and

$$\lim_{\Delta\to 0} p_r(\Delta\gamma) = 1; \quad \forall r \in \{1, 2, \ldots\} \tag{5.38}$$

Proof The result in (5.36) is obtained directly from the definition

$$p_r(\Delta\gamma) = \det \begin{bmatrix} 1 & \frac{\Delta}{2!} & \cdots & \frac{\Delta^{r-2}}{(r-1)!} & \frac{\Delta^{r-1}}{r!} \\ -\gamma & 1 & & \cdots & \frac{\Delta^{r-2}}{(r-1)!} \\ & \ddots & \ddots & & \vdots \\ \vdots & & -\gamma & 1 & \frac{\Delta}{2} \\ 0 & \cdots & & -\gamma & 1 \end{bmatrix} \tag{5.39}$$

Computing the determinant across the last row leads to

$$p_r(\Delta\gamma) = \det \begin{bmatrix} 1 & \frac{\Delta}{2!} & \cdots & \frac{\Delta^{r-2}}{(r-1)!} \\ -\gamma & 1 & \cdots & \frac{\Delta^{r-3}}{(r-2)!} \\ & \ddots & \ddots & \vdots \\ 0 & & -\gamma & 1 \end{bmatrix} + \gamma \det \begin{bmatrix} 1 & \cdots & \frac{\Delta^{r-3}}{(r-2)!} & \frac{\Delta^{r-1}}{r!} \\ -\gamma & \cdots & \frac{\Delta^{r-4}}{(r-3)!} & \frac{\Delta^{r-2}}{(r-1)!} \\ & \ddots & \ddots & \vdots \\ 0 & & -\gamma & \frac{\Delta}{2} \end{bmatrix} \tag{5.40}$$

where the first determinant corresponds to $p_{r-1}(\Delta\gamma)$, and the second determinant can be computed again across the last row to obtain

$$\begin{aligned} p_r(\Delta\gamma) &= p_{r-1}(\Delta\gamma) + \gamma \left(\Delta p_{r-2}(\Delta\gamma) + \gamma \det \begin{bmatrix} 1 & \cdots & \frac{\Delta^{r-4}}{(r-3)!} & \frac{\Delta^{r-1}}{r!} \\ -\gamma & \ddots & \vdots & \frac{\Delta^{r-2}}{(r-1)!} \\ & \ddots & \ddots & \vdots \\ 0 & & -\gamma & \frac{\Delta^2}{3!} \end{bmatrix} \right) \\ &= p_{r-1}(\Delta\gamma) + \frac{\Delta\gamma}{2} p_{r-2}(\Delta\gamma) \\ &\quad + \gamma^2 \left(\frac{\Delta^2}{3!} p_{r-3}(\Delta\gamma) + \gamma \det \begin{bmatrix} 1 & \cdots & \frac{\Delta^{r-5}}{(r-4)!} & \frac{\Delta^{r-1}}{r!} \\ -\gamma & \ddots & \vdots & \frac{\Delta^{r-2}}{(r-1)!} \\ & \ddots & \ddots & \vdots \\ 0 & & -\gamma & \frac{\Delta^3}{4!} \end{bmatrix} \right) \\ &\ \ \vdots \\ &= p_{r-1}(\Delta\gamma) + \frac{\Delta\gamma}{2} p_{r-2}(\Delta\gamma) + \cdots + \gamma^{r-2} \det \begin{bmatrix} 1 & \frac{\Delta^{r-1}}{r!} \\ -\gamma & \frac{\Delta^{r-2}}{(r-1)!} \end{bmatrix} \\ &= p_{r-1}(\Delta\gamma) + \frac{\Delta\gamma}{2} p_{r-2}(\Delta\gamma) + \cdots + \frac{(\Delta\gamma)^{r-2}}{(r-1)!} + \frac{(\Delta\gamma)^{r-1}}{r!} \end{aligned} \tag{5.41}$$

which, in fact, corresponds to (5.37).

Finally, (5.38) readily follows from the recursion (5.37), on noting that

$$\lim_{\Delta \to 0} p_r(\Delta\gamma) = \lim_{\Delta \to 0} p_{r-1}(\Delta\gamma) = \cdots = \lim_{\Delta \to 0} p_1(\Delta\gamma) = 1 \tag{5.42}$$

□

5.2 General Linear Systems

Next consider the case of a general single-input single-output (SISO) linear continuous-time system. The focus here is on the corresponding discrete-time model when a ZOH input is applied. The relationship between the continuous-time poles and those of the discrete-time model can be easily determined. However, the relationship between the zeros in the continuous- and discrete-time domains is more involved. The asymptotic case (as the sampling period decreases) is emphasised in the sequel. A frequency-domain development is first presented. Later in Chap. 9, an alternative time-domain development will be presented.

Lemma 5.10 *Let $G(s)$ be a rational function:*

$$G(s) = \frac{F(s)}{E(s)} = K\frac{(s-z_1)(s-z_2)\cdots(s-z_m)}{(s-p_1)(s-p_2)\cdots(s-p_n)} \tag{5.43}$$

and $G_q(z)$ the corresponding discrete-time transfer function with ZOH input. Assume that $m < n$, i.e., $G(s)$ is strictly proper. Then, as the sampling period Δ goes to 0, m zeros of $G_q(z)$ converge to 1 as $e^{z_i\Delta}$, and the remaining $n-m-1$ zeros of $G_q(z)$ converge to the zeros of the polynomial $B_{n-m}(z)$ defined in Lemma 5.2*, i.e.,*

$$G_q(z) \xrightarrow{\Delta \approx 0} K\frac{\Delta^{n-m}(z-1)^m B_{n-m}(z)}{(n-m)!(z-1)^n} \tag{5.44}$$

Proof Consider the representation (3.28) on p. 26, i.e.,

$$\begin{aligned} G_q(z) &= \left(1-z^{-1}\right)\frac{1}{2\pi j}\int_{\gamma-j\infty}^{\gamma+j\infty} \frac{e^{s\Delta}}{z-e^{s\Delta}}\frac{G(s)}{s}\,ds \\ &= \left(1-z^{-1}\right)\frac{1}{2\pi j}\int_{\gamma\Delta-j\infty}^{\gamma\Delta+j\infty} \frac{e^{w}}{z-e^{w}}G\left(\frac{w}{\Delta}\right)\frac{dw}{w} \end{aligned} \tag{5.45}$$

From (5.43), it follows that

$$G_q\left(\frac{w}{\Delta}\right) = K\left(\frac{\Delta}{w}\right)^{n-m} \frac{(1 - \frac{z_1\Delta}{w})\cdots(1 - \frac{z_m\Delta}{w})}{(1 - \frac{p_1\Delta}{w})\cdots(1 - \frac{p_n\Delta}{w})} \tag{5.46}$$

Hence,

$$\begin{aligned} \lim_{\Delta\to 0} \Delta^{m-n} G_q(z) &= \left(1 - z^{-1}\right)\frac{1}{2\pi j}\int \frac{e^w}{z - e^w}\frac{K}{w^{n-m}}\frac{dw}{w} \\ &= K\frac{B_{n-m}(z)}{(n-m)!(z-1)^{n-m}} \end{aligned} \tag{5.47}$$

where the integration path is the imaginary axis with a small detour to the right around the origin, and where Lemma 5.2 on p. 48 has been used. □

The above result has been expressed in terms of the shift operator. It can also be re-expressed in the delta operator framework as follows.

Lemma 5.11 *Consider a SISO linear continuous-time system described by the transfer function* (5.43). *Given a sampling period* Δ, *the corresponding discrete-time sampled-data model in the delta domain, for a ZOH input, is given by a rational transfer function of the form*

$$G_i(\gamma) = \frac{F_i(\gamma)}{E_i(\gamma)} \tag{5.48}$$

As the sampling period Δ *goes to zero*:

$$F_i(\gamma) \longrightarrow F(\gamma)p_{n-m}(\Delta\gamma) \tag{5.49}$$

$$E_i(\gamma) = \prod_{\ell=1}^{n}\left(\gamma - \frac{e^{p_\ell\Delta} - 1}{\Delta}\right) \longrightarrow E(\gamma) \tag{5.50}$$

Remark 5.12 Comparing the results in Lemmas 5.10 and 5.11, we notice that there is an important difference between shift operator models and incremental (or δ-operator) models:

- For shift operator models, the sampling zeros converge to fixed locations in the complex plane. However, the intrinsic poles and zeros all converge to $z = 1$.
- In incremental models, the sampling zeros converge to locations which depend upon the sampling period Δ. Moreover, as $\Delta \to 0$, the sampling zeros converge to $-\infty$. The intrinsic poles and zeros all converge to their corresponding continuous-time locations.

Shift operator models are potentially problematic when using fast sampling since all intrinsic poles and zeros converge to 1 (in the shift domain) as $\Delta \to 0$, whereas for incremental (or δ-domain) models, the intrinsic poles and zeros converge to their continuous locations as $\Delta \to 0$, thus preserving information about the underlying dynamics of the process.

5.3 Summary

The key points covered in this chapter are:

- The exact sampled-data model corresponding to an rth order integrator, assuming a ZOH input:

$$G_q(z) = \frac{\Delta^r}{r!} \frac{B_r(z)}{(z-1)^r} \tag{5.51}$$

- The properties of the Euler–Frobenius polynomials in Remark 5.3; also, the first few polynomials given in (5.19)–(5.22).
- The alternative expression for the Euler–Frobenius polynomials given by

$$\sum_{k=-\infty}^{\infty} \frac{1}{(\log z + j2\pi k)^r} = \frac{z B_{r-1}(z)}{(r-1)!(z-1)^r}, \quad r \geq 2 \tag{5.52}$$

- The incremental model for the sampled rth order integrator given by

$$G_i(\gamma) = \frac{p_r(\Delta\gamma)}{\gamma^r} \tag{5.53}$$

- The first few Euler–Frobenius polynomials in incremental form given in (5.33)–(5.35).
- The recursive formula for Euler–Frobenius polynomials in incremental form:

$$p_0(\Delta\gamma) \triangleq 1 \tag{5.54}$$

$$p_r(\Delta\gamma) = \sum_{\ell=1}^{r} \frac{(\Delta\gamma)^{\ell-1}}{\ell!} p_{r-\ell}(\Delta\gamma); \quad r \geq 1 \tag{5.55}$$

- The asymptotic sampled-data models for the general linear system

$$G(s) = \frac{F(s)}{E(s)} = K \frac{(s-z_1)(s-z_2)\cdots(s-z_m)}{(s-p_1)(s-p_2)\cdots(s-p_n)} \tag{5.56}$$

  in shift form

$$G_q(z) \xrightarrow{\Delta\approx 0} K \frac{\Delta^{n-m}(z-1)^m B_{n-m}(z)}{(n-m)!(z-1)^n} \tag{5.57}$$

and in incremental form

$$G_i(\gamma) \xrightarrow{\Delta \approx 0} \frac{F(\gamma) p_{n-m}(\Delta\gamma)}{E(\gamma)} \tag{5.58}$$

- The observation that:
  - In shift operator models, the sampling zeros converge to fixed locations in the complex plane. However, the intrinsic poles and zeros all converge to $z = 1$ and thus reveal nothing about the underlying continuous-time dynamics.
  - In incremental models, the sampling zeros converge to locations which depend upon the sampling period Δ. However, intrinsic poles and zeros all converge to their corresponding continuous-time locations.

Further Reading

Asymptotic sampling zeros were first described in

Åström KJ, Hagander P, Sternby J (1984) Zeros of sampled systems. Automatica 20(1):31–38
Mårtensson B (1982) Zeros of sampled systems. Master's thesis, Department of Automatic Control, Lund University, Lund, Sweden. Report TFRT-5266

Further results were later presented in

Blachuta MJ (1999) On zeros of pulse transfer functions. IEEE Trans Autom Control 44(6):1229–1234
Hagiwara T, Yuasa T, Araki M (1993) Stability of the limiting zeros of sampled-data systems with zero- and first-order holds. Int J Control
Weller SR, Moran W, Ninness B, Pollington AD (2001) Sampling zeros and the Euler–Fröbenius polynomials. IEEE Trans Autom Control 46(2):340–343

Chapter 6
Generalised Hold Devices

In previous chapters, it has been shown that the poles of a sampled-data model depend only on the continuous-time poles and the sampling period. However, the zeros also depend on the *artefacts* of the sampling process, namely, how the continuous-time input is generated and how the output samples are obtained. For example, different models arise when using a zero order hold (ZOH) or a first order hold (FOH) to generate the continuous-time input.

In this chapter, the effect of the hold device is studied in a more general setting. In particular, hold devices are characterised by their *impulse response*. In this framework, ZOH and FOH are particular cases of what are called *generalised hold functions* (GHFs).

It is shown that, given a continuous-time system and a sampling period Δ, a GHF can be used to shift the zeros of the corresponding sampled-data model. In particular, a hold design is proposed that shifts the sampling zeros in the discrete-time model.

For the stochastic case, a similar result holds. This will be treated in Chap. 15.

The design procedures presented in this chapter are independent of the particular system, and depend only on the system relative degree. Furthermore, the results obtained are asymptotic as the sampling period goes to zero.

6.1 Generalised Hold Functions

Here the role of the hold device is examined with respect to its role in obtaining sampled-data models for deterministic systems. A more general setting is considered than that which was used in Chap. 3, where only ZOHs and FOHs were studied.

The starting point here is that a linear hold device can be completely characterised by its *impulse response*, $h_g(t)$. This function is the continuous-time signal generated by the hold device when its (discrete-time) input is a Kronecker delta

J.I. Yuz, G.C. Goodwin, *Sampled-Data Models for Linear and Nonlinear Systems*, Communications and Control Engineering, DOI 10.1007/978-1-4471-5562-1_6,

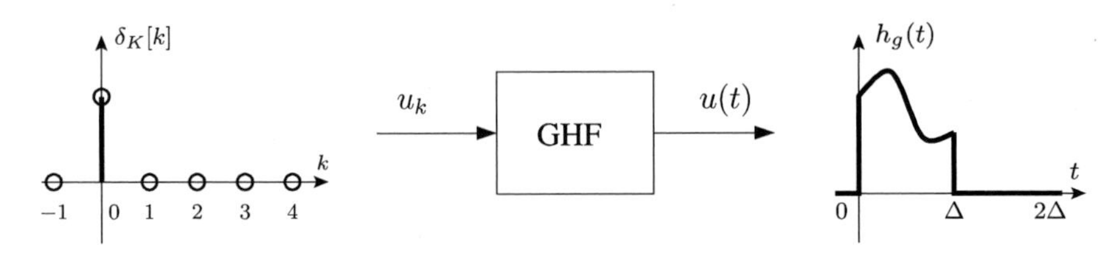

Fig. 6.1 Schematic representation of a generalised hold function (GHF)

function, i.e.,

$$u_k = \delta_K[k] = \begin{cases} 1 & k = 0 \\ 0 & k \neq 0 \end{cases} \quad \Rightarrow \quad u(t) = h_g(t) \tag{6.1}$$

Figure 6.1 schematically represents a GHF and its impulse response. ZOHs and FOHs can be understood as particular cases of GHFs. Indeed, their impulse responses were shown previously in Fig. 3.1 on p. 25.

Note that, given any input sequence u_k, the continuous-time signal generated by the hold is given by:

$$u(t) = \sum_{k=-\infty}^{\infty} h_g(t - k\Delta)u_k \tag{6.2}$$

Assumption 6.1 For simplicity, the analysis will be restricted (for the moment) to the class of GHFs whose impulse response has support on one sampling interval, i.e., $h_g(t) = 0$, for all $t \notin [0, \Delta)$.

The previous assumption excludes from the analysis some holds, for example, the FOH in Fig. 6.1(b). However, the following results show that the class of GHFs considered above provides sufficient *freedom* to arbitrarily assign the zeros of the sampled-data model.

Lemma 6.2 *Consider the continuous-time state-space model* (3.1)–(3.2). *If a GHF is used (with impulse response* $h_g(t)$*) to generate the input* $u(t)$*, then the equivalent discrete-time model is given by:*

$$x_{k+1} = A_q x_k + B_g u_k \tag{6.3}$$

$$y_k = C x_k \tag{6.4}$$

where $A_q = e^{A\Delta}$*, and*

$$B_g = \int_0^{\Delta} e^{A(\Delta-\tau)} B h_g(\tau)\, d\tau \tag{6.5}$$

Proof From Eq. (3.24), it follows that

$$x_{k+1} = e^{A\Delta} x_k + \int_{k\Delta}^{k\Delta+\Delta} e^{A(k\Delta+\Delta-\eta)} B u(\eta)\, d\eta \tag{6.6}$$

Assumption 6.1 allows one to simplify the continuous-time input (6.2). Within a single sampling period, it follows that

$$u(t) = h_g(t - k\Delta) u_k; \quad k\Delta \le t < k\Delta + \Delta \tag{6.7}$$

Thus,

$$\int_{k\Delta}^{k\Delta+\Delta} e^{A(k\Delta+\Delta-\eta)} B u(\eta)\, d\eta = \left[\int_{k\Delta}^{k\Delta+\Delta} e^{A(k\Delta+\Delta-\eta)} B h_g(\eta - k\Delta)\, d\eta \right] u_k \tag{6.8}$$

where, changing variables or simply considering $k = 0$, the last integral is shown to be equal to B_g in (6.5). □

Note that the previous result coincides with Lemma 3.4 on p. 25, when restricting to the ZOH case. The impulse response of the ZOH appears in Fig. 3.1(a) and is defined by

$$h_g(t) = h_{ZOH}(t) = \begin{cases} 1; & t \in [0, \Delta) \\ 0; & t \notin [0, \Delta) \end{cases} \tag{6.9}$$

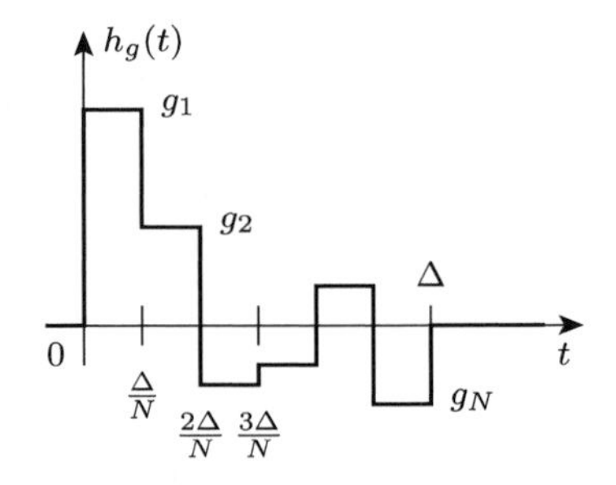

Fig. 6.2 Impulse response of a piecewise constant GHF

Corollary 6.3 *The zeros of the discrete-time system* (6.3)–(6.4) *are given by the solutions of the equation*

$$C\,\mathrm{adj}(zI_n - A_q)B_g = 0 \tag{6.10}$$

where $\mathrm{adj}(\cdot)$ *denotes the adjoint matrix.*

Proof This result is a direct consequence of expressing the state-space model (6.3)–(6.4) in transfer function form:

$$G_q(z) = C(zI_n - A_q)^{-1}B_g = \frac{C\,\mathrm{adj}(zI_n - A_q)B_g}{\det(zI_n - A_q)} \tag{6.11}$$

□

Remark 6.4 The sequel will be focused on designing a hold function such that the asymptotic zeros are arbitrarily assigned. Indeed, controllability of the pair (A, B) is enough to arbitrarily assign the roots of (6.10), i.e., to arbitrarily assign all discrete zeros.

Lemma 6.2 highlights the fact that sampled-data model characteristics depend both on the continuous-time system and the sampling process. Indeed, Eq. (6.11) shows that the poles of the system depend only on A and Δ, but the zeros are functions of B_g and, thus, of the GHF impulse response $h_g(t)$.

Note that a simpler expression can be obtained for the matrix B_g in (6.5) if a GHF defined by a piecewise constant impulse response like the type shown in Fig. 6.2 is considered:

$$h_g(t) = f_N(t) = \begin{cases} g_1; & 0 \le t < \frac{\Delta}{N} \\ g_2; & \frac{\Delta}{N} \le t < \frac{2\Delta}{N} \\ \vdots & \\ g_N; & \frac{(N-1)\Delta}{N} \le t < \Delta \end{cases} \tag{6.12}$$

Substituting (6.12) into (6.5) leads to the expression for B_g in terms of the *weights* g_ℓ, $\ell = 1, \ldots, N$. These coefficients will be used later, in Sect. 6.2, as design parameters to assign the sampling zeros. Note that the matrix B_g is linearly parameterised:

$$B_g = \sum_{\ell=1}^{N} g_\ell \int_{\frac{(\ell-1)\Delta}{N}}^{\frac{\ell\Delta}{N}} e^{A(\Delta-\tau)} B \, d\tau \tag{6.13}$$

Lemma 6.2 provides a state-space sampled-data model corresponding to a continuous-time system, also expressed in state-space form. However, the sampled-data model can also be obtained directly from the continuous-time transfer function $G(s)$. Specifically, the discrete-time transfer function can be obtained by computing the $\mathcal{Z}$-transform of the response of the combined GHF and continuous-time system, when the hold input is a Kronecker delta function. This line of reasoning leads to

$$G_q(z) = \mathcal{Z}\{\mathcal{L}^{-1}\{H_g(s)G(s)\}|_{t=k\Delta}\} \tag{6.14}$$

where $G(s)$ is the transfer function of the continuous-time system, and $H_g(s)$ is the Laplace transform of the GHF impulse response $h_g(t)$. This expression coincides with the result in Chap. 3, for the ZOH case. In fact, if $H_g(s)$ is replaced by the ZOH *transfer function* (3.41), then (6.14) can be rewritten as in expression (3.27).

6.2 Asymptotic Sampling Zeros for Generalised Holds

Here, the asymptotic sampling zeros that arise in sampled-data models are investigated when the continuous-time input is generated by a GHF. For simplicity, consider a piecewise constant GHF with impulse response of the form illustrated in Fig. 6.2.

The following preliminary result is first established.

Lemma 6.5 *Using the polynomials defined in* (5.7)–(5.8), *and defining* $B_0(z) = z^{-1}$, *it follows that*

$$\sum_{k=1}^{\infty} k^p z^{-k} = \frac{z B_p(z)}{(z-1)^{p+1}}; \quad \forall p \ge 0 \tag{6.15}$$

Proof Use induction. First note that for $p = 0$, the result is straightforward. For $p = 1$, it follows that

$$\sum_{k=1}^{\infty} k z^{-k} = \mathcal{Z}\{k\} = -z \frac{d}{dz} \mathcal{Z}\{1\} = \frac{z B_1(z)}{(z-1)^2} \tag{6.16}$$

Assuming that (6.15) holds for p, it is shown below that it also holds for $p+1$. It follows that

$$\sum_{k=1}^{\infty} k^{p+1} z^{-k} = \mathcal{Z}\{k^{p+1}\} = -z \frac{d}{dz} \mathcal{Z}\{k^p\} = \frac{z[z(1-z)B_p'(z) + (pz+1)B_p(z)]}{(z-1)^{p+2}} \tag{6.17}$$

The result then follows from the recursion (5.18) satisfied by the polynomials $B_p(z)$:

$$z(1-z)B_p'(z) + (pz+1)B_p(z) = B_{p+1}(z) \tag{6.18}$$

for all $p \geq 0$. □

Using the previous result, Lemma 5.2 can be extended to the GHF case. In particular, the sampling zeros of the sampled-data model of an r-th order integrator can be characterised when the input is generated by a piecewise constant GHF.

Lemma 6.6 *Consider the r-th order integrator $G(s) = s^{-r}$. If the continuous-time input $u(t)$ is generated by a piecewise constant GHF, defined as in* (6.12), *with r different sub-intervals, then the corresponding discrete-time transfer function is given by*

$$G_q(z) = \frac{\Delta^r}{r!(z-1)^r} \sum_{p=0}^{r-1} z B_p(z)(z-1)^{r-p-1} C_{r,p} \tag{6.19}$$

where the polynomials $B_p(z)$ are defined in (5.7)–(5.8), *and*

$$C_{r,p} = \binom{r}{p} \left(\frac{-1}{r} \right)^{r-p} \sum_{\ell=1}^{r} g_\ell \left[(\ell-1)^{r-p} - \ell^{r-p} \right] \tag{6.20}$$

Proof The sampled-data model will be obtained from Eq. (6.14). The Laplace transform of the impulse response of the GHF (6.12) is first obtained. The latter is a piecewise constant function defined in r sub-intervals. Thus,

$$f_r(t) = \sum_{\ell=1}^{r} g_\ell \left[\mu\left(t - \frac{(\ell-1)\Delta}{r}\right) - \mu\left(t - \frac{\ell\Delta}{r}\right) \right] \tag{6.21}$$

$$F_r(s) = \sum_{\ell=1}^{r} g_\ell F_{r\ell}(s) \tag{6.22}$$

where $\mu(\cdot)$ is the unitary step function, and

$$F_{r\ell}(s) = \frac{1}{s}\left(e^{-s\frac{(\ell-1)\Delta}{r}} - e^{-s\frac{\ell\Delta}{r}}\right); \quad \ell = 1, \ldots, r \tag{6.23}$$

Next consider the impulse response of the combined continuous-time model:

$$G(s)F_r(s) = \sum_{\ell=1}^{r} g_\ell H_\ell(s) \tag{6.24}$$

The inverse Laplace transform of each element in the sum can be readily computed as:

$$H_\ell(s) = G(s)F_{r\ell}(s) = \frac{e^{-s\frac{(\ell-1)\Delta}{r}} - e^{-s\frac{\ell\Delta}{r}}}{s^{r+1}} \tag{6.25}$$

$$h_\ell(t) = \frac{(t - \frac{\ell-1}{r}\Delta)^r}{r!}\mu\left(t - \frac{(\ell-1)\Delta}{r}\right) - \frac{(t - \frac{\ell}{r}\Delta)^r}{r!}\mu\left(t - \frac{\ell\Delta}{r}\right) \tag{6.26}$$

Consider this signal at the sampling instants $h_\ell[k] = h_\ell(k\Delta)$. Note that $h_\ell[0] = 0$ and, for $k \geq 1$, the *binomial theorem* can then be utilised to obtain

$$h_\ell[k] = \frac{\Delta^r}{r!}\sum_{p=0}^{r-1} k^p \binom{r}{p}\left(\frac{-1}{r}\right)^{r-p}\left[(\ell-1)^{r-p} - \ell^{r-p}\right] \tag{6.27}$$

The $\mathcal{Z}$-transform of this signal is then given by

$$H_\ell(z) = \frac{\Delta^r}{r!}\sum_{p=0}^{r-1}\left[\binom{r}{p}\left(\frac{-1}{r}\right)^{r-p}\left[(\ell-1)^{r-p} - \ell^{r-p}\right]\sum_{k=1}^{\infty} k^p z^{-k}\right] \tag{6.28}$$

Hence, applying the result in Lemma 6.5,

$$H_\ell(z) = \frac{\Delta^r}{r!}\sum_{p=0}^{r-1}\left[\binom{r}{p}\left(\frac{-1}{r}\right)^{r-p}\left[(\ell-1)^{r-p} - \ell^{r-p}\right]\frac{z B_p(z)}{(z-1)^{p+1}}\right] \tag{6.29}$$

Finally, the result is obtained by substituting (6.29) into the linear combination obtained from (6.24):

$$G_q(z) = \mathcal{Z}\left\{\mathcal{L}^{-1}\left\{G(s)F_r(s)\right\}_{t=k\Delta}\right\} = \mathcal{Z}\left\{\sum_{\ell=1}^{r} g_\ell h_l[k]\right\} = \sum_{\ell=1}^{r} g_\ell H_\ell(z) \tag{6.30}$$

□

Remark 6.7 Note that (6.19) establishes the fact that the sampled-data model of an r-th order integrator has all r poles at $z = e^{0 \cdot \Delta} = 1$ (as expected). The discrete-time model also has $r - 1$ *sampling zeros*. In fact, the numerator of the corresponding sampled-data model can be rewritten as a polynomial of order $r - 1$, i.e.,

$$G_q(z) = \frac{\Delta^r}{r!(z-1)^r} \sum_{p=0}^{r-1} \alpha_p z^p \tag{6.31}$$

Next consider a more general system. Theorem 5.10 on p. 55 is extended to the case when a piecewise constant GHF is used to generate the input.

Theorem 6.8 *Let $G(s)$ be a rational function as in* (5.43), *with relative degree $r = n - m$. Let $G_q(z)$ be the corresponding sampled transfer function obtained using a piecewise constant GHF with r stages.*

Assume $m < n$ (or, equivalently, $r > 0$). Then, as the sampling period Δ goes to 0, m zeros of $G_q(z)$ go to 1 as $e^{z_i \Delta}$, and the remaining $r - 1$ zeros of $G_q(z)$ (the sampling zeros*) go to the roots of the polynomial:*

$$\sum_{p=0}^{r-1} z B_p(z)(z-1)^{r-p-1} C_{r,p} = \sum_{p=0}^{r-1} \alpha_p z^p \tag{6.32}$$

Proof The proof of this result is similar to the proof of Theorem 5.10. First obtain the Laplace transform of the piecewise constant GHF with r sub-intervals. Proceeding as in the proof of Lemma 6.6,

$$H_g(s) = \frac{1}{s} \sum_{\ell=1}^{r} g_\ell \left(e^{-s\frac{(\ell-1)\Delta}{r}} - e^{-s\frac{\ell\Delta}{r}} \right) \tag{6.33}$$

Using the definition of the Laplace and $\mathcal{Z}$-transforms (and their inverse transforms, respectively), Eq. (6.14) can be rewritten as:

$$\begin{aligned} G_q(z) &= \sum_{k=0}^{\infty} \frac{1}{2\pi j} \int_{\gamma - j\infty}^{\gamma + j\infty} G(s) H_g(s) e^{sk\Delta} \, ds \, z^{-k} \\ &= \frac{1}{2\pi j} \int_{\gamma - j\infty}^{\gamma + j\infty} G(s) H_g(s) \left(\sum_{k=1}^{\infty} e^{sk\Delta} z^{-k} \right) ds \\ &= \frac{1}{2\pi j} \int_{\gamma - j\infty}^{\gamma + j\infty} G(s) H_g(s) \frac{e^{s\Delta}}{z - e^{s\Delta}} \, ds \end{aligned} \tag{6.34}$$

where γ is such that $G(s)/s$ has all its poles to the left of $\Re\{s\} = \gamma$. Substituting the system transfer function (5.43) and the GHF (6.33) into the expression leads to the following by changing variables in the integral, using $w = s\Delta$:

$$\lim_{\Delta\to 0} \Delta^{-r} G_q(z) = \frac{K}{2\pi j} \int_{\gamma\Delta-j\infty}^{\gamma\Delta+j\infty} \frac{e^w \sum_{\ell=1}^{r} g_\ell (e^{-\frac{(\ell-1)w}{r}} - e^{-\frac{\ell w}{r}})}{w^{r+1}(z - e^w)} \, dw \tag{6.35}$$

It is readily shown that this expression corresponds to replacing $G(s)$ by an r-th order integrator in (6.14). Thus,

$$\lim_{\Delta\to 0} \frac{1}{\Delta^r} G_q(z) = \frac{K \sum_{p=0}^{r-1} z B_p(z)(z-1)^{r-p-1} C_{r,p}}{r!(z-1)^r} = \frac{K(z-1)^m}{r!(z-1)^n} \sum_{p=0}^{r-1} \alpha_p z^p \tag{6.36}$$

□

Based on the previous results, a procedure is next presented to design a GHF such that the *sampling zeros* of the discrete-time model are asymptotically assigned to the origin, as the sampling period Δ goes to 0.

Theorem 6.9 *The coefficients g_ℓ, $\ell = 1, \ldots, r$ of the GHF in* (6.33) *can be chosen in such a way that the sampling zeros of the discrete-time model* (6.32) *converge asymptotically to $z = 0$.*

Proof To assign the sampling zeros to the origin, it follows from (6.32) that the following condition must hold:

$$\alpha_p = 0, \quad \forall p = 0, \ldots, n-2 \tag{6.37}$$

This is equivalent to having $r - 1$ linear equations in the coefficients $C_{r,p}$, and thus for the weights g_ℓ. Moreover, the GHF must satisfy an extra condition to ensure unitary gain at zero frequency, i.e.,

$$\frac{1}{r} \sum_{\ell=1}^{r} g_\ell = 1 \tag{6.38}$$

Equations (6.37) and (6.38) define r conditions on the coefficients g_ℓ, $\ell = 1, \ldots, r$, which are (generically) linearly independent provided (A, B) is controllable (see Remark 6.4). □

A key observation is that the GHF obtained by solving (6.37)–(6.38) does not depend on the particular continuous-time system. Theorem 6.8 ensures that the sampling zeros, and, thus, the GHF design procedure, depend only on the system relative degree.

Remark 6.10 In Theorem 6.9, the asymptotic sampling zeros have been designed to the origin. This implies that, by a continuity argument, there exists a sampling period $\Delta_\varepsilon > 0$ such that, for every sampling period $\Delta < \Delta_\varepsilon$, all the sampling zeros are stable, i.e., they lie *inside* the unit circle in the complex plane z. Indeed, for Δ_ε small enough, all the sampling zeros will be inside a circle of radius $r_\varepsilon \ll 1$.

Theorem 6.9 assigns the asymptotic sampling zeros to the origin and hence ensures that the sampled-data model is minimum phase. However, a different set of conditions can be imposed on the weighting coefficients if one wants to assign the sampling zeros to any other location in the complex plane.

The following example illustrates the GHF design procedure described above for a particular system.

Example 6.11 Consider the third order system

$$G(s) = \frac{1}{(s+1)^3} \tag{6.39}$$

By Theorem 5.10, if a ZOH is used to generate the input, then, as the sampling period Δ tends to zero, the associated sampled-data transfer function is given by

$$G_q(z) \xrightarrow[\text{(ZOH)}]{\Delta\approx 0} \frac{\Delta^3(z+3.732)(z+0.268)}{(z-1)^3} \tag{6.40}$$

Note that the resulting discrete-time model has a non-minimum phase (NMP) sampling zero, even though the continuous-time system has no finite zeros.

On the other hand, using (6.32), (6.37), and (6.38), a GHF is obtained:

$$h_g(t) = \begin{cases} 29/2; & 0 \le t < \frac{\Delta}{3} \\ -17; & \frac{\Delta}{3} \le t < \frac{2\Delta}{3} \\ 11/2; & \frac{2\Delta}{3} \le t < \Delta \end{cases} \tag{6.41}$$

Note that this assigns the limiting sampling zeros asymptotically to the origin; i.e., the combined hold and plant discrete-time model is, as Δ goes to 0,

$$G_q(z) \xrightarrow[\text{(GHF)}]{\Delta\approx 0} \frac{\Delta^3 z^2}{(z-1)^3} \tag{6.42}$$

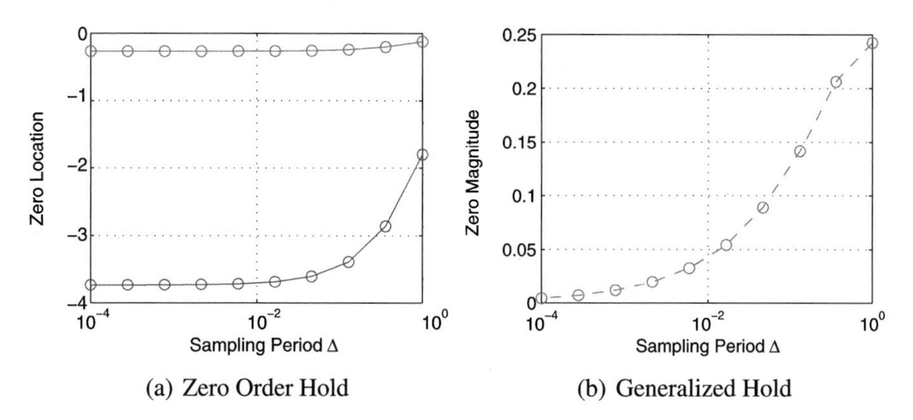

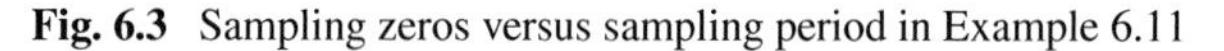
Fig. 6.3 Sampling zeros versus sampling period in Example 6.11

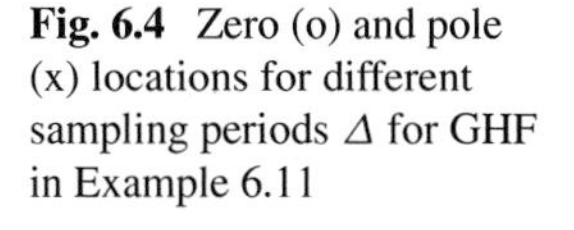
Fig. 6.4 Zero (o) and pole (x) locations for different sampling periods Δ for GHF in Example 6.11

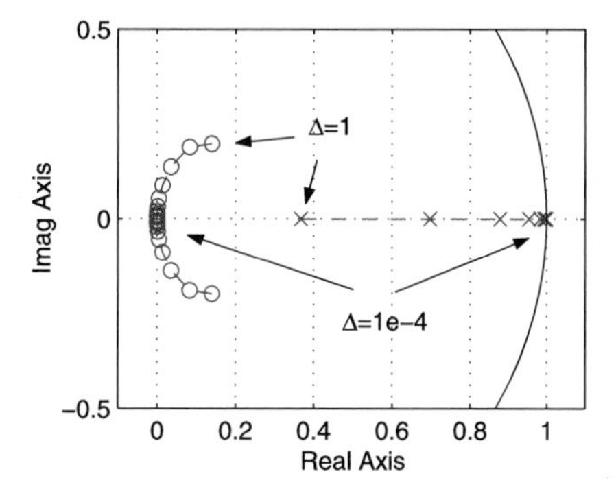

Figure 6.3(a) shows the exact sampling zeros as a function of Δ for the ZOH case. Note that the sampling zeros converge to their asymptotic locations as $\Delta \to 0$. Indeed, the sampling zeros are very close to the asymptotic location, provided the sampling frequency is approximately one decade above the fastest system pole. Figure 6.3(b) shows the magnitude of the (complex) sampling zeros obtained when the GHF (6.41) is used. It can be seen that the sampling zeros converge to the origin as intended by the design procedure.

Furthermore, Fig. 6.4 shows the zero and pole locations for the sampled version of the system using the fixed GHF (6.41), for sampling periods from 1 to 10^{-4}. Note that all of the resulting discrete-time models are minimum phase.

6.3 Summary

The key points covered in this chapter are:

- The characterisation of generalised hold functions by their impulse response

$$u(t) = \sum_{k=-\infty}^{\infty} h_g(t - k\Delta)u_k \tag{6.43}$$

- The exact sampled-data model in state-space form given in Lemma 6.2, i.e.,

$$x_{k+1} = A_q x_k + B_g u_k \tag{6.44}$$

$$y_k = C x_k \tag{6.45}$$

where $A_q = e^{A\Delta}$, and

$$B_g = \int_0^{\Delta} e^{A(\Delta-\tau)} B h_g(\tau)\, d\tau \tag{6.46}$$

- The observation that sampled-data models depend, not only on the underlying continuous-time system and sampling period, but also on the hold device used to generate the continuous-time input.
- The generalised hold having a piecewise constant impulse response (Fig. 6.2 on p. 62):

$$h_g(t) = f_N(t) = \begin{cases} g_1; & 0 \le t < \frac{\Delta}{N} \\ g_2; & \frac{\Delta}{N} \le t < \frac{2\Delta}{N} \\ \vdots & \\ g_N; & \frac{(N-1)\Delta}{N} \le t < \Delta \end{cases} \tag{6.47}$$

- The characterisation of the sampling zeros of an r-th order integrator in Lemma 6.6.
- The characterisation of the asymptotic sampling zeros of a general linear system of relative degree $r = n - m$ in Theorem 6.8.
- The equation to assign the asymptotic sampling zeros in Theorem 6.9.

Further Reading

Further results on using generalised hold functions can be found in

Arriagada I, Yuz JI (2008) On the relationship between splines, sampling zeros and numerical integration in sampled-data models for linear systems. In: American control conference, ACC 2008, Seattle, WA, USA

Feuer A, Goodwin GC (1994) Generalized sample hold functions: frequency domain analysis of robustness, sensitivity, and intersample difficulties. IEEE Trans Autom Control 39(5):1042–1047

Kabamba P (1987) Control of linear systems using generalized sampled-data hold functions. IEEE Trans Autom Control 32(9):772–783

Yuz JI, Goodwin GC, Garnier H (2004) Generalised hold functions for fast sampling rates. In: 43rd IEEE conference on decision and control, Nassau, Bahamas

Zhang J, Zhang C (1994) Robustness analysis of control systems using generalized sample hold functions. In: 33th IEEE conference on decision and control, Lake Buena, Vista, FL, USA

Chapter 7
Robustness

The reader will have noticed that all of the sampled-data models described previously capture the effect of folding of high frequency aspects of the continuous-time model back to lower frequencies. Thus, the models depend upon hypotheses regarding the nature of the high frequency behaviour of the system. For example, the formulae for asymptotic sampling zeros follow by applying the assumption that, when the sampling frequency is sufficiently high, the model behaves *above the Nyquist frequency* as $1/s^r$ (where r is the relative degree). Clearly, this begs the question about the impact of unmodelled high frequency poles or zeros. If these are present, they will clearly affect the validity of discrete-time models based on the (possibly false) assumption that the continuous-time model is behaving as $1/s^r$.

Thus, it is necessary to be careful about the (frequency) range of validity of models. In particular, sampling zeros correspond to very precise assumptions about how the system behaves in the region directly above the Nyquist frequency. This chapter explores these robustness issues.

7.1 Robustness of Asymptotic Sampled-Data Models

Here the robustness of sampled-data models with zero order holds (ZOHs) is examined with respect to the presence of high frequency undermodelling. A simple example is used to illustrate the core ideas.

Example 7.1 Consider a system which has two dominant poles. This model is viewed here as the nominal model of the system. The effect of an unmodelled fast pole is next considered. Thus, let the true system be given by

$$G(s) = \frac{\beta_o}{(s^2 + \alpha_1 s + \alpha_o)(\frac{s}{\omega_u} + 1)} = \frac{G_o(s)}{(\frac{s}{\omega_u} + 1)} \tag{7.1}$$

where the term $(\frac{s}{\omega_u} + 1)$ denotes an 'unmodelled' pole.

J.I. Yuz, G.C. Goodwin, *Sampled-Data Models for Linear and Nonlinear Systems*, Communications and Control Engineering, DOI 10.1007/978-1-4471-5562-1_7,

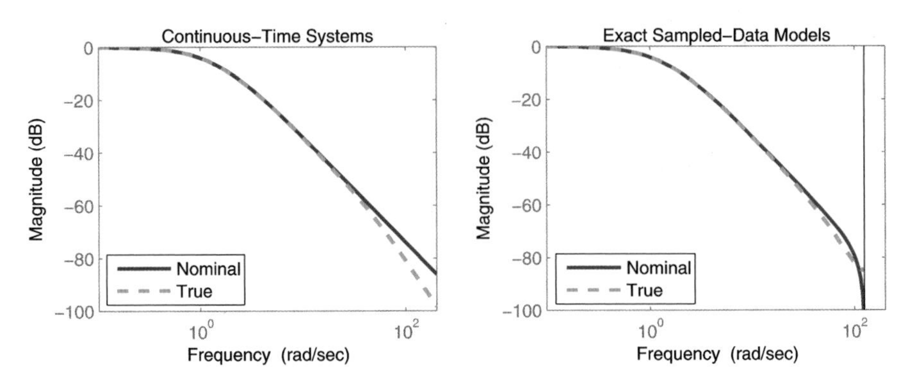

Fig. 7.1 Frequency response for nominal and true models

Figure 7.1 compares the frequency response of nominal and true models, both for the continuous-time system and the corresponding sampled-data models. The nominal poles of the system are at $s = -1$ and $s = -2$, the sampling frequency is $\omega_s = 250$ [rad/s], and the unmodelled pole is at $s = -50$.

Note that the true system has relative degree 3, and thus, the corresponding true discrete-time model will have two sampling zeros. As a consequence, the *true* model will yield different asymptotic sampling zeros as Δ goes to zero. Thus, the nominal model satisfies

$$G_{o,\delta}(\gamma) \to \frac{b_o(1 + \frac{\Delta}{2}\gamma)}{\gamma^2 + a_1\gamma + a_o} \tag{7.2}$$

whereas the true model satisfies

$$G_{\delta}(\gamma) \to \frac{b_o(1 + \Delta\gamma + \frac{\Delta^2}{6}\gamma^2)}{(\gamma^2 + a_1\gamma + a_o)(\frac{\gamma}{\omega_u} + 1)} \tag{7.3}$$

The differences between (7.2) and (7.3) are reflected in the frequency responses shown in Fig. 7.1.

The previous example illustrates the fact that asymptotic sampling zeros depend upon the exact nature of the model above the Nyquist frequency. In the example, the sampling frequency was chosen well above the nominal poles of the system—in fact, two decades above. In theory, this should allow one to use the asymptotic characterisation of the sampled-data model. However, it can be seen that, if there are unmodelled dynamics (in this case, one decade above the nominal fastest pole), then there will also be undermodelling in the sampled-data description. Moreover, even though the sampling zeros go to infinity for the nominal and true models, their precise characterisation depends significantly on high frequency aspects of the model, as shown in (7.2) and (7.3).

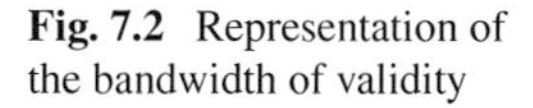

Fig. 7.2 Representation of the bandwidth of validity

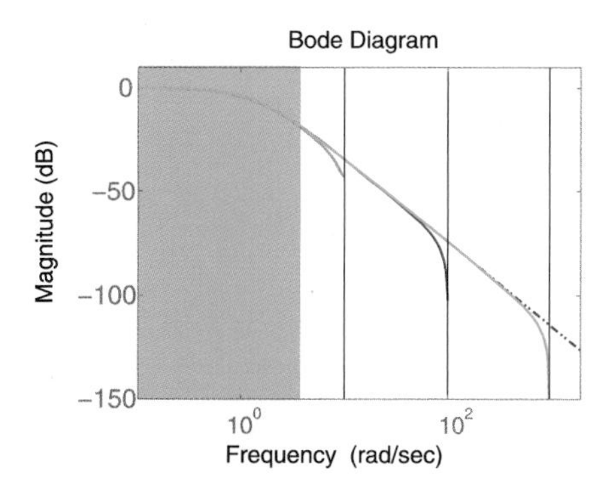

Remark 7.2 The above discussion highlights the issues that must be taken into account when using sampled-data models. Specifically:

- Any method, including asymptotic sampling zeros, that relies on high frequency system characteristics depends upon high frequency model behaviour. Hence,
- Models should be considered within a *bandwidth of validity*, to avoid the folded effect of high frequency modelling errors. The bandwidth of validity is illustrated by the shaded area in Fig. 7.2.

7.2 Robustness of Generalised Hold Designs

The robustness issues discussed above are next highlighted in the context of generalised hold design as described in Sect. 6.2. The ideas are again illustrated via an example.

Example 7.3 (Deterministic Systems with GHFs) Consider again the system in Example 6.11 on p. 68. Assume that the system includes an *unmodelled* fast pole, i.e.,

$$G(s) = \frac{1}{(s+1)^3(0.01s+1)} \tag{7.4}$$

For the ZOH case, Theorem 5.10 predicts that the asymptotic sampling zeros will be $\{-3.732, -0.268\}$, based on a *nominal model* of relative degree 3, and $\{-9.899, -1, -0.101\}$ for the *true model* of relative degree 4. Indeed, it is shown in Fig. 7.3(a) that, as Δ decreases, the sampling zeros first approach those corresponding to the *nominal* model (of relative degree 3), but then move to those corresponding to the *true* model (of relative degree 4). For this case, it can be seen that the nominal discrete-time model (6.40) is basically reached for a sampling period $\Delta \approx 0.2$ but is not valid for $\Delta < 0.1$.

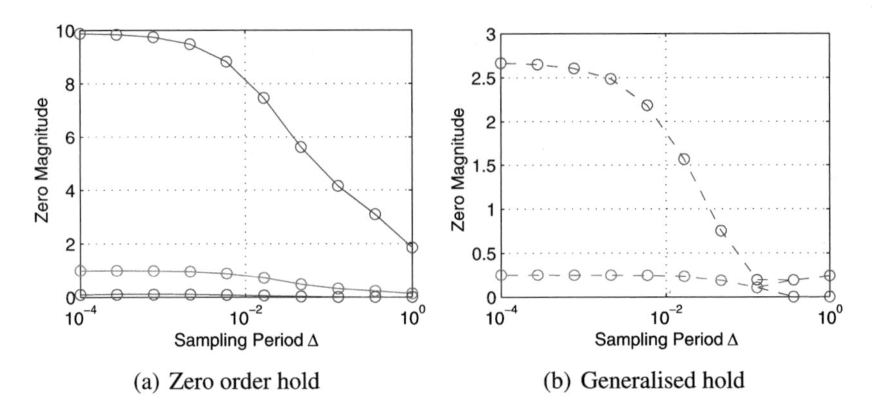

(a) Zero order hold (b) Generalised hold

Fig. 7.3 Magnitudes of the sampling zeros in Example 7.3: fast pole at $s = -10^2$

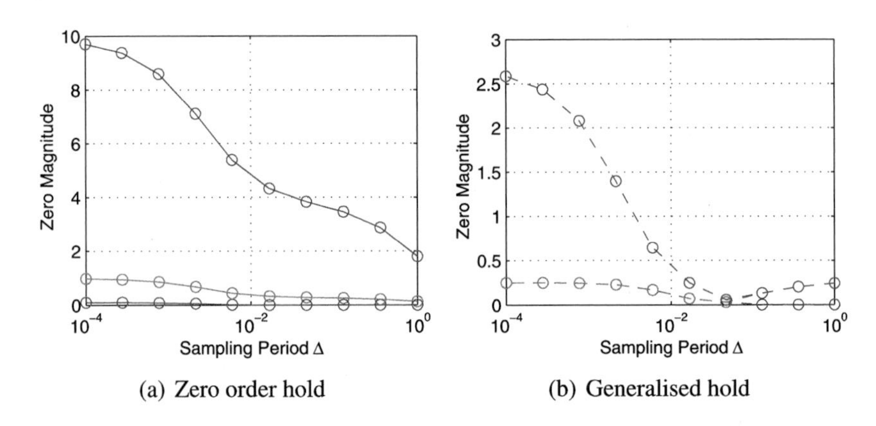

(a) Zero order hold (b) Generalised hold

Fig. 7.4 Magnitudes of the sampling zeros in Example 7.3: fast pole at $s = -10^3$

Similarly, it is shown in Fig. 7.3(b) that the zeros obtained with the fixed GHF (6.41) are *close* to the origin for $\Delta > 0.1$. However, when the sampling period is reduced further, the unmodelled pole at $s = -10^2$ in (7.4) begins to impact the result, and the zeros clearly depart from the prescribed nominal values.

The plots in Fig. 7.4 correspond to the plots in Fig. 7.3, save that now the unmodelled pole appears further from the origin, namely at $s = -10^3$. Comparing the plots in Fig. 7.4 with those in Fig. 7.3, it can be seen that the analysis and design procedures, based on the nominal model, now hold for smaller values of the sampling period δ, i.e., for faster sampling rates. This is to be expected since the unmodelled pole now impacts the system at higher frequencies. The plots in Fig. 7.5 correspond to those in Figs. 7.3 and 7.4 save that now the unmodelled pole appears much closer to the origin, namely at $s = -10$. Comparing the plots in Fig. 7.5 with those in Figs. 7.3 and 7.4, it can be seen that the analysis and design procedures, based on the nominal model, now only hold for relative large values of the sampling period δ (of the order of 1). Again this is to be expected since the unmodelled pole now impacts the system at relatively low frequencies.

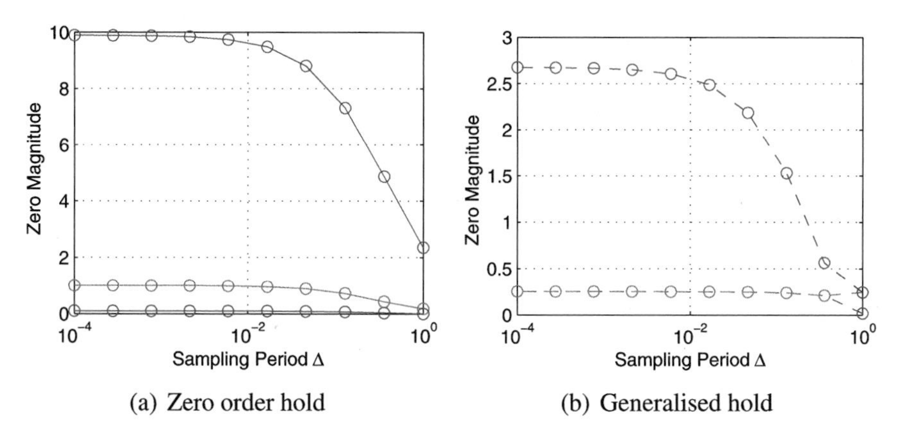

Fig. 7.5 Magnitudes of the sampling zeros in Example 7.3: fast pole at $s = -10$

7.3 Summary

The key points covered in this chapter are the following:

- Accurate sampled-data models capture the folding effect of high frequency characteristics reflected to the band $[0, \frac{\pi}{\Delta}]$.
- As a consequence, any control or identification method, or, indeed, any algorithm, that relies on high frequency system characteristics is sensitive to high frequency behaviour.
- Models, both continuous and sampled-data, should be considered within a *bandwidth of validity*, to avoid high frequency modelling errors having a negative impact on design procedures.

Further Reading

Further background on robustness of sampled-data models can be found in

Feuer A, Goodwin GC (1994) Generalized sample hold functions: frequency domain analysis of robustness, sensitivity, and intersample difficulties. IEEE Trans Autom Control 39(5):1042–1047

Goodwin GC, Yuz JI, Garnier H (2005) Robustness issues in continuous-time system identification from sampled data. In: Proceedings of 16th IFAC world congress, Prague, Czech Republic

Goodwin GC, Agüero JC, Welsh JS, Yuz JI, Adams GJ, Rojas CR (2008) Robust identification of process models from plant data. J Process Control 18(9):810–820

Yuz JI, Goodwin GC (2008) Robust identification of continuous-time systems from sampled data. In: Garnier H, Wang L (eds) Continuous-time model identification from sampled data. Springer, Berlin

Zhang J, Zhang C (1994) Robustness analysis of control systems using generalized sample hold functions. In: 33th IEEE conference on decision and control, Lake Buena Vista, FL, USA

Chapter 8
Approximate Models for Linear Deterministic Systems

This chapter develops various approximate discrete-time models for general *linear* deterministic systems. Approximate models are not necessary for linear systems, because it is always possible to obtain an *exact* description. However, approximate models are treated here for three principal reasons:

(i) They give further insight into the structure of discrete-time models. For example, they more clearly capture the relationship between discrete-time and continuous-time parameters.
(ii) Some of these models are easier to obtain than the corresponding exact sampled-data models, which involve, amongst other things, the computation of matrix exponentials (see Chap. 3).
(iii) Some of the methods that are used here to construct approximate discrete-time models will be directly applicable to *nonlinear systems*, as shown in Chap. 9.

8.1 Approximate Models in the Frequency Domain

One initial question that may occur to the reader is: Why bother with discrete-time models at all? Indeed, it has been argued that sampled-data models converge (in a sense that has been made precise) to the underlying continuous-time system. Hence, why not simply sample quickly and use the continuous-time model? The following example illustrates the fact that, no matter how fast one samples, there is always a difference between the continuous-time frequency response and the discrete-time frequency response.

Example 8.1 Consider a second order deterministic system, described by

$$\frac{d^2}{dt^2}y(t) + \alpha_1 \frac{d}{dt}y(t) + \alpha_o y(t) = \beta_o u(t) \tag{8.1}$$

J.I. Yuz, G.C. Goodwin, *Sampled-Data Models for Linear and Nonlinear Systems*, Communications and Control Engineering, DOI 10.1007/978-1-4471-5562-1_8,

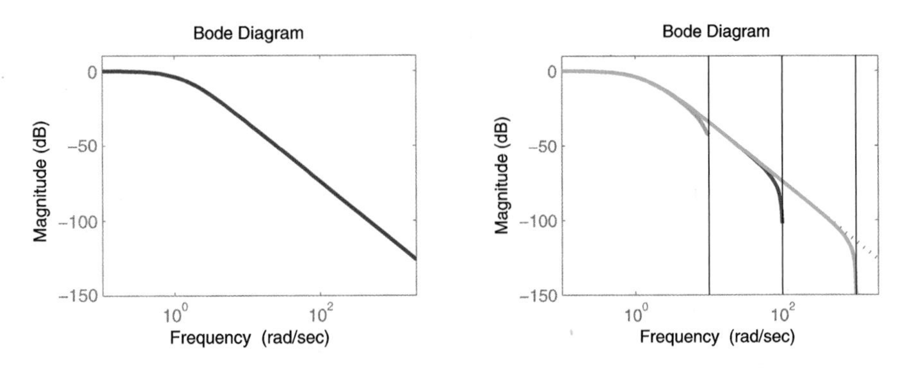

Fig. 8.1 Continuous-time (*left plot*) and discrete-time (*right plot*) frequency response magnitudes

The exact discrete-time model based on the use of a zero order hold (ZOH) has a sampling zero. Indeed the discrete-time model has the following form:

$$\delta^2 y_k + a_1 \delta y_k + a_o y_k = b_0 u_k + b_1 \delta u_k \tag{8.2}$$

Moreover, as the sampling period Δ goes to zero, the continuous-time coefficients are recovered, i.e., $a_1 \to \alpha_1, a_0 \to \alpha_0, b_0 \to \beta_0$. Also, the sampling zero can be readily characterised (see Theorem 5.11 on p. 56) leading to an asymptotic model of the form

$$\delta^2 y_k + \alpha_1 \delta y_k + \alpha_o y_k = \beta_0 \left(1 + \frac{\Delta}{2}\delta\right) u_k \tag{8.3}$$

Figure 8.1 shows a comparison of the Bode magnitude diagrams corresponding to a continuous-time system as in (8.4) (on the left-hand side) and the exact sampled-data model (8.2), obtained for different sampling frequencies (on the right):

$$G(s) = \frac{1}{(s+1)^2} \tag{8.4}$$

The figure clearly illustrates the fact that, *no matter how high the sampling rate*, there is always a difference (near the folding frequency) between the continuous-time model and the discretised models.

8.1.1 Error Quantification in the Frequency Domain

The reader may have noticed that the plots given in Fig. 8.1 use a logarithmic scale for the magnitude of the frequency response. This was not an arbitrary choice! A logarithmic (i.e., dB) scale emphasises relative errors. Had a linear scale been used in Fig. 8.1, the difference between the responses would have been too small to show up. Thus, in the sequel, relative errors will be emphasised. The focus on relative errors is reinforced by the observation that an absolute error of 0.1 in a small quantity (say 0.01) would be catastrophically large, whereas the same absolute error of 0.1 in a large quantity (say 100) would be irrelevant.

Different approximate sampled-data models $\hat{G}_q^i(z)$ will be compared, where the superscript i refers to the particular type of approximate sampled-data model considered. Comparisons are also made to the exact sampled-data (ESD) model obtained in Sect. 19.5.2, i.e.,

$$G_q(z) = G_q^{\mathrm{ESD}}(z) = \mathcal{Z}\left\{\frac{1-e^{s\Delta}}{s}G(s)\right\} \tag{8.5}$$

Relative errors are defined in two ways, depending upon the normalisation factor used.

Definition 8.2 The relative error of type 1 between a transfer function $G_q(z)$ and an approximate transfer function $\hat{G}_q(z)$ is defined as

$$RE_1 = \sup_{\omega\in[0,\frac{\pi}{\Delta}]} R_1(\omega) \tag{8.6}$$

where

$$R_1(\omega) = \left|\frac{G_q(e^{j\omega\Delta}) - \hat{G}_q(e^{j\omega\Delta})}{G_q(e^{j\omega\Delta})}\right|; \quad 0 \le \omega \le \frac{\pi}{\Delta} \tag{8.7}$$

Definition 8.3 The relative error of type 2 between a transfer function $G_q(z)$ and an approximate transfer function $\hat{G}_q(z)$ is defined as

$$RE_2 = \sup_{\omega\in[0,\frac{\pi}{\Delta}]} R_2(\omega) \tag{8.8}$$

where

$$R_2(\omega) = \left|\frac{\hat{G}_q(e^{j\omega\Delta}) - G_q(e^{j\omega\Delta})}{\hat{G}_q(e^{j\omega\Delta})}\right|; \quad 0 \le \omega \le \frac{\pi}{\Delta} \tag{8.9}$$

The error function $R_2(\omega)$ is closely related to those in control system analysis, where relative errors of this type appear when considering the robustness of a design to modelling errors.

In the next sections RE_1 and RE_2 will be characterised in terms of the order of convergence as the sampling period Δ goes to 0. In order to precisely measure the accuracy, the *big-$\mathcal{O}$* notation for asymptotic analysis will be utilised. This notation is defined below.

Definition 8.4 A function $f(\Delta)$ is said to be of the order of $g(\Delta)$ if and only if

$$|f(\Delta)| < C|g(\Delta)| \tag{8.10}$$

for all $\Delta < \Delta_o$, for some $\Delta_o > 0$, and where $C > 0$ is a constant. This is denoted $f(\Delta) \in \mathcal{O}(g(\Delta))$ or, with a slight abuse of notation, simply as $f(\Delta) = \mathcal{O}(g(\Delta))$.

In particular, consider $g(\Delta) = \Delta^\ell$, where ℓ is an integer. Comparisons will be made between $f(\Delta)$ and powers of the sampling period Δ.

The first approximate model to be considered will be based on Euler integration.

8.1.2 Approximate Models Based on Euler Integration

Consider a linear continuous-time deterministic system

$$G(s) = \frac{K\prod_{i=1}^{m}(s - c_i)}{\prod_{i=1}^{n}(s - p_i)} \tag{8.11}$$

The system can equivalently be described in state-space form (see (3.1)–(3.2) on p. 21). Then, an easy way to construct an approximate sampled-data model is to use a first order Taylor series expansion for the state (equivalent to Euler integration), i.e.,

$$\hat{x}(k\Delta + \tau) = \hat{x}(k\Delta) + \tau \left.\frac{dx}{dt}\right|_{x=\hat{x}(k\Delta)} \tag{8.12}$$

Using (3.1), the state derivative can be replaced as above to obtain an approximate description of the state evolution at some instant $t = k\Delta + \tau$, where $\tau > 0$:

$$\hat{x}(k\Delta + \tau) = \hat{x}(k\Delta) + \tau\left[A\hat{x}(k\Delta) + Bu_k\right] \tag{8.13}$$

From (8.13), an approximate model can be obtained by evaluating the expression at $\tau = \Delta$, i.e.,

$$\hat{x}_{k+1} = (I + A\Delta)\hat{x}_k + B\Delta u_k \tag{8.14}$$

$$\hat{z}_k = C\hat{x}_k \tag{8.15}$$

where the discrete-time sub-index has been used as a compact notation to indicate the sampling instant:

$$\hat{x}_k = \hat{x}(k\Delta) \tag{8.16}$$

In incremental form, the above model can equivalently be expressed as:

$$d\hat{x}_k^+ = A\hat{x}_k\Delta + Bu_k\Delta \tag{8.17}$$

$$\hat{z}_k = C\hat{x}_k \tag{8.18}$$

The model (8.17) is particularly simple. It is sometimes called a *simple derivative replacement* (SDR) model, because the associated discrete-time transfer function from the input u to the output $\hat{z}$ can be obtained by replacing the complex Laplace variable s (in the continuous-time transfer function (3.6)) by $\frac{z-1}{\Delta}$, i.e.,

$$\begin{aligned}\hat{G}_q(z) = \hat{G}_q^{\mathrm{SDR}}(z) &= C\big(zI - (I + \Delta A)\big)^{-1} B\Delta \\ &= C\left(\frac{z-1}{\Delta}I - A\right)^{-1} B \\ &= G(s)\big|_{s=\frac{z-1}{\Delta}}\end{aligned} \tag{8.19}$$

Notice that, in the time domain, this corresponds to replacing the derivative operator by the forward Euler approximation of the time derivative, i.e., $\frac{d}{dt} \approx \frac{q-1}{\Delta}$.

The equivalent model in incremental form is similarly obtained by replacing the complex Laplace variable s by γ (the complex variable corresponding to the δ-operator), i.e.,

$$\hat{G}_\delta^{\mathrm{SDR}}(\gamma) = G(\gamma) = C(\gamma I - A)^{-1} B = \frac{K\prod_{i=1}^m(\gamma - c_i)}{\prod_{i=1}^n(\gamma - p_i)} \tag{8.20}$$

Note that this model does not include any sampling zeros. Also note that this model is *not* equivalent to the exact sampled-data model expressed in delta form.

The confusion arises because, in delta form, Euler integration leads to a discrete-time model which closely resembles the continuous-time model. However, it is important to note that any model (including some which do not resemble the continuous-time model) can be expressed in delta form.

Since a first order Taylor series has been used in (8.12), it follows that this model has errors of the order of Δ^2 in all states and output. This fact is established below.

Lemma 8.5 *Say that the error in all states associated with* (8.14) *is of the order of* Δ^2 *at time* $k\Delta$. *Then the error at time* $(k+1)\Delta$ *is also of the order of* Δ^2.

Proof The proof follows by using a higher order Taylor expansion of the state

$$x(k\Delta + \tau) = x(k\Delta) + \tau \frac{dx}{dt}\bigg|_{t=k\Delta} + \frac{\tau^2}{2}\frac{d^2x}{dt^2}\bigg|_{t=k\Delta} + \cdots \tag{8.21}$$

Using (8.13) and substituting $\tau = \Delta$, it follows that

$$x(k\Delta + \Delta) = x(k\Delta) + \Delta\big[Ax(k\Delta) + Bu_k\big] + \mathcal{O}\big(\Delta^2\big) \tag{8.22}$$

From (8.14),

$$\begin{aligned} x(k\Delta + \Delta) - \hat{x}_{k+1} &= (I + \Delta A)x(k\Delta) + \Delta Bu_k \\ &\quad - \big[(I + \Delta A)\hat{x}_k + \Delta Bu_k\big] + \mathcal{O}\big(\Delta^2\big) \\ &= (I + \Delta A)\big[x(k\Delta) - \hat{x}_k\big] + \mathcal{O}\big(\Delta^2\big) \end{aligned} \tag{8.23}$$

Thus, if $\|x(k\Delta) - \hat{x}_k\|$ is of the order of Δ^2, then $\|x(k\Delta + \Delta) - \hat{x}_{k+1}\|$ is also of order Δ^2. □

Lemma 8.5 shows that the error associated with an Euler model is of the order of Δ^2. This may appear to be perfectly adequate for all practical purposes because, to obtain an accurate model, it seems that is only necessary to choose the sampling period Δ small enough. However, it will be shown in the sequel that there is a fundamental difficulty with the SDR model when *relative* errors are considered.

Example 8.6 Consider again the continuous-time system presented in Example 3.7 on p. 32 having transfer function

$$G(s) = \frac{6(s+5)}{(s+2)(s+3)(s+4)} \tag{8.24}$$

The associated SDR model is given by

$$\hat{G}_q^{\mathrm{SDR}}(z) = \frac{6(\frac{z-1}{\Delta} + 5)}{(\frac{z-1}{\Delta} + 2)(\frac{z-1}{\Delta} + 3)(\frac{z-1}{\Delta} + 4)} \tag{8.25}$$

or, equivalently, in the δ-domain,

$$\hat{G}_\delta^{\mathrm{SDR}}(\gamma) = \frac{6(\gamma+5)}{(\gamma+2)(\gamma+3)(\gamma+4)} \tag{8.26}$$

The associated relative errors $R_1(\omega)$ and $R_2(\omega)$, as defined in (8.7) and (8.9), are plotted in Fig. 8.2 as a function of frequency, for three different sampling periods $\Delta = 0.1$, 0.01, and 0.001. The relative errors do not go to zero near the Nyquist frequency for each sampling period. Indeed, for $R_1(\omega)$, the error becomes very large near the Nyquist frequency and for $R_2(\omega)$, the error approaches 1.

The above example points to the fact that the SDR model is not sufficiently accurate, in general, to produce an approximate model whose *relative* errors converge to zero as $\Delta \to 0$. Of course, the SDR model does not include the sampling zeros. This leads to the question as to whether or not a better result is obtained if the asymptotic sampling zeros are selected in the model.

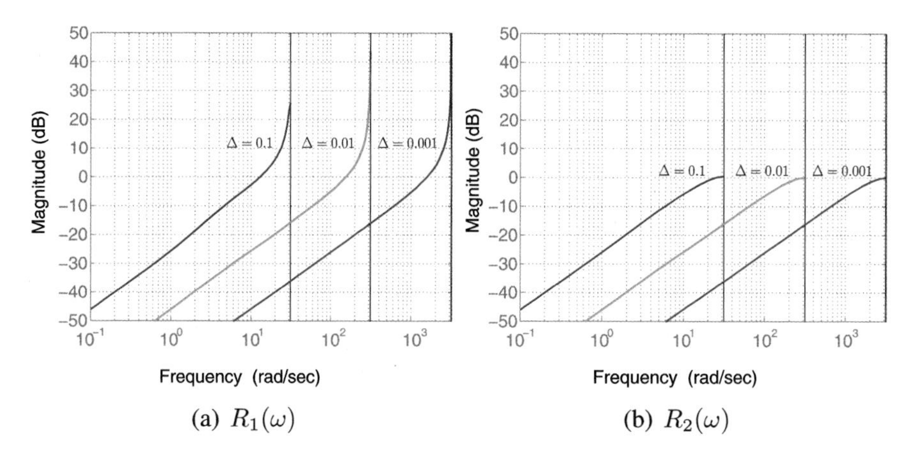

Fig. 8.2 Bode magnitude of the relative errors $R_1(\omega)$ and $R_2(\omega)$ associated with SDR models (see Example 8.6)

8.1.3 Approximate Models Using Asymptotic Sampling Zeros

The asymptotic results described in Chap. 5 lead to the following question: Is it possible to obtain useful approximate models for any value of Δ by exploiting the results on asymptotic sampling zeros?

In this section, approximate sampled models are constructed by appending sampling zeros located at their asymptotic values to the SDR model. The intrinsic poles and zeros are mapped via the SDR approach. This idea leads to the following equivalent approximate models:

$$G_q^{\mathrm{ASZ}}(z) = \frac{B_r(z)\prod_{i=1}^{m}(\frac{z-1}{\Delta} - c_i)}{r!\prod_{i=1}^{n}(\frac{z-1}{\Delta} - p_i)} = \frac{B_r(z)}{r!} G_q^{\mathrm{SDR}}(z) \tag{8.27}$$

$$G_\delta^{\mathrm{ASZ}}(\gamma) = \frac{p_r(\gamma\Delta)\prod_{i=1}^{m}(\gamma - c_i)}{\prod_{i=1}^{n}(\gamma - p_i)} = p_r(\gamma\Delta) G_\delta^{\mathrm{SDR}}(\gamma) \tag{8.28}$$

Note that, by using the fact that $B_r(1) = r!$, the dc gain of this model can be determined so as to match the continuous-time dc gain.

Equations (8.27)–(8.28) define an *asymptotic sampling zeros model* (ASZ model).

Example 8.7 Consider again the continuous-time system presented in Example 8.6 on p. 84 with transfer function (8.24). The system has relative degree 2. Thus, an asymptotic sampling zero exists at $z = -1$. Therefore, the associated ASZ model is given by

$$\hat{G}_q^{\mathrm{ASZ}}(z) = \frac{3(z+1)(\frac{z-1}{\Delta} + 5)}{(\frac{z-1}{\Delta} + 2)(\frac{z-1}{\Delta} + 3)(\frac{z-1}{\Delta} + 4)} \tag{8.29}$$

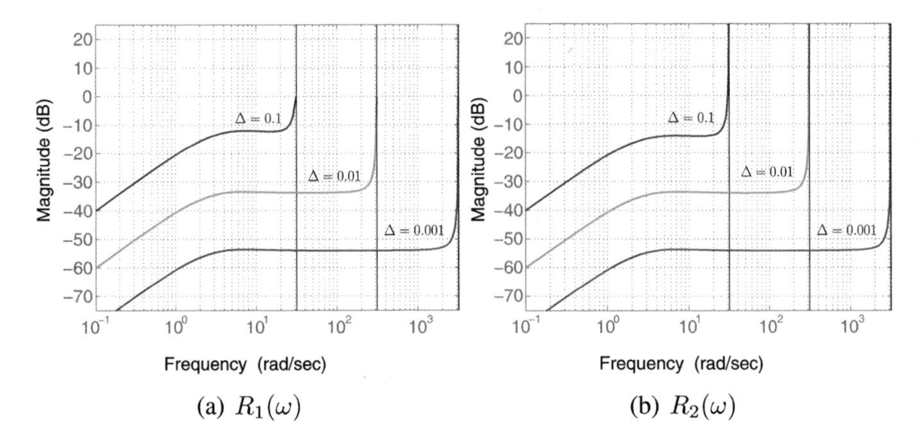

Fig. 8.3 Bode magnitude of the relative errors $R_1(\omega)$ and $R_2(\omega)$ associated with ASZ models (see Example 8.7)

or, equivalently, in the δ-domain,

$$\hat{G}_\delta^{\mathrm{ASZ}}(\gamma) = \frac{6(1+\frac{\gamma\Delta}{2})(\gamma+5)}{(\gamma+2)(\gamma+3)(\gamma+4)} \tag{8.30}$$

The associated relative errors $R_1(\omega)$ and $R_2(\omega)$, as defined in (8.7) and (8.9), are plotted on Fig. 8.3 as a function of frequency, for three different sampling periods $\Delta = 0.1$, 0.01, and 0.001. Comparing Fig. 8.3(a) with Fig. 8.2(a) shows that inclusion of the asymptotic sampling zeros has made a major improvement to the model in terms of relative errors. Indeed, it is clear from Fig. 8.3(a) that the relative error $R_1(\omega)$ for the ASZ model decreases with the sampling period. However, Fig. 8.3(b) contains a further surprise. In particular, the relative error, $R_2(\omega)$, increases (even becoming unbounded) at the Nyquist frequency for each sampling period.

The above example shows that the relative errors (as measured by $R_1(\omega)$) are greatly reduced by including asymptotic sampling zeros into the approximate sampled-data model. However, different behaviours are observed for $R_1(\omega)$ and $R_2(\omega)$.

Surprisingly, $R_2(\omega)$ goes to infinity in the vicinity of the Nyquist frequency $\frac{\pi}{\Delta}$. There is a simple reason for this behaviour. In particular, when the continuous-time relative degree is an even number, there exists an asymptotic sampling zero at $z = -1$ (or, equivalently, at $\gamma = -\frac{2}{\Delta}$ in the δ-operator model). This point corresponds, in fact, to the Nyquist frequency $\frac{\pi}{\Delta}$. Hence, normalising by the approximate model $\hat{G}_q^{\mathrm{ASZ}}(z)$, as in Definition 8.3, leads to an infinite value for $R_2(\frac{\pi}{\Delta})$.

A more refined value for the asymptotic sampling zero at $z = -1$ (when this zero exists) is needed to overcome the above difficulty. Hence, a modified approximate model is next defined which includes a refined approximation to the sampling zero (known as the corrected sampling zero, CSZ) near -1 (if one exists). The other sampling zeros and the intrinsic poles and zeros are located as previously. Also the dc gain is matched to that of the continuous-time model.

Lemma 8.8 *Consider the continuous-time system* (8.11). *An approximate sampled-data model including a more refined (corrected) approximation of the asymptotic sampling zero at* $z=-1$ *is given by*

$$G_q^{\text{CSZ}}(z) = \frac{P(z)}{P(1)} G_q^{\text{ASZ}}(z) \tag{8.31}$$

where

$$P(z) = \frac{z+1+\sigma_\Delta}{z+1} \tag{8.32}$$

For r *odd, we have* $\sigma_\Delta = 0$. *For* r *even,* σ_Δ *is selected as*

$$\sigma_\Delta = \frac{\Delta}{r+1}\left\{\sum_{i=1}^{n} p_i - \sum_{i=1}^{m} c_i\right\} \tag{8.33}$$

where r *is the continuous-time relative degree and* p_i, c_i *denote the continuous poles and zeros, respectively.*

Proof The system transfer function (8.11) can be rewritten as

$$G(s) = \frac{K}{s^r} + \Theta \frac{K}{s^{r+1}} + \cdots \tag{8.34}$$

where

$$\Theta = b_{m-1} - a_{n-1} = -\sum_{i=1}^{m} c_i + \sum_{i=1}^{n} p_i \tag{8.35}$$

Then, the exact sampled-data model can be expressed as

$$G_q^{\text{ESD}}(z) = K\frac{\Delta^r B_r(z)}{r!(z-1)^r} + K\Theta \frac{\Delta^{r+1} B_{r+1}(z)}{(r+1)!(z-1)^{r+1}} + \mathcal{O}(\Delta^{r+2}) \tag{8.36}$$

where Lemma 5.2 on p. 48 has been used.

On the other hand, from (8.27), the ASZ model can be expressed as

$$\begin{aligned} G_q^{\text{ASZ}}(z) &= \frac{B_r(z)}{r!} G\left(\frac{z-1}{\Delta}\right) \\ &= K\frac{\Delta^r B_r(z)}{r!(z-1)^r} + K\frac{\Theta \Delta^{r+1} B_r(z)}{r!(z-1)^{r+1}} + \cdots \end{aligned} \tag{8.37}$$

Comparing (8.36) and (8.37), we notice that the ASZ model is accurate up to order Δ^r. To refine the ASZ model, higher order terms in Δ must be included. Thus, it is required that

$$\frac{P(z)}{P(1)} = 1 + \mathcal{O}(\Delta) \tag{8.38}$$

In particular, when r is an even number, $P(z)$ should have a pole at $z = -1$ (and a zero). The simplest form is given by (8.32), where, from (8.38), σ_Δ is chosen to be a function of the order of Δ. Using (8.32), this procedure yields

$$\begin{aligned}\frac{P(z)}{P(1)} &= \left(\frac{z+1+\sigma_\Delta}{z+1}\right)\left(\frac{2}{2+\sigma_\Delta}\right)\\ &= \left(1+\frac{\sigma_\Delta}{z+1}\right)\left(\frac{1}{1+\frac{\sigma_\Delta}{2}}\right)\\ &= 1-\left(\frac{z-1}{z+1}\right)\left(\frac{\sigma_\Delta}{2}\right)+\mathcal{O}(\Delta^2)\end{aligned} \tag{8.39}$$

Equations (8.37) and (8.39) can be substituted into (8.31), yielding

$$\begin{aligned}G_q^{\mathrm{CSZ}}(z) = K\frac{\Delta^r B_r(z)}{r!(z-1)^r} &+ K\frac{\Theta\Delta^{r+1} B_r(z)}{r!(z-1)^{r+1}}\\ &-\left(\frac{z-1}{z+1}\right)\left(\frac{\sigma_\Delta}{2}\right)K\frac{\Delta^r B_r(z)}{r!(z-1)^r}+\cdots\end{aligned} \tag{8.40}$$

Finally, σ_Δ is chosen such that the term corresponding to Δ^{r+1} in $G_q^{\mathrm{CSZ}}(z)$ and in $G_q^{\mathrm{ESD}}(z)$ (8.36) are the same, i.e.,

$$K\frac{\Theta\Delta^{r+1} B_r(z)}{r!(z-1)^{r+1}} - \left(\frac{z-1}{z+1}\right)\left(\frac{\sigma_\Delta}{2}\right)K\frac{\Delta^r B_r(z)}{r!(z-1)^r} = K\Theta\frac{\Delta^{r+1} B_{r+1}(z)}{(r+1)!(z-1)^{r+1}} \tag{8.41}$$

This leads to

$$\Theta\Delta B_r(z) - B_r(z)\frac{(z-1)^2}{(z+1)}\frac{\sigma_\Delta}{2} = \Theta\frac{\Delta B_{r+1}(z)}{r+1} \tag{8.42}$$

A particular (fixed) value for σ_Δ is obtained from the last equation to ensure that the identity holds at $z = -1$. Notice that $B_r(-1) = 0$ when r is an even number. Then, from (8.42) the following expression is obtained:

$$0 - \left(\lim_{z\to -1}\frac{B_r(z)}{z+1}\right)(-2)^2\frac{\sigma_\Delta}{2} = \Theta\frac{\Delta B_{r+1}(-1)}{r+1} \tag{8.43}$$

The limit is obtained applying *l'Hôpital*'s rule and the Euler-Frobenius polynomial property (5.18) on p. 50. This yields

$$\lim_{z\to -1}\frac{B_r(z)}{z+1} = \lim_{z\to -1} B_r'(z) = \left.\frac{B_{r+1}(z)-(rz+1)B_r(z)}{z(1-z)}\right|_{z=-1}$$
$$= -\frac{B_{r+1}(-1)}{2} \tag{8.44}$$

Substituting into (8.43) yields the result (8.33), i.e.,

$$\sigma_\Delta = \frac{\Theta\Delta}{r+1} = \frac{\Delta}{r+1}\left\{\sum_{i=1}^{n} p_i - \sum_{i=1}^{m} c_i\right\} \tag{8.45}$$

□

Special cases of the result in Lemma 8.8 are

$$(r=2) \qquad \sigma_\Delta = \frac{\Delta}{3}\left(\sum_{i=1}^{n} p_i - \sum_{i=1}^{m} c_i\right) \tag{8.46}$$

$$(r=4) \qquad \sigma_\Delta = \frac{\Delta}{5}\left(\sum_{i=1}^{n} p_i - \sum_{i=1}^{m} c_i\right) \tag{8.47}$$

Example 8.9 Consider again the continuous-time system presented in Examples 8.6 and 8.7 with transfer function (8.24). The system has relative degree 2. Thus, an asymptotic sampling zero exists at $z=-1$. The associated CSZ model is given by

$$\hat{G}_q^{\text{CSZ}}(z) = \frac{6(\frac{z-1}{\Delta}+5)}{(\frac{z-1}{\Delta}+2)(\frac{z-1}{\Delta}+3)(\frac{z-1}{\Delta}+4)}\,\frac{(z+1+\sigma_\Delta)}{(2+\sigma_\Delta)} \tag{8.48}$$

where

$$\sigma_\Delta = \frac{\Delta}{3}(-2-3-4+5) = -\frac{4}{3}\Delta \tag{8.49}$$

Equivalently, in the δ-domain, the CSZ model is

$$\hat{G}_\delta^{\text{CSZ}}(\gamma) = \frac{6(\gamma+5)}{(\gamma+2)(\gamma+3)(\gamma+4)}\left(1+\frac{\gamma\Delta}{2+\sigma_\Delta}\right) \tag{8.50}$$

The associated relative errors $R_1(\omega)$ and $R_2(\omega)$, as defined in (8.7) and (8.9), are plotted in Fig. 8.4 as a function of frequency, for three different sampling periods $\Delta = 0.1$, 0.01, and 0.001. It can be seen that the relative error for both $R_1(\omega)$ and $R_2(\omega)$ now decreases with the sampling period for high frequencies, including those near the Nyquist frequency.

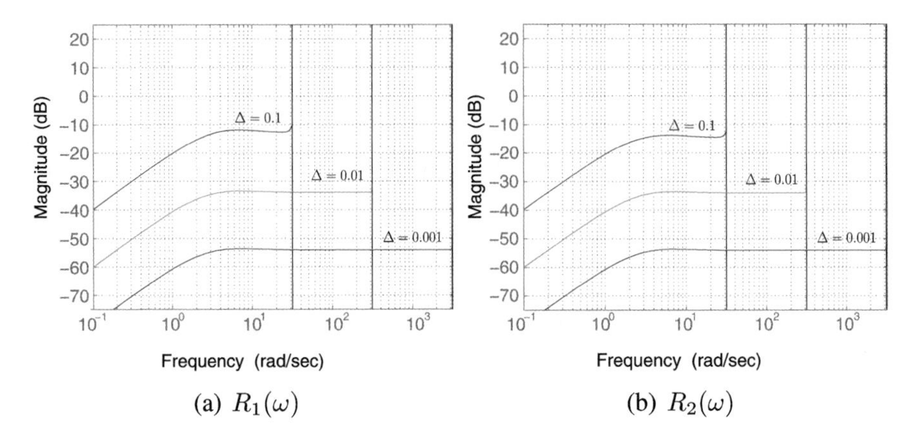

Fig. 8.4 Bode magnitude of the relative errors $R_1(\omega)$ and $R_2(\omega)$ associated with CSZ models (see Example 8.9)

8.2 Approximate Models in the Time Domain

The ideas presented above show that there are fundamental differences between continuous-time models and their sampled equivalents. These differences are of particular significance when inputs having frequency content near the folding frequency are considered. This leads to the following question: Given the importance of the differences between continuous-time and sampled-data models, can insight into the results on sampling zeros be gained using a purely time-domain perspective? A key motivation for doing this is that the time-domain ideas have the potential to be extended to the nonlinear case. These extensions will be explained in Chap. 9.

Two ideas are presented, namely the use of up-sampling and the use of the truncated Taylor series.

8.2.1 Approximate Models Based on Up-Sampling

A first idea to obtain a sampled-data model would be to use Euler integration. However, Example 8.1 has pointed to deficiencies in this simple model. The explanation of the problem is as follows.

Recall that Euler integration models have truncated errors of order Δ^2. However, systems of relative degree r behave as $\frac{1}{s^r}$ at high frequencies. Thus, at the folding frequency, the continuous-time model has a frequency response with a magnitude of the order of $\frac{1}{(\frac{\pi}{\Delta})^r}$, i.e., $\frac{\Delta^r}{\pi^r}$. However, the Euler model only has errors of order Δ^2. This suggests that an Euler model will be inadequate if a small *relative* error is desired in the vicinity of the folding frequency.

One simple way to retain the simplicity of Euler integration but to have a model of increased accuracy is to up-sample. To illustrate, say that it is desired to ultimately use a sampling period Δ with a ZOH input. Then it is possible to run an Euler model at some period $\Delta' = \frac{\Delta}{m}$, $m > 1$, as follows:

$$x_{k+1} = (I + A\Delta')x_k + B\Delta' u_k, \quad k = 1, \ldots, m \tag{8.51}$$

Iterating this model and noting the ZOH nature of the input yields

$$\bar{x}_{k+m} = \left(I + A\frac{\Delta}{m}\right)^m \bar{x}_k + \sum_{i=1}^{m}\left(I + A\frac{\Delta}{m}\right)^{i-1} B\frac{\Delta}{m}u_k \tag{8.52}$$

It is easily seen that as $m \to \infty$, the above model converges to the exact discrete-time model at period Δ. Hence, the model will necessarily capture the sampling zeros as $m \to \infty$. In practice, it often suffices to use a relatively small m to obtain acceptable approximations to the sampling zeros.

To illustrate the idea of up-sampling, a simple example is presented.

Example 8.10 Consider the third order integrator

$$\dot{x}_1 = x_2$$

$$\dot{x}_2 = x_3$$

$$\dot{x}_3 = u$$

The following up-sampled models are immediate:

$m = 1$

$$\begin{bmatrix} x_1 \\ x_2 \\ x_3 \end{bmatrix}_{k+1} = \begin{bmatrix} 1 & \Delta & 0 \\ 0 & 1 & \Delta \\ 0 & 0 & 1 \end{bmatrix} \begin{bmatrix} x_1 \\ x_2 \\ x_3 \end{bmatrix}_k + \begin{bmatrix} 0 \\ 0 \\ \Delta \end{bmatrix} u_k \tag{8.53}$$

$m = 2$

$$\begin{bmatrix} x_1 \\ x_2 \\ x_3 \end{bmatrix}_{k+2} = \begin{bmatrix} 1 & \Delta & \frac{\Delta^2}{4} \\ 0 & 1 & \Delta \\ 0 & 0 & 1 \end{bmatrix} \begin{bmatrix} x_1 \\ x_2 \\ x_3 \end{bmatrix}_k + \begin{bmatrix} 0 \\ \frac{\Delta^2}{4} \\ \Delta \end{bmatrix} u_k \tag{8.54}$$

$m = 3$

$$\begin{bmatrix} x_1 \\ x_2 \\ x_3 \end{bmatrix}_{k+3} = \begin{bmatrix} 1 & \Delta & \frac{\Delta^2}{3} \\ 0 & 1 & \Delta \\ 0 & 0 & 1 \end{bmatrix} \begin{bmatrix} x_1 \\ x_2 \\ x_3 \end{bmatrix}_k + \begin{bmatrix} \frac{\Delta^3}{9} \\ \frac{\Delta^2}{3} \\ \Delta \end{bmatrix} u_k \tag{8.55}$$

These results can be compared with the true sampled-data model, which is

$$\begin{bmatrix} x_1 \\ x_2 \\ x_3 \end{bmatrix}^+ = \begin{bmatrix} 1 & \Delta & \frac{\Delta^2}{2} \\ 0 & 1 & \Delta \\ 0 & 0 & 1 \end{bmatrix} \begin{bmatrix} x_1 \\ x_2 \\ x_3 \end{bmatrix}_k + \begin{bmatrix} \frac{\Delta^3}{6} \\ \frac{\Delta^2}{2} \\ \Delta \end{bmatrix} u_k \tag{8.56}$$

Note, in particular, the convergence of the model to the exact sampled-data model.

8.2.2 Approximate Models Based on Truncated Taylor Series

The method described in Sect. 8.2.1 builds an approximate model by iterating a simpler model. Another form of iterative refinement is achieved by utilising a truncated Taylor series. The key idea is to use a higher order approximation to $e^{A\Delta}$. This suggests an alternative approximate discrete-time model at period Δ, i.e.,

$$x_{k+1} = A_T x_k + B_T u_k \tag{8.57}$$

where

$$A_T = \sum_{i=1}^{m+1} \frac{A^{i-1}\Delta^{i-1}}{(i-1)!}$$

$$B_T = \Delta \left\{ \sum_{i=1}^{m} \frac{A^{i-1}}{i!} \right\} B$$

The error in this model is of order Δ^{m+1}. Again it is clear that the approximate model converges (as $m \to \infty$) to the true discrete-time model.

8.2.3 Approximate Models Based on Near Euler Integration

Here, another approach for obtaining an approximate discrete-time model is explored. The motivation for introducing this third approach is that whilst the methods in Sects. 8.2.1 and 8.2.2 can, in principle, be extended to the nonlinear case, there is a clear advantage to the method of Sect. 8.2.1 as it uses only an Euler approximation. This leads to a further question: Is there some way that we can retain the simplicity of Euler integration but, at the same time, obtain a sampled-data model that captures the essence of the sampling zero dynamics? The answer to this question is a definite yes! To explain the idea, it is first necessary to transform the model to a new format in which the relative degree becomes explicit. This special form is called the 'normal form.' The key idea is presented in the next section.

8.2.4 Normal Forms for Linear Systems

Consider a linear continuous-time system having relative degree $r = n - m$ expressed in transfer function form as

$$G(s) = \frac{b_m s^m + b_{m-1} s^{m-1} + \cdots + b_1 s + b_0}{s^n + a_{n-1} s^{n-1} + \cdots + a_1 s + a_0} \tag{8.58}$$

where $b_m \neq 0$ and $a_{n-1} = -\sum_{i=1}^{n} p_i$. The transfer function (8.58) can be expressed in state-space form as

$$\dot{x} = Ax + Bu \tag{8.59}$$

$$y = Cx \tag{8.60}$$

where

$$A = \left[\begin{array}{c|c} 0_{n-1\times 1} & I_{n-1} \\ \hline -a_0 & \dots - a_{n-1} \end{array}\right], \qquad B = \left[\begin{array}{c} 0_{n-1\times 1} \\ \hline 1 \end{array}\right] \tag{8.61}$$

$$C = \begin{bmatrix} b_0 & b_1 & \cdots & b_{m-1} & b_m & 0 & \cdots & 0 \end{bmatrix} \tag{8.62}$$

The above state-space representation can be re-expressed in normal form:

$$\begin{bmatrix} \dot{\xi}_1 \\ \vdots \\ \dot{\xi}_{r-1} \\ \dot{\xi}_r \\ \dot{\eta} \end{bmatrix} = \begin{bmatrix} \xi_2 \\ \vdots \\ \xi_r \\ Q_{11}\xi + Q_{12}\eta + u \\ Q_{21}\xi + Q_{22}\eta \end{bmatrix} \tag{8.63}$$

$$y = \xi_1 \tag{8.64}$$

where $\xi = [\xi_1, \dots, \xi_r]^T$, $\eta = [\xi_{r+1}, \dots, \xi_n]^T$.

The process of transforming a general linear state-space model to normal form is described in the next result.

Lemma 8.11 *The conversion of the state-space model* (8.59)–(8.62) *to the normal form* (8.63)–(8.64) *is achieved via a similarity transformation of the state model given by*

$$z = \begin{bmatrix} \xi \\ \eta \end{bmatrix} = \Phi x \tag{8.65}$$

where

$$\Phi = \left[\begin{array}{c|c} \multicolumn{2}{c}{C} \\ \multicolumn{2}{c}{\vdots} \\ \multicolumn{2}{c}{CA^{r-1}} \\ \hline I_m & \vec{0}_{m\times r} \end{array}\right] \tag{8.66}$$

Proof Note that $\xi_1 = y$, which establishes (8.64). As a consequence,

$$\dot{\xi}_1 = \dot{y} = C[Ax + Bu] = CAx = \xi_2$$

since $CA^k B = 0$ for $k = 0, \ldots, r-1$ according to the relative degree definition in (3.65). Similarly,

$$\dot{\xi}_\ell = \xi_{\ell+1}; \quad \ell = 1, \ldots, r-1 \tag{8.67}$$

and

$$\begin{aligned}\dot{\xi}_r &= CA^r x + CA^r Bu \\ &= CA^r \Phi^{-1} \begin{bmatrix} \xi \\ \eta \end{bmatrix} + b_m u \\ &= Q_{11}\xi + Q_{12}\eta + b_m u\end{aligned} \tag{8.68}$$

Also,

$$\begin{aligned}\dot{\eta} &= \begin{bmatrix} I_m & 0 \end{bmatrix} \dot{x} = \begin{bmatrix} I_m & 0 \end{bmatrix} Ax \\ &= \begin{bmatrix} I_m & 0 \end{bmatrix} A\Phi^{-1} \begin{bmatrix} \xi \\ \eta \end{bmatrix} \\ &= Q_{21}\xi + Q_{22}\eta\end{aligned} \tag{8.69}$$

Combining (8.67)–(8.69) yields (8.63). □

The remarkable thing about the model (8.61) is that the states $\xi_1, \ldots, \xi_r$ have differing relative degrees, namely $r, \ldots, 1$, respectively. This key observation will be exploited in the next section.

8.2.5 Variable Truncated Taylor Series Model

A key observation concerning the normal form presented in the previous subsection is that the first r states (where r is the relative degree) are described by a simple chain of integrators. Indeed, this is consistent with the *smoothness* of the output resulting from the relative degree of the system. Also note that the *first* state ξ_1 corresponds to the output y, the *second* state ξ_2 to $\dot{y}$, etc. This observation motivates the use of a Taylor series of different orders for the states. In particular, a truncated Taylor series expansion of order $r, r-1, \ldots, 1$ will be used for the states $\xi_1, \ldots, \xi_r$ and a first order expansion for all the components in η. Performing this calculation shows that the states, at time $k\Delta + \Delta$, can be exactly expressed as

$$\xi_1(k\Delta+\Delta)=\xi_1(k\Delta)+\Delta\xi_2(k\Delta)+\cdots+\frac{\Delta^r}{r!}[Q_{11}\xi+Q_{12}\eta+b_m u]_{t=t_1} \tag{8.70}$$

$$\xi_2(k\Delta+\Delta)=\xi_2(k\Delta)+\cdots+\frac{\Delta^{r-1}}{(r-1)!}[Q_{11}\xi+Q_{12}\eta+b_m u]_{t=t_2} \tag{8.71}$$

$$\vdots$$

$$\xi_r(k\Delta+\Delta)=\xi_r(k\Delta)+\Delta[Q_{11}\xi+Q_{12}\eta+b_m u]_{t=t_r} \tag{8.72}$$

and

$$\eta(k\Delta+\Delta)=\eta(k\Delta)+\Delta[Q_{21}\xi+Q_{22}\eta]_{t=t_{r+1}} \tag{8.73}$$

for some (unknown) time instants $k\Delta < t_\ell < k\Delta+\Delta$; $\ell=1,\dots,r+1$.

Note that the model (8.70)–(8.73) is exact. An approximate sampled-data model is then obtained by replacing the unknown time instants t_ℓ by $k\Delta$, i.e.,

$$\hat{\xi}_1^+=\hat{\xi}_1+\Delta\hat{\xi}_2+\cdots+\frac{\Delta^r}{r!}[Q_{11}\hat{\xi}+Q_{12}\hat{\eta}+b_m u_k] \tag{8.74}$$

$$\hat{\xi}_2^+=\hat{\xi}_2+\cdots+\frac{\Delta^{r-1}}{(r-1)!}[Q_{11}\hat{\xi}+Q_{12}\hat{\eta}+b_m u_k] \tag{8.75}$$

$$\vdots$$

$$\hat{\xi}_r^+=\hat{\xi}_r+\Delta[Q_{11}\hat{\xi}+Q_{12}\hat{\eta}+b_m u_k] \tag{8.76}$$

and

$$\hat{\eta}^+=\hat{\eta}+\Delta[Q_{21}\hat{\xi}+Q_{22}\hat{\eta}] \tag{8.77}$$

where $\hat{\xi}_\ell=\hat{\xi}_\ell(k\Delta)$, $\hat{\xi}_\ell^+=\hat{\xi}_\ell(k\Delta+\Delta)$, and u_k is the ZOH input.

The model (8.74)–(8.77) is called the truncated Taylor series (TTS) model.

The TTS model can be expressed compactly as:

$$q\zeta=A_r^q\zeta+B_r^q\left(CA^r\Phi^{-1}\begin{bmatrix}\zeta\\ \eta\end{bmatrix}+u\right) \tag{8.78}$$

$$q\eta=\Delta Q_{12}\zeta+(I_m+\Delta Q_{22})\eta \tag{8.79}$$

where q is the forward shift operator and where:

$$A_r^q = e^{A_r \Delta} = \begin{bmatrix} 1 & \Delta & \cdots & \frac{\Delta^{r-1}}{(r-1)!} \\ 0 & & \ddots & \vdots \\ \vdots & \ddots & & \Delta \\ 0 & \cdots & 0 & 1 \end{bmatrix} \tag{8.80}$$

$$B_r^q = \int_0^\Delta e^{A_r \eta} B_r \, d\eta = \begin{bmatrix} \frac{\Delta^r}{r!} \\ \vdots \\ \Delta \end{bmatrix} \tag{8.81}$$

Equations (8.78) and (8.79) can also be expressed in incremental form as:

$$\delta\zeta = \frac{q-1}{\Delta}\zeta = A_r^\delta \zeta + B_r^\delta \left(CA^r \Phi^{-1} \begin{bmatrix} \zeta \\ \eta \end{bmatrix} + u \right) \tag{8.82}$$

$$\delta\eta = \frac{q-1}{\Delta}\eta = Q_{21}\zeta + Q_{22}\eta \tag{8.83}$$

where:

$$A_r^\delta = \frac{A_r^q - I}{\Delta} = A_r + \mathcal{O}(\Delta) = \begin{bmatrix} 0 & 1 & \cdots & \frac{\Delta^{r-2}}{(r-1)!} \\ \vdots & \ddots & \ddots & \vdots \\ \vdots & \ddots & \ddots & 1 \\ 0 & \cdots & \cdots & 0 \end{bmatrix} \tag{8.84}$$

$$B_r^\delta = \frac{B_r^q}{\Delta} = B_r + \mathcal{O}(\Delta) = \begin{bmatrix} \frac{\Delta^{r-1}}{r!} & \cdots & 1 \end{bmatrix}^T \tag{8.85}$$

Example 8.12 Consider again the ongoing example of the continuous-time system given by the transfer function (8.24). This system can be expressed in the normal form as follows:

$$\dot{x} = \begin{bmatrix} \dot{x}_1 \\ \dot{x}_2 \\ \dot{x}_3 \end{bmatrix} = \begin{bmatrix} x_2 \\ a(x) \\ c(x) \end{bmatrix} + \begin{bmatrix} 0 \\ b(x) \\ 0 \end{bmatrix} u \tag{8.86}$$

$$y = x_1 \tag{8.87}$$

where

$$a(x) = \begin{bmatrix} -6 & -4 & 6 \end{bmatrix} x \tag{8.88}$$

$$b(x) = 6 \tag{8.89}$$

$$c(x) = \begin{bmatrix} 1 & 0 & -5 \end{bmatrix} x \tag{8.90}$$

The corresponding TTS model is then given by

$$\hat{x}_{1,k+1} = \hat{x}_{1,k} + \Delta \hat{x}_{2,k} + \frac{\Delta^2}{2}\big[a(\hat{x}_k) + b(\hat{x}_k)u_k\big] \tag{8.91}$$

$$\hat{x}_{2,k+1} = \hat{x}_{2,k} + \Delta\big[a(\hat{x}_k) + b(\hat{x}_k)u_k\big] \tag{8.92}$$

$$\hat{x}_{3,k+1} = \hat{x}_{3,k} + \Delta\big[c(\hat{x}_k)\big] \tag{8.93}$$

$$y_k = \hat{x}_{1,k} \tag{8.94}$$

Substituting (8.88)–(8.90) into (8.91)–(8.94) shows that the transfer function associated with the approximate sampled-data model is:

$$G_q^{\mathrm{TTS}}(z) = \frac{3\Delta^2(z+1)(z-1+5\Delta)}{(z-1+4\Delta)\,\bar{E}(z)} \tag{8.95}$$

$$\bar{E}(z) = z^2 - 2z + 1 + 3\Delta^2 z + 3\Delta^2 + 5\Delta z - 5\Delta \tag{8.96}$$

A key observation regarding the above model is that the model (8.95) contains a sampling zero exactly at the asymptotic location $z = -1$. The associated incremental form corresponding to (8.95) is given by

$$G_\delta^{\mathrm{TTS}}(\gamma) = \frac{6(1 + \frac{\Delta}{2}\gamma)(\gamma+5)}{(\gamma+4)(\gamma^2 + 5\gamma + 6 + 3\Delta\gamma)} \tag{8.97}$$

The above example suggests that the TTS idea may, indeed, capture the asymptotic sampling zeros. This is further illustrated below.

Example 8.13 Consider a more complex system which includes a slightly damped resonant mode

$$G(s) = \frac{p_1\omega_n^2}{(s+p_1)(s^2 + 2\zeta\omega_n s + \omega_n^2)} \tag{8.98}$$

where $p_1 = 1$, $\zeta = 0.1$, $\omega_n = 4$. In this case, the asymptotic sampling zeros are the roots of $B_3(z) = z^2 + 4z + 1$.

Indeed, Fig. 8.5 shows that the TTS model provides similar relative errors to those of the ASZ model. This happens because the TTS model, inter alia, includes the asymptotic sampling zeros.

Moreover, it can be seen that the TTS model provides the highest accuracy for both low and high frequencies, and it is only slightly outperformed by the ASZ model in a band around $\omega = 1$.

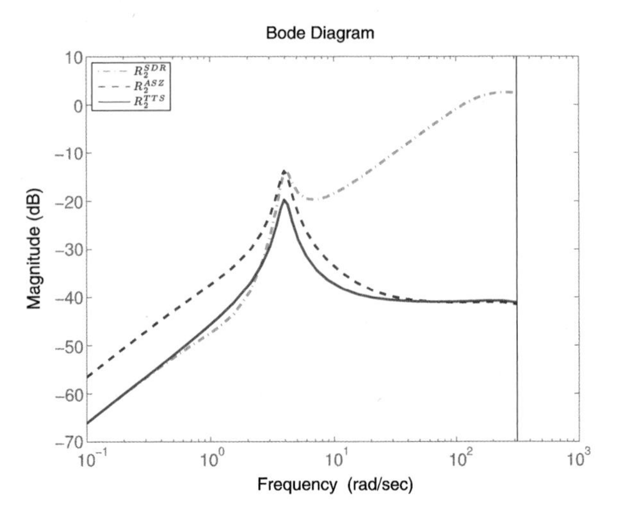

Fig. 8.5 Relative errors $R_2^{\mathrm{SDR}}(\omega)$, $R_2^{\mathrm{ASZ}}(\omega)$, and $R_2^{\mathrm{TTS}}(\omega)$, for a sampling period $\Delta = 0.01$ (Example 8.13)

8.3 Summary

The key points covered in this chapter are the following:

- For linear systems, exact models can be obtained. Approximate models can be obtained, for example,
  - Using derivative approximation, such as the Euler method
  - Including the presence of (asymptotic) sampling zeros
  - Performing a truncated Taylor series expansion
- The error associated with approximate sampled-data models can be quantified in the time domain, for example, by considering truncation errors, or in the frequency domain, for example, by considering the maximum relative error.

Further Reading

Further information regarding corrected asymptotic sampling zeros can be found in

Blachuta MJ (1999a) On approximate pulse transfer functions. IEEE Trans Autom Control 44(11):2062–2067

Blachuta MJ (1999b) On zeros of pulse transfer functions. IEEE Trans Autom Control 44(6):1229–1234

An application of up-sampled models can be found in

Cea M, Goodwin GC (2010) Up-sampling strategies in sampled data nonlinear filtering. In: 49th IEEE conference on decision and control

Further information regarding relative errors for approximate linear sampled-data models is given in

Goodwin GC, Yuz JI, Agüero JC (2008) Relative error issues in sampled data models. In: 17th IFAC world congress, Seoul, Korea
Yucra E, Yuz JI (2011) Frequency domain accuracy of approximate sampled-data models. In: 18th IFAC world congress, Milan, Italy

Normal forms for linear (and nonlinear) systems are described in detail in

Isidori A (1995) Nonlinear control systems, 3rd edn. Springer, Berlin

Chapter 9
Approximate Models for Nonlinear Deterministic Systems

This chapter extends the key ideas of Chap. 8 to develop discrete-time models for *nonlinear* deterministic systems.

9.1 Approximate Models Based on Up-Sampling

The ideas of Sect. 8.2.1 can be readily extended to the nonlinear case. Specifically, consider a nonlinear continuous-time model

$$\dot{x} = f(x, u) \tag{9.1}$$

$$y = h(x) \tag{9.2}$$

Say that it is desired that an accurate discrete-time model be developed at period Δ. An up-sampled Euler model can be used at period $\Delta' = \frac{\Delta}{m}$, i.e.,

$$x_{k+i} = x_{k+i-1} + \frac{\Delta}{m} f(x_{k+i-1}, u_k); \quad i = 1, \ldots, m \tag{9.3}$$

$$y_{k+m} = h(x_{k+m}) \tag{9.4}$$

This is a simple yet powerful idea. However, it fails to give insight into the resultant model. In Sects. 8.2.3, 8.2.4, and 8.2.5 other ideas were developed which led, in the linear case, to approximate discrete-time models containing the deterministic asymptotic sampling zero dynamics. Remarkably, as shown below, these same ideas readily extend to the nonlinear case. As in the linear case, a key ingredient is the conversion of the model to normal form. Thus, normal forms for nonlinear systems are next reviewed.

J.I. Yuz, G.C. Goodwin, *Sampled-Data Models for Linear and Nonlinear Systems*, Communications and Control Engineering, DOI 10.1007/978-1-4471-5562-1_9,

9.2 Normal Forms for Nonlinear Systems

Consider the class of nonlinear continuous-time systems where the model is affine in the input:

$$\dot{x}(t) = f\big(x(t)\big) + g\big(x(t)\big)u(t) \tag{9.5}$$

$$y(t) = h\big(x(t)\big) \tag{9.6}$$

where $x(t)$ is the state evolving in an open subset $\mathcal{M} \subset \mathbb{R}^n$, and where the vector fields $f(\cdot)$ and $g(\cdot)$ and the output function $h(\cdot)$ are analytic.

The appropriate extension of the notion of relative degree to the nonlinear case is captured in the following definition.

Definition 9.1 The nonlinear system (9.5)–(9.6) is said to have *relative degree* r at a point x_o if:

(i) $L_g L_f^k h(x) = 0$ for x in a neighbourhood of x_o and for $k = 0, \dots, r-2$, and

(ii) $L_g L_f^{r-1} h(x_o) \neq 0$,

where L_g and L_f correspond to Lie derivatives. For example, $L_g h(x) = \frac{\partial h}{\partial x} g(x)$.

Intuitively, the relative degree, as defined above, corresponds to the number of times that we need to differentiate the output $y(t)$ to make the input $u(t)$ appear explicitly. This is the same idea used in the linear case—see Chap. 4.

Next the nonlinear version of Lemma 8.11 (Sect. 8.2.4 on p. 92) is developed. The reader is referred to the sources given at the end of this chapter for proof of the next two results.

Lemma 9.2 *Suppose that the system has relative degree r at x_o. Consider the new coordinate defined as:*

$$z_1 = \phi_1(x) = h(x) \tag{9.7}$$

$$z_2 = \phi_2(x) = L_f h(x) \tag{9.8}$$

$$\vdots$$

$$z_r = \phi_r(x) = L_f^{r-1} h(x) \tag{9.9}$$

Furthermore, if $r < n$, *it is always possible to define* $z_{r+1} = \phi_{r+1}(x), \dots,$ $z_n = \phi_n(x)$ *such that:*

$$z = \begin{bmatrix} z_1 \\ \vdots \\ z_n \end{bmatrix} = \begin{bmatrix} \phi_1(x) \\ \vdots \\ \phi_n(x) \end{bmatrix} = \Phi(x) \tag{9.10}$$

has a nonsingular Jacobian at x_o. *Then,* $\Phi(\cdot)$ *is a* local coordinate transformation *in a neighbourhood of* x_o. *Moreover, it is always possible to define* $z_{r+1} = \phi_{r+1}(x), \dots, z_n = \phi_n(x)$ *in such a way that:*

$$L_g \phi_i(x) = 0 \tag{9.11}$$

in a neighbourhood of x_o, *for all* $i = r+1, \dots, n$.

Lemma 9.3 *The state-space description of the nonlinear system* (9.5)–(9.6) *in the new coordinate defined by Lemma* 9.2 *is given by the* normal form:

$$\dot{\xi} = \left[\begin{array}{c|c} 0 & \\ \vdots & I_{r-1} \\ 0 & \\ \hline 0 & 0 \ \dots \ 0 \end{array}\right] \xi + \begin{bmatrix} 0 \\ \vdots \\ 0 \\ 1 \end{bmatrix} \big(a(\xi, \eta) + b(\xi, \eta) u(t)\big) \tag{9.12}$$

$$\dot{\eta} = c(\xi, \eta) \tag{9.13}$$

where the output is $z_1 = h(x) = y$.

The state vector in (9.12), (9.13) *is:*

$$z(t) = \begin{bmatrix} \xi(t) \\ \eta(t) \end{bmatrix} \quad \begin{cases} \xi(t) = [z_1(t), z_2(t), \dots, z_r(t)]^T \\ \eta(t) = [z_{r+1}(t), z_{r+2}(t), \dots, z_n(t)]^T \end{cases} \tag{9.14}$$

and

$$a(\xi, \eta) = a(z) = L_f^r h\big(\Phi^{-1}(z)\big) \tag{9.15}$$

$$b(\xi, \eta) = b(z) = L_g L_f^{r-1} h\big(\Phi^{-1}(z)\big) \tag{9.16}$$

$$c(\xi, \eta) = c(z) = \begin{bmatrix} L_f \phi_{r+1}(\Phi^{-1}(z)) \\ \vdots \\ L_f \phi_n(\Phi^{-1}(z)) \end{bmatrix} \tag{9.17}$$

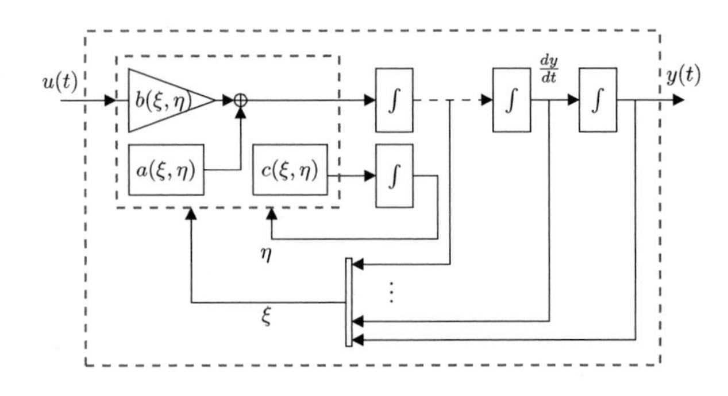

Fig. 9.1 Block diagram representation of the normal form for nonlinear systems

Remark 9.4 The connection between the linear and nonlinear cases will be apparent to the reader. Also, note that the state variables contained in $\xi(t)$, defined in (9.7)–(9.9), correspond to the output $y(t)$ and its first $r-1$ derivatives:

$$z_\ell(t) = z_1^{(\ell-1)}(t) = y^{(\ell-1)}(t); \quad \ell = 1, \ldots, r \tag{9.18}$$

The normal form is represented as a block diagram in Fig. 9.1.

9.3 Variable Truncated Taylor Series Model

Based on the nonlinear version of the normal form, the ideas presented in Sect. 8.2.5 can now be extended to the nonlinear case. Note that, when the input signal $u(t)$ is generated by a zero order hold (ZOH), the rth derivative, $y^{(r)}(t) = \dot{z}_r(t) = b(z) + a(z)u(t)$, is well defined but is, in general, discontinuous at the sampling instants $t = k\Delta$ (when the ZOH control signal (3.19) is updated). This observation allows a Taylor series expansion for $y(t)$ and for each one of its $r-1$ derivatives to be developed. Thus, the state variables z_ℓ at $t = k\Delta + \Delta$ can be *exactly* expressed by:

$$z_1(k\Delta + \Delta) = z_1(k\Delta) + \Delta z_2(k\Delta) + \cdots + \frac{\Delta^r}{r!}\big[a(\xi, \eta) + b(\xi, \eta)\, u\big]_{t=\xi_1} \tag{9.19}$$

$$z_2(k\Delta + \Delta) = z_2(k\Delta) + \cdots + \frac{\Delta^{r-1}}{(r-1)!}\big[a(\xi, \eta) + b(\xi, \eta)\, u\big]_{t=\xi_2} \tag{9.20}$$

$$\vdots$$

$$z_r(k\Delta + \Delta) = z_r(k\Delta) + \Delta\big[a(\xi, \eta) + b(\xi, \eta)\, u\big]_{t=\xi_r} \tag{9.21}$$

and

$$\eta(k\Delta + \Delta) = \eta(k\Delta) + \Delta\big[c(\xi, \eta)\big]_{t=\xi_{r+1}} \tag{9.22}$$

for some time instants $k\Delta < \xi_\ell < k\Delta + \Delta$, $\ell = 1, \ldots, r+1$. Note that the sampled model above is *exact* within the sampling interval $k\Delta \le \xi_\ell < k\Delta + \Delta$ (for some unknown time instants ξ_ℓ) where the input u is continuous (in fact, it is constant due to the ZOH). The reader will note that (9.19)–(9.22) represent the nonlinear extension of the equivalent linear version given earlier in (8.70)–(8.73).

Proceeding as in Chap. 8, an approximate discrete-time model is obtained by replacing the unknown time instants by $k\Delta$, i.e.,

$$\hat{z}_1^+ = \hat{z}_1 + \Delta\hat{z}_2 + \cdots + \frac{\Delta^r}{r!}\big[a(\hat{\xi},\hat{\eta}) + b(\hat{\xi},\hat{\eta})u_k\big] \tag{9.23}$$

$$z_2^+ = z_2 + \cdots + \frac{\Delta^{r-1}}{(r-1)!}\big[a(\hat{\xi},\hat{\eta}) + b(\hat{\xi},\hat{\eta})\,u_k\big] \tag{9.24}$$

$$\vdots$$

$$z_r^+ = z_r + \Delta\big[a(\hat{\xi},\hat{\eta}) + b(\hat{\xi},\hat{\eta})u_k\big] \tag{9.25}$$

and

$$\hat{\eta}^+ = \hat{\eta} + \Delta c(\hat{\xi},\hat{\eta}) \tag{9.26}$$

where $\hat{z}_\ell = \hat{z}_\ell(k\Delta)$ and $\hat{z}_\ell^+ = \hat{z}_\ell(k\Delta + \Delta)$, similarly for $\hat{\xi}$ and $\hat{\eta}$, and u_k is the ZOH input.

The approximate model (9.23)–(9.26) is the nonlinear extension of the equivalent linear version given earlier in (8.74)–(8.77). As for the linear case, the model (9.23)–(9.26) is called the truncated Taylor series (TTS) model.

9.4 Approximation Errors for Nonlinear Sampled Models

In the previous chapter it made sense to define errors in the frequency domain. In particular, this gave insight into the interaction between the rate of change of the input (i.e., frequency) and the associated model errors. Perhaps not surprisingly, it was found that the most demanding errors occurred at high frequency (i.e., in the vicinity of the Nyquist rate). This led to the notion of relative errors which were defined on a frequency-by-frequency basis.

In the case of nonlinear systems, it is difficult to relate the frequency content of the input to the model accuracy. For example, a single frequency input can lead to a multifrequency output. Thus, alternative measures of model accuracy are needed in the time domain. This topic will be explored in the sequel.

9.4.1 Links to Numerical Analysis

There exists substantial material on approximate sampled-data models in the numerical analysis literature (see the citations at the end of the chapter). This literature will form the basis of the presentation here. However, to describe the errors in the TTS model, it is necessary to go beyond the standard results available in the numerical analysis literature. In particular, three new elements, inspired by system theory, are injected:

1. Special emphasis is placed on the *output* of the model.
2. The concept of relative degree is used to inspire a special class of state-space model (the normal form).
3. Separate truncation errors are used for each element of the state-space model.

9.4.2 Local and Global Truncation Errors

The analysis presented below is based on a general nonlinear state-space model having n states of the form

$$\dot{x} = f(x, u) \tag{9.27}$$

$$y = h(x) \tag{9.28}$$

and an associated (and, in general, approximate) discrete-time model

$$\hat{x}_{k+1} = \hat{f}(\hat{x}_k, u_k) \tag{9.29}$$

$$\hat{y}_k = \hat{h}(\hat{x}_k) \tag{9.30}$$

To measure the accuracy of an approximate sampled-data model of the type given in (9.29), (9.30), different options can be used. For example, the states of the model can be compared with the true system states in one sampling interval, after a fixed number of sampling instants, or at the end of a fixed continuous-time interval. This will lead to three different error measures, as described below. These errors are measured in terms of the order of convergence to zero as a function of powers of the sampling period (see Definition 8.4 on p. 82).

The following definition captures one aspect of the error between the approximate model (9.29), (9.30) and the true sampled-data response of the system (9.27), (9.28).

Definition 9.5 Consider a dynamical system with states $(x_1, \ldots, x_n)$ and an associated approximate model with states $(\hat{x}_1, \ldots, \hat{x}_n)$. Then, the *local vector truncation error* of the approximate model is said to be of the order of $(\Delta^{m_1}, \ldots, \Delta^{m_n})$ if and only if

$$\begin{cases} \hat{x}_1[k] - x_1[k] = 0 \\ \vdots \\ \hat{x}_n[k] - x_n[k] = 0 \end{cases} \Rightarrow \begin{cases} \hat{x}_1[k+1] - x_1[k+1] \in \mathcal{O}(\Delta^{m_1}) \\ \vdots \\ \hat{x}_n[k+1] - x_n[k+1] \in \mathcal{O}(\Delta^{m_n}) \end{cases} \tag{9.31}$$

where $m_1, \ldots, m_n$ are integers.

Definition 9.5 embeds the usual definition for local truncation error used in the numerical analysis literature. In the latter body of work, the integers $m_1, \ldots, m_n$ are all chosen as the same integer m. Here the usual definition has been generalised to characterise the accuracy for individual states (and the system output) when applying a discretisation strategy such as the TTS model proposed in Sect. 9.3.

The following definition captures the accumulation of errors when iterating an approximate discrete model over several steps.

Definition 9.6 The *local vector fixed-steps truncation error* of the approximate model is said to be of the order of $(\Delta^{m_1}, \ldots, \Delta^{m_n})$ if and only if, for initial state errors

$$\begin{cases} \hat{x}_1[k] - x_1[k] \in \mathcal{O}(\Delta^{\bar{m}_1}) \\ \vdots \\ \hat{x}_n[k] - x_n[k] \in \mathcal{O}(\Delta^{\bar{m}_n}) \end{cases} \tag{9.32}$$

for any $\bar{m}_i \geq m_i, i = 1, \ldots, n$, after N steps, where N is a finite *fixed* number, it follows that

$$\begin{cases} \hat{x}_1[k+N] - x_1[k+N] \in \mathcal{O}(\Delta^{m_1}) \\ \vdots \\ \hat{x}_n[k+N] - x_n[k+N] \in \mathcal{O}(\Delta^{m_n}) \end{cases} \tag{9.33}$$

Finally, consider the case where the sampled-data model is used at a fixed continuous-time interval. The following definition captures the associated errors.

Definition 9.7 The *global vector fixed-time truncation error* of an approximate model is said to be of the order of $(\Delta^{\tilde{m}_1}, \ldots, \Delta^{\tilde{m}_n})$ if and only if, for initial state errors

$$\begin{cases} \hat{x}_1[k] - x_1[k] \in \mathcal{O}(\Delta^{\bar{m}_1}) \\ \vdots \\ \hat{x}_n[k] - x_n[k] \in \mathcal{O}(\Delta^{\bar{m}_n}) \end{cases} \tag{9.34}$$

for any $\bar{m}_i \geq \tilde{m}_i, i = 1, \ldots, n$, after a *fixed* (continuous) time T, i.e., after $N = \lfloor T/\Delta \rfloor$ steps, it follows that

$$\begin{cases} \hat{x}_1[k+N] - x_1[k+N] \in \mathcal{O}(\Delta^{\tilde{m}_1}) \\ \vdots \\ \hat{x}_n[k+N] - x_n[k+N] \in \mathcal{O}(\Delta^{\tilde{m}_n}) \end{cases} \tag{9.35}$$

The key difference between Definition 9.6 and Definition 9.7 is that, in the latter, the number of steps is chosen to be a function of the sampling time Δ so that the same fixed continuous-time interval T is covered. As Δ decreases, then in Definition 9.7, the number of iteration steps N increases since T must remain fixed. This is in contrast with Definition 9.6, where independent of the sampling time, the error bound is always measured after the same fixed number of iterations.

The next section applies these ideas to the TTS model.

9.4.3 Truncation Errors for the TTS Model

Consider the truncated Taylor series (TTS) model. This model was proposed in Sects. 8.2.5 and 9.3 for linear and nonlinear models, respectively. The following result describes the error properties of the model.

Theorem 9.8 *Assume that the continuous-time nonlinear model* (9.5)–(9.6) *has uniform relative degree $r \leq n$ in the open subset $\mathcal{M} \subset \mathbb{R}^n$, and, thus, it can be expressed in the normal form* (9.12)–(9.14). *Then, the approximate TTS model* (9.23)–(9.26) *has*

1. *Local vector truncation error of the order of* $(\Delta^{r+1}, \Delta^r, \ldots, \Delta^2, \Delta^2)$,
2. *Local vector fixed-steps truncation error of the order of* $(\Delta^{r+1}, \Delta^r, \ldots, \Delta^2, \Delta^2)$, *and*
3. *Global vector fixed-time truncation error of the order of* $(\Delta, \ldots, \Delta)$.

Proof

1. From the definition of local (vector) truncation error, it is assumed that

$$\hat{z}_1[k] - z_1[k] = 0$$
$$\vdots$$
$$\hat{z}_r[k] - z_r[k] = 0$$
$$\hat{\eta}[k] - \eta[k] = 0$$

and, because of this, the approximate states $(\hat{\xi}, \hat{\eta}) = (\xi, \eta)$, and therefore

$$\widehat{F} = F = \{b(\xi, \eta) + a(\xi, \eta) \cdot u_k\}\big|_{t=k\Delta}$$

Hence, the one-step-ahead errors are:

$$\hat{z}_1^+ - z_1^+ = \frac{\Delta^r}{r!} \cdot (F - F_1)$$
$$\vdots$$
$$\hat{z}_r^+ - z_r^+ = \Delta \cdot (F - F_r)$$
$$\hat{\eta}^+ - \eta^+ = \Delta \cdot \big(c(\xi, \eta)|_{t=k\Delta} - c(\xi, \eta)|_{t=\xi_{r+1}}\big)$$

Define the errors $e_i[p] = \hat{z}_i[p] - z_i[p]$, $i = 1, \ldots, r$ and $e_\eta[p] = \hat{\eta}[p] - \eta[p]$. In order to show that the approximate sampled-data model given in (9.23)–(9.26) has a local vector truncation error of the order of $(\Delta^{r+1}, \Delta^r, \ldots, \Delta^3, \Delta^2, \Delta^2)$, it suffices to show that

$$|F - F_i| \in \mathcal{O}(\Delta), \quad i = 1, \ldots, r \tag{9.36}$$

$$\big|c(\xi, \eta)|_{t=k\Delta} - c(\xi, \eta)|_{t=\xi_{r+1}}\big| \in \mathcal{O}(\Delta) \tag{9.37}$$

From (9.23)–(9.26) it follows that

$$\begin{aligned} |F - F_i| &= \big|\{b(\xi, \eta) + a(\xi, \eta) \cdot u_k\}\big|_{t=k\Delta} - \{b(\xi, \eta) + a(\xi, \eta) \cdot u_k\}\big|_{t=\xi_i}\big| \\ &\le L \cdot \big\|(\xi, \eta)|_{t=k\Delta} - (\xi, \eta)|_{t=\xi_i}\big\| \\ &= L \cdot \big\|z(k\Delta) - z(\xi_i)\big\| \end{aligned}$$

where the existence of the Lipschitz constant L is guaranteed by the analyticity of $f(\cdot)$, $g(\cdot)$, and $h(\cdot)$. In addition, it guarantees that the state trajectory $z(t) = [\xi^T(t)\ \eta^T(t)^T]$ is bounded in the following form:

$$\big\|z(k\Delta) - z(\xi_i)\big\| \le C \cdot \frac{e^{L|k\Delta - \xi_i|} - 1}{L}$$

Therefore,

$$\begin{aligned} |F - F_i| &\leq L \cdot C \cdot \frac{e^{L|k\Delta - \xi_i|} - 1}{L} \\ &< C \cdot \left(e^{L\Delta} - 1\right) \\ &\in \mathcal{O}(\Delta) \end{aligned}$$

which shows (9.36). The proof for e_η is analogous, i.e.,

$$\left|c(\xi,\eta)|_{t=k\Delta} - c(\xi,\eta)|_{t=\xi_{r+1}}\right| \leq L \cdot \left\|(\xi,\eta)|_{t=k\Delta} - (\xi,\eta)|_{t=\xi_{r+1}}\right\|$$

from which (9.37) follows.

2. For the local vector fixed-steps truncation error, the cumulative effect of errors is evaluated when the approximate model is iterated.

   Define $e_i[p] = \hat{z}_i[p] - z_i[p], i = 1, \ldots, r$. It then follows that

$$\begin{bmatrix} e_1[k+1] \\ e_2[k+1] \\ e_3[k+1] \\ \vdots \\ e_{r-1}[k+1] \\ e_r[k+1] \end{bmatrix} \leq \begin{bmatrix} 1 & \Delta & \frac{\Delta^2}{2} & \cdots & \frac{\Delta^{r-2}}{(r-2)!} & \frac{\Delta^{r-1}}{(r-1)!} \\ 0 & 1 & \Delta & \cdots & \frac{\Delta^{r-3}}{(r-3)!} & \frac{\Delta^{r-2}}{(r-2)!} \\ 0 & 0 & 1 & \cdots & \frac{\Delta^{r-4}}{(r-4)!} & \frac{\Delta^{r-3}}{(r-3)!} \\ \vdots & \vdots & & \ddots & \vdots & \vdots \\ 0 & 0 & 0 & \cdots & 1 & \Delta \\ 0 & 0 & 0 & \cdots & 0 & 1 \end{bmatrix} \begin{bmatrix} e_1[k] \\ e_2[k] \\ e_3[k] \\ \vdots \\ e_{r-1}[k] \\ e_r[k] \end{bmatrix} + \begin{bmatrix} \frac{\Delta^r}{r!} \\ \frac{\Delta^{r-1}}{(r-1)!} \\ \frac{\Delta^{r-2}}{(r-2)!} \\ \vdots \\ \frac{\Delta^2}{2} \\ \Delta \end{bmatrix} \cdot E \tag{9.38}$$

   where, as shown in the proof of Theorem 9.8, $E = \max_{i,k} |F - F_i|_{t=k\Delta} \in \mathcal{O}(\Delta), \forall k \in \mathbb{N}, \forall i = 1, \ldots, r, \eta$. Since the system (9.38) has the form

$$\theta[k+1] = A \cdot \theta[k] + \Phi \cdot E \tag{9.39}$$

   then, after N steps,

$$\theta[k+N] = A^N \cdot \theta[k] + \sum_{i=0}^{N-1} A^i \cdot \Phi \cdot E \tag{9.40}$$

It is easily seen that the matrices are as follows:

$$A^N = \begin{bmatrix} 1 & N\Delta & \frac{N^2\Delta^2}{2} & \cdots & \frac{N^{r-2}\Delta^{r-2}}{(r-2)!} & \frac{N^{r-1}\Delta^{r-1}}{(r-1)!} \\ 0 & 1 & N\Delta & \cdots & \frac{N^{r-3}\Delta^{r-3}}{(r-3)!} & \frac{N^{r-2}\Delta^{r-2}}{(r-2)!} \\ 0 & 0 & 1 & \cdots & \frac{N^{r-4}\Delta^{r-4}}{(r-4)!} & \frac{N^{r-3}\Delta^{r-3}}{(r-3)!} \\ \vdots & \vdots & & \ddots & \vdots & \vdots \\ 0 & 0 & 0 & \cdots & 1 & N\Delta \\ 0 & 0 & 0 & \cdots & 0 & 1 \end{bmatrix}$$

$$\sum_{i=0}^{N-1} A^i = \begin{bmatrix} N & \frac{N(N-1)}{2}\Delta & \frac{N(N-1)(2N-1)}{6}\frac{\Delta^2}{2} & \cdots & S_{r-2}^{N-1}\frac{\Delta^{r-2}}{(r-2)!} & S_{r-1}^{N-1}\frac{\Delta^{r-1}}{(r-1)!} \\ 0 & N & \frac{N(N-1)}{2}\Delta & \cdots & S_{r-3}^{N-1}\frac{\Delta^{r-3}}{(r-3)!} & S_{r-2}^{N-1}\frac{\Delta^{r-2}}{(r-2)!} \\ 0 & 0 & N & \cdots & S_{r-4}^{N-1}\frac{\Delta^{r-4}}{(r-4)!} & S_{r-3}^{N-1}\frac{\Delta^{r-3}}{(r-3)!} \\ \vdots & \vdots & & \ddots & \vdots & \vdots \\ 0 & 0 & 0 & \cdots & N & \frac{N(N-1)}{2}\Delta \\ 0 & 0 & 0 & \cdots & 0 & N \end{bmatrix}$$

where

$$S_r^{N-1} = \sum_{i=0}^{N-1} i^r = \frac{N^{r+1}}{r+1} + \sum_{k=1}^{r} \frac{B_k}{r-k+1}\binom{r}{k} N^{r-k+1} \tag{9.41}$$

and where B_k is the kth Bernoulli number.

Finally, if N is a fixed number, then both $A^N \cdot \theta[k]$ and $\sum_{i=0}^{N-1} A^i \cdot \Phi \cdot E$ preserve the order of $\theta[k]$ element-wise.

The proof for the remaining states η is analogous to the proof for z_r.

3. Based on the proof of part (ii), note that the sum $\sum_{i=0}^{N-1} i^r$ is a polynomial in N of order $r+1$ (see the references at the end of the chapter). Therefore, since $N = T/\Delta$ with T fixed, every element in the last column of the matrix $\sum_{i=0}^{N-1} A^i$ is of order $\mathcal{O}(\Delta^{-1})$. Then, given that the last row of Φ is Δ, there will always be a term of order $\mathcal{O}(1)$ multiplying E for every state. Finally, it has already been established that $E \in \mathcal{O}(\Delta)$. Thus, it is straightforward to show that

$$\sum_{i=0}^{N-1} A^i \cdot \Phi \cdot E \in \mathcal{O}(\Delta) \tag{9.42}$$

On the other hand, note that all the elements of A^N are of order $\mathcal{O}(1)$, and since the initial errors are at least $\mathcal{O}(\Delta)$ in all states by assumption ($\bar{m}_i \geq \tilde{m}_i$), we can conclude that

$$e_i[k+N] \in \mathcal{O}(\Delta), \quad i = 1, \ldots, r. \tag{9.43}$$

The proof for the remaining states η is analogous to the proof for z_r.

□

Remark 9.9 Parts (i) and (ii) of the above result are very promising. For example, recall that Euler integration gives errors of order Δ^2 in every state. The TTS model is only marginally more complex than the Euler model, but it has significantly better local error properties.

Remark 9.10 Part (iii) of the above theorem may, at first, seem disappointing. The result in part (iii) has its roots in the fact that the lower order errors in the lower part of the chain of integrators, e.g., the r-th state, appear (and accumulate) in the upper part of the chain, e.g., the first state. This behaviour is the phenomenon that has led to the use of the term global truncation error in the numerical analysis literature. Typically only one order of Δ is lost from local errors when global errors are considered. However, because of the structure of the normal model, instead of dropping only one order from the local to the global truncation error, the TTS model drops all the way down to Δ for all states.

The following example illustrates the above results.

Example 1 Consider the following second order nonlinear system:

$$\dot{x}_1(t) = x_2(t) \tag{9.44}$$

$$\dot{x}_2(t) = -\alpha_1 x_2(t) - \alpha_0 x_1(t)\big(1 + \varepsilon_1 x_1(t)^2\big) + \beta_0\big(1 + \varepsilon_2 x_1(t)\big)u(t) \tag{9.45}$$

The system is already in the normal form and has relative degree equal to 2. The corresponding approximate, TTS, sampled model is:

$$x_1^+ = x_1 + \Delta x_2 + \frac{\Delta^2}{2}\big(-\alpha_1 x_2 - \alpha_0 x_1\big(1 + \varepsilon_1 x_1^2\big) + \beta_0(1 + \varepsilon_2 x_1)u\big) \tag{9.46}$$

$$x_2^+ = x_2 + \Delta\big(-\alpha_1 x_2 - \alpha_0 x_1\big(1 + \varepsilon_1 x_1^2\big) + \beta_0(1 + \varepsilon_2 x_1)u\big) \tag{9.47}$$

The system parameters are chosen as $\alpha_0 = 2$, $\alpha_1 = 3$, $\beta_0 = 2$, $\varepsilon_1 = 0.5$, $\varepsilon_2 = 0.5$. According to Theorem 9.8, this system has local vector truncation error and local vector fixed-steps truncation error of order (Δ^3, Δ^2) and global vector fixed-time truncation error of order (Δ, Δ). The exact discrete-time model is impossible to obtain. However, in order to accurately describe the errors, the continuous-time system is simulated using Simulink with ode45 (Dormand-Prince) as a solver. This gives high accuracy. The state signals are then sampled at the corresponding sampling instants.

The behaviour of the truncation errors is analysed as the sampling period is varied, specifically as it tends to zero as $\Delta = T/2^n$, where $n = 3, \ldots, 10$ and $T = 1[s]$. A step input and zero initial conditions are considered.

Figure 9.2 shows the discretisation errors obtained after one iteration of the approximate model, for both states of the system, for different sampling times. It can be seen that, for the TTS model, the slope is 30 dB/decade for the state x_1. This confirms that the error is of order Δ^3. Also, the figure shows that the error in the

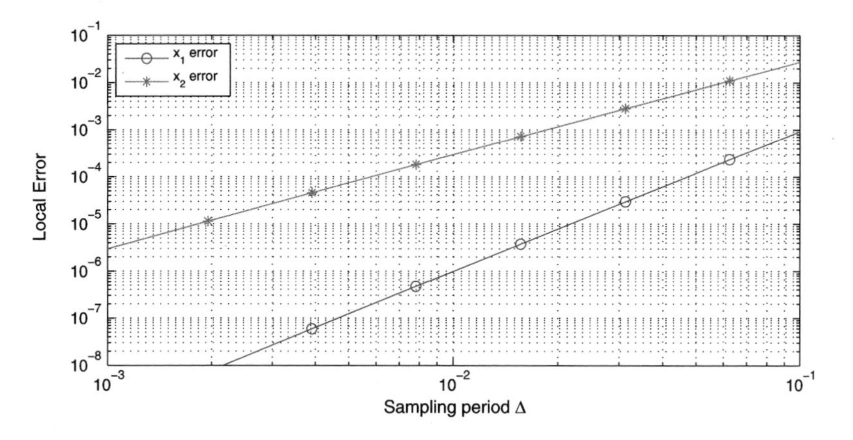

Fig. 9.2 Local truncation errors for x_1 and x_2

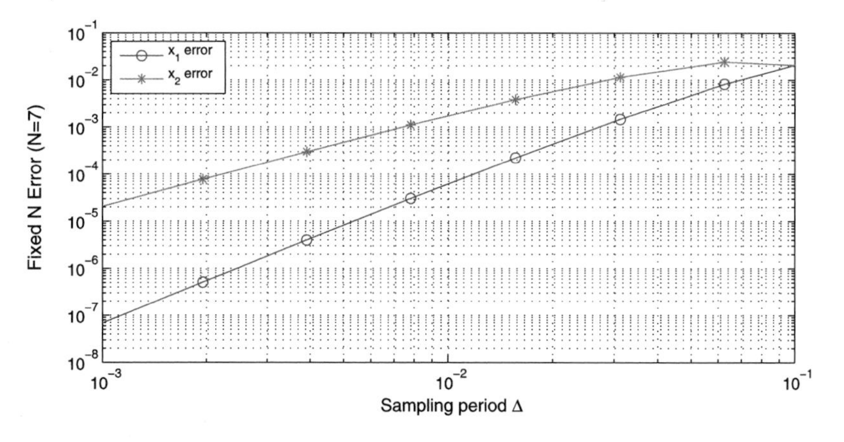

Fig. 9.3 Local fixed-steps truncation errors with $N = 7$

state x_2 is of the order of Δ^2. These results are in accord with the conclusions of part (i) of Theorem 9.8.

Figure 9.3 shows the errors after a fixed number of steps, $N = 7$. It can be seen that, for a sufficiently small sampling period, the accumulated errors still preserve the behaviour of the local truncation error. This result is in accord with the conclusions of part (ii) of Theorem 9.8.

Finally, Fig. 9.4 shows the error for a fixed continuous-time interval, i.e., after N steps, where $N = T/\Delta$ with $T = 1$. It can be seen that the errors in both states are now of the order of Δ. This result is in accord with the conclusions of part (iii) of Theorem 9.8.

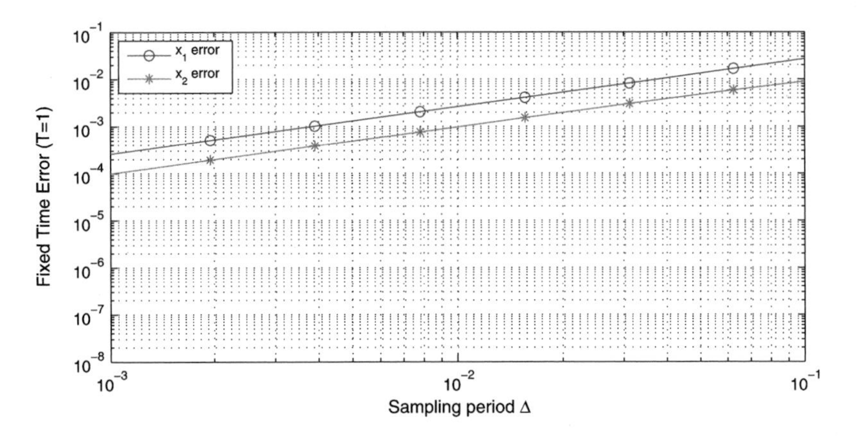

Fig. 9.4 Global fixed-time truncation errors ($N = T/\Delta$)

9.5 Summary

The key points covered in this chapter are as follows:

- Exact sampled-data models for nonlinear systems are generally not available.
- An approximate discretisation strategy based on a (variable) truncated Taylor series (TTS) expansion can be applied to a class of nonlinear systems expressed in the normal form.
- The accuracy of the associated approximate sampled model can be characterised in terms of time-domain measures. Specifically, the TTS model has:
  1. Local vector truncation error of the order of $(\Delta^{r+1}, \Delta^r, \ldots, \Delta^2; \Delta^2)$,
  2. Local vector fixed-steps truncation error of the order of $(\Delta^{r+1}, \Delta^r, \ldots, \Delta^2; \Delta^2)$,
  3. Global vector fixed-time truncation error of the order of $(\Delta, \ldots; \Delta)$.

Further Reading

A general introduction to the accuracy of numerical integration schemes is given in

Butcher JC (2008) Numerical methods for ordinary differential equations, 2nd edn. Wiley, New York

Normal forms for nonlinear systems are described in detail in

Isidori A (1995) Nonlinear control systems, 3rd edn. Springer, Berlin

Discrete-time normal forms are discussed in

Califano C, Monaco S, Normand-Cyrot D (1998) On the discrete-time normal form. IEEE Trans Autom Control 43(11):1654–1658

State-space models for nonlinear systems, in both continuous time and discrete time, are discussed in

Belikov J, Kotta U, Tonso M (2012) State-space realization of nonlinear control systems: unification and extension via pseudo-linear algebra. Kybernetika 48(6):1100–1113

Approximate models for nonlinear systems based on variable truncated Taylor series were first described in

Yuz JI, Goodwin GC (2005) On sampled-data models for nonlinear systems. IEEE Trans Autom Control 50(10):1477–1489

The definitions of vector truncation errors and an associated accuracy study of truncated Taylor series models were first presented in

Carrasco DS, Goodwin GC, Yuz JI (2013) Vector measures of accuracy for sampled data models of nonlinear systems. IEEE Trans Autom Control 58(1):224–230

Sampled-data models for Hamiltonian systems have been proposed in

Laila D, Astolfi A (2006) Construction of discrete-time models for port-controlled Hamiltonian systems with applications. Syst Control Lett 55:673–680
Monaco S, Normand-Cyrot D, Tiefensee F (2011) Sampled-data stabilization; a PBC approach. IEEE Trans Autom Control 56(4):907–912

The problem of sampled-data control for nonlinear systems has been extensively investigated. See, for example

Laila D, Nesic D, Astolfi A (2006) Sampled-data control of nonlinear systems. In: Advanced topics in control systems theory II, pp 91–137
Nesic D, Teel AR (2004) A framework for stabilization of nonlinear sampled-data systems based on their approximate discrete-time models. IEEE Trans Autom Control 49(7):1103–1122

Chapter 10
Applications of Approximate Sampled-Data Models in Estimation and Control

This chapter investigates the application of the approximate sampled-data models described in earlier chapters to problems arising in estimation and control. Both linear and nonlinear systems are considered. It is demonstrated, via several examples, that including sampling zero dynamics can have a significant impact on achievable performance.

10.1 When Are Deterministic Sampling Zeros Important?

Before delving into specific case studies, it is relevant to ask, at a heuristic level, under what circumstances deterministic sampling zeros are likely to be important.

The answer to the question posed above is related to how the sampled-data model is used. Two possible scenarios (amongst many) are considered below.

1. Design bandwidth is small relative to the sampling frequency.
   Consider the situation where the goal is to achieve a fixed (continuous) response time. In this case, the design can be thought of as achieving a 'design bandwidth' which is independent of Δ. Next say that the sampling period is selected in such a way that the Nyquist frequency (π/Δ) is well above the chosen design bandwidth. In this case, it is heuristically reasonable that the sampling zeros will play a diminishing role as the sampling frequency increases, because the sampling zeros impact the discrete-time response near the Nyquist frequency. Specifically, provided the sampling frequency is greater (say by an order of magnitude) than the design bandwidth, it seems reasonable to expect that the derivative replacement model obtained by means of Euler integration (which does not include the sampling zero dynamics) should be adequate.
2. Design bandwidth is comparable to the sampling frequency.
   A different situation arises when the desired response time is comparable to the sampling period. Then, the 'design bandwidth' can be said to be comparable to the Nyquist frequency (π/Δ). In this case, it seems reasonable to expect that one needs to accurately capture the behaviour of the model in the vicinity of

J.I. Yuz, G.C. Goodwin, *Sampled-Data Models for Linear and Nonlinear Systems*, Communications and Control Engineering, DOI 10.1007/978-1-4471-5562-1_10,

the Nyquist frequency. This, in turn, implies that a discrete-time model which includes the sampling zero dynamics may be needed.

These ideas are illustrated in the case studies presented below.

10.2 State Feedback Control for Linear Systems

Consider a *high-gain* controller applied to a linear system. Based on the discussion in Sect. 10.1, such a controller is likely to be sensitive to high frequency model errors, because the 'design bandwidth' is comparable to the Nyquist frequency. This conclusion is borne out in the following example.

Example 10.1 Consider a third order continuous-time system represented by the transfer function

$$G(s) = \frac{1}{(s-1)^3} \tag{10.1}$$

The corresponding state-space model in the normal form is

$$\begin{cases} \dot{x} = \begin{bmatrix} 0 & 1 & 0 \\ 0 & 0 & 1 \\ 1 & -3 & 3 \end{bmatrix} x + \begin{bmatrix} 0 \\ 0 \\ 1 \end{bmatrix} u \\ y = [1 \quad 0 \quad 0] x \end{cases} \tag{10.2}$$

Note that the system is *unstable* and has relative degree 3.

The sampling period is chosen as $\Delta = 0.01$. The exact sampled-data model for the system, assuming a zero order hold (ZOH) input, can be readily obtained and is given by

$$G_q(z) = \frac{1.6792 \times 10^{-7}(z+3.76)(z+0.27)}{(z-1.01)^3} \tag{10.3}$$

However, the goal here is to explore the potential impact of using approximate models in the design process. Two models are examined: the SDR and TTS models.

The simple derivative replacement (SDR) model is given by

$$G_q^{\mathrm{SDR}}(z) = \frac{10^{-6}}{(z-1.01)^3} \tag{10.4}$$

The equivalent state-space SDR model for the system in the normal form (10.2) is:

$$\begin{cases} (\frac{q-1}{\Delta}) x_k = \begin{bmatrix} 0 & 1 & 0 \\ 0 & 0 & 1 \\ 1 & -3 & 3 \end{bmatrix} x_k + \begin{bmatrix} 0 \\ 0 \\ 1 \end{bmatrix} u_k \\ y_k = [1 \quad 0 \quad 0] x_k \end{cases} \tag{10.5}$$

The alternative model obtained by the truncated Taylor series (TTS) method has transfer function

$$G_q^{\mathrm{TTS}}(z) = \frac{1.6667 \times 10^{-7}(z+3.732)(z+0.2679)}{(z-1.009)(z^2-2.021z+1.021)} \tag{10.6}$$

Note that this model is extremely close to the true model (10.3).

Assume that all the states (i.e., the output y and its derivatives) are sampled at period $\Delta = 0.01$. This is physically reasonable given that the system has relative degree 3, and, hence, the output trajectories are *smooth*.

Design a state variable feedback, using the two approximate sampled-data models (10.4), (10.6), to place the closed-loop poles at the origin in the shift domain.

For the SDR model, the state feedback gain is found to be

$$K_d^{\mathrm{SDR}} = 10^6 \begin{bmatrix} 1.000001 & 0.029997 & 0.000303 \end{bmatrix} \tag{10.7}$$

whereas for the TTS model, the state feedback gain is found to be

$$K_d^{\mathrm{TTS}} = 10^6 \begin{bmatrix} 1.000001 & 0.019997 & 0.0001863 \end{bmatrix} \tag{10.8}$$

When applied to the true plant, it is found that K_d^{TTS} gives excellent results. The true closed-loop poles are located at $\{-0.2902, 0.1257 \pm j0.1497\}$. These are very close to the design values at the origin. Certainly the true system is stable with this control law. However, when using K_d^{SDR}, the state feedback strategy fails to stabilise the true system. In fact, the closed-loop poles are now located at $\{-2.7079, 0.4897 \pm j0.1157\}$.

Thus, for this example, good performance can be achieved when the design is based on an approximate model, but it is crucial that the model include key aspects of the true system, e.g., the sampling zero dynamics.

10.3 State Estimation for Linear Systems

Next consider the design of a *high-gain* observer so that state estimates are obtained in a few samples. Specifically, the observer error poles are placed at the origin (in the shift operator domain). The reason for examining high-gain observers is that there is a natural nonlinear extension to these ideas. For this scenario, the 'design bandwidth' is comparable again to the sampling frequency, and hence it is reasonable to anticipate (based on the discussion in Sect. 10.1) that sampling zero dynamics will be important.

Example 10.2 (Example 10.1 Continued) Design an observer having error dynamics with poles at the origin. For the SDR model the following observer gains are obtained:

$$L_d^{\text{SDR}} = \begin{bmatrix} 3.03 \\ 309.06 \\ 10918.1 \end{bmatrix} \tag{10.9}$$

This observer recovers the true states in three samples when applied to the approximate system.

An alternative observer design is carried out based on the TTS model (assigning the error poles to the origin). The observer gain is

$$L_d^{\text{TTS}} = \begin{bmatrix} 3.029 \\ 255.487 \\ 10601.9 \end{bmatrix} \tag{10.10}$$

10.4 Output Feedback Control for Linear Systems

Example 10.3 (Example 10.1 Continued) Finally, the two elements (observer plus state estimate feedback) designed in the previous two sections are combined to generate an output feedback controller. Not surprisingly, given the results in Sect. 10.2, the output feedback controller based on the SDR model fails to stabilise the *true* plant. In fact, the closed-loop poles are now located at

$$\{-2.5167, 0.1002 \pm j2.5682, 0.7982 \pm j0.3139, 0.7201\} \tag{10.11}$$

On the other hand, the output feedback controller based on the approximate TTS model gives excellent results and the closed-loop poles are located at

$$\{-0.4849 \pm j0.0353, 0.0402 \pm j0.6392, 0.4451 \pm j0.1950\} \tag{10.12}$$

Example 10.4 (Example 10.1 Continued) It is also of interest to see if it is possible to use the SDR model to design the controller but use the more accurate TTS model to design the observer. Testing the idea, it is found that the combination $\{K^{\text{SDR}}, L^{\text{TTS}}\}$ fails to stabilise the true plant.

Similarly, if the controller design is based on the TTS model and the observer design on the SDR model, then the resulting combination $\{K^{\text{TTS}}, L^{\text{SDR}}\}$ again fails to stabilise the true plant.

Thus, the only combination that leads to satisfactory results when applied to the true plant is the combination $\{K^{\text{TTS}}, L^{\text{TTS}}\}$.

We conclude from Examples 10.1 through 10.4 that the asymptotic sampling zeros play a crucial role in obtaining satisfactory designs when:

(i) the sampling rate is high, and
(ii) when rapid responses are sought (i.e., the design bandwidth is near the Nyquist frequency).

Interestingly, the above designs can be carried out to a high degree of accuracy if the true system dynamics are replaced by a pure integrator. This observation will be exploited in the next two sections dealing with nonlinear systems.

10.5 Sampled-Data State Feedback Control for Nonlinear Systems

Here the ideas of Sects. 10.1 and 10.2 are extended to the case of nonlinear systems. The goal is to again demonstrate the importance of having a good approximate sampled-data model.

The discussion is based on an underlying continuous-time nonlinear system in the normal form. Assume that the system has relative degree r and that the input appears in an affine fashion. Then the corresponding model can be described in the following form:

$$\begin{aligned}
\dot{\xi}_1 &= \xi_2 \\
\dot{\xi}_2 &= \xi_3 \\
&\vdots \\
\dot{\xi}_r &= a(\xi, \eta) + b(\xi, \eta)u \\
\dot{\eta} &= c(\xi, \eta) \\
y &= \xi_1
\end{aligned} \tag{10.13}$$

It is assumed in the sequel that the intrinsic zero dynamics are stable.

A design procedure is developed for the corresponding sampled-data system. The design will be based on the approximate TTS sampled-data model, which takes the form:

$$\begin{aligned}
\delta\bar{\xi}_1 &= \bar{\xi}_2 + \frac{\Delta}{2}\bar{\xi}_3 + \cdots + \frac{\Delta^{r-1}}{r!}\left[a(\bar{\xi}, \bar{\eta}) + b(\bar{\xi}, \bar{\eta})u\right] \\
&\vdots \\
\delta\bar{\xi}_r &= a(\bar{\xi}, \bar{\eta}) + b(\bar{\xi}, \bar{\eta})u \\
\delta\bar{\eta} &= c(\bar{\xi}, \bar{\eta}) \\
y &= \bar{\xi}_1
\end{aligned} \tag{10.14}$$

It is assumed that $b(\bar{\xi}, \bar{\eta})$ is non-zero for all $\bar{\xi}, \bar{\eta}$ in the region of interest. Also, it is assumed that samples of all states are available to the controller.

The design strategy described below is inspired by feedback linearisation (see the references given at the end of the chapter). In this framework, a new input variable υ is defined via

$$\upsilon = a(\bar{\xi}, \bar{\eta}) + b(\bar{\xi}, \bar{\eta})u \tag{10.15}$$

Note that, for the purpose of design, the difference between the exact and approximate model is ignored. Substituting (10.15) into the TTS model yields:

$$\begin{aligned}
\delta\bar{\xi}_1 &= \bar{\xi}_2 + \frac{\Delta}{2}\bar{\xi}_3 + \cdots + \frac{\Delta^{r-1}}{r!}\upsilon \\
&\vdots \\
\delta\bar{\xi}_r &= \upsilon \\
\delta\bar{\eta} &= c(\bar{\xi}, \bar{\eta})
\end{aligned} \tag{10.16}$$

This model is interesting. It is linear but contains unstable zero dynamics arising from the asymptotic sampling zero dynamics.

Using the above approximate model, state variable feedback is designed to place the closed-loop poles at any desired location leading to a closed-loop system of the form:

$$\begin{aligned}
\delta\bar{\xi}_1 &= \bar{\xi}_2 + \frac{\Delta}{2}\bar{\xi}_3 + \cdots + \frac{\Delta^{r-1}}{r!}(-K \cdot \bar{\xi}) \\
\delta\bar{\xi}_2 &= \bar{\xi}_3 + \cdots + \frac{\Delta^{r-2}}{(r-1)!}(-K \cdot \bar{\xi}) \\
&\vdots \\
\delta\bar{\xi}_r &= (-K \cdot \bar{\xi}) \\
\delta\bar{\eta} &= c(\bar{\xi}, \bar{\eta})
\end{aligned} \tag{10.17}$$

Note that 'high-gain' feedback is utilised to ensure that the intrinsic zero dynamics become (approximately) part of the closed-loop system. In particular, the closed-loop poles for the (linear) sampled data (10.17) are assigned to the point $-\frac{1}{\Delta}$ in the δ-domain (equivalently the origin in the shift domain).

Stability and performance issues are explored via a sequence of examples.

Example 10.5 Consider the continuous-time nonlinear system

$$\dot{x}_1 = x_2 \tag{10.18}$$

$$\dot{x}_2 = \big(1 + |x_1|\big)u \tag{10.19}$$

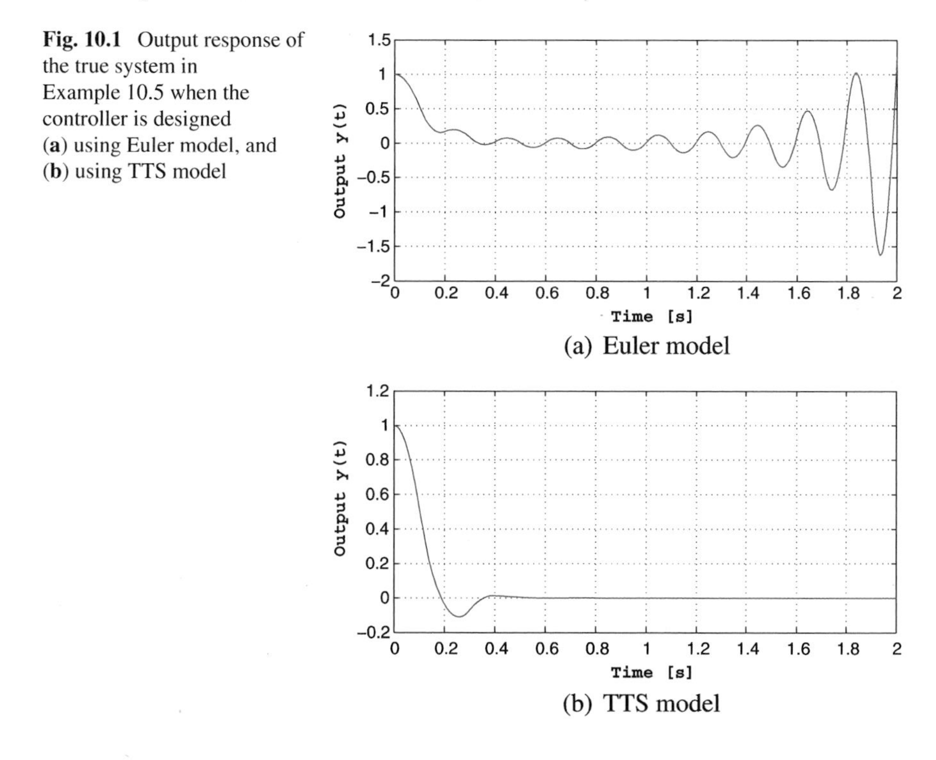

Fig. 10.1 Output response of the true system in Example 10.5 when the controller is designed (**a**) using Euler model, and (**b**) using TTS model

A sampling period of $\Delta = 0.1$ is chosen. The designs are based on two approximate models: (i) the Euler model and (ii) the TTS model. Linearisation via the substitution (10.15) is first employed. The discrete-time closed-loop poles are then assigned to the origin in the shift domain. The resultant controllers are applied to the true system (simulated using a high order numerical integration routine). The initial state was chosen to be $(1, 0)$.

The results are shown in Fig. 10.1. It can be seen from Fig. 10.1(a) that the design based on the Euler model is unstable when applied to the true system. However, the design based on the TTS model gives excellent performance when applied to the true system.

Next a more demanding example is explored.

Example 10.6 Consider the continuous-time system

$$\dot{x}_1 = x_2 \tag{10.20}$$

$$\dot{x}_2 = \big(1 + 0.2|x_2|\big)u \tag{10.21}$$

A sampling period of $\Delta = 0.1$ is chosen. The designs are again based on two approximate models, (i) the Euler model and (ii) the TTS model. Linearisation via the substitution (10.15) is first employed. The discrete-time closed-loop poles are then assigned to the origin in the shift domain. The resultant controllers are applied

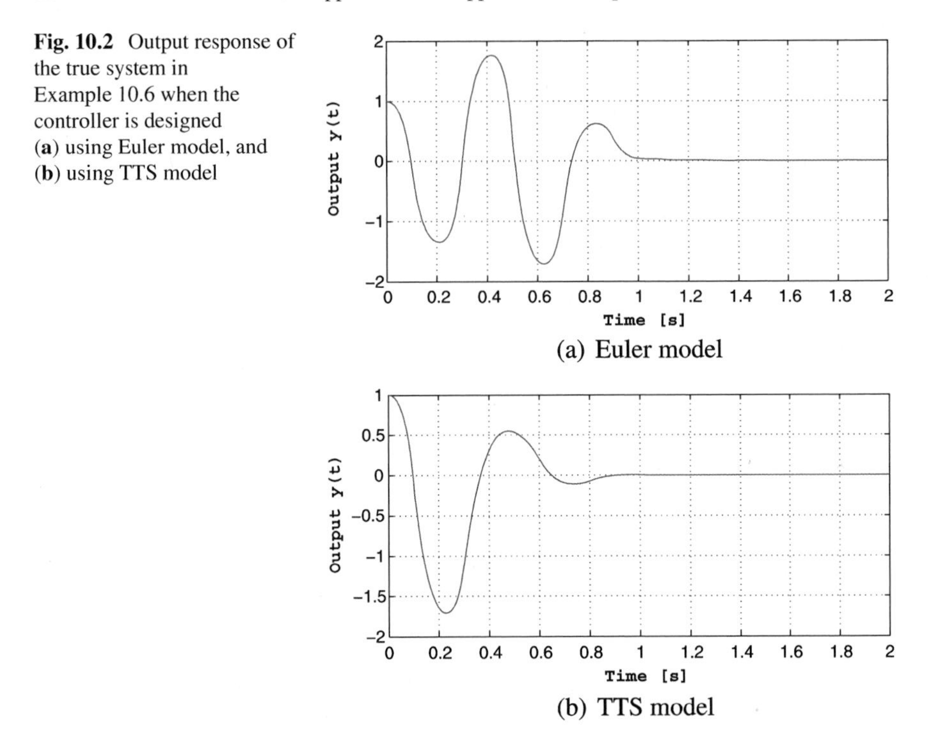

Fig. 10.2 Output response of the true system in Example 10.6 when the controller is designed (**a**) using Euler model, and (**b**) using TTS model

to the true system (simulated using a high order numerical integration routine). The initial state is chosen as $(1, 0)$.

The results are shown in Fig. 10.2. It can be seen from Fig. 10.2(a) that the design based on the Euler model gives poor performance, whereas, as seen in Fig. 10.2(b), the design based on the TTS model gives improved performance.

Finally, a very demanding case is examined.

Example 10.7 Consider the following continuous-time system:

$$\dot{x}_1 = x_2 \tag{10.22}$$

$$\dot{x}_2 = \left(1 + kx_2^2\right)u \tag{10.23}$$

A sampling period of $\Delta = 0.1$ is chosen. The designs are based on two approximate models, (i) the Euler model and (ii) the TTS model. Linearisation via the substitution (10.15) is first employed. The discrete-time closed-loop poles are again assigned to the origin in the shift domain. The resultant controllers are applied to the true system (simulated using a high order numerical integration routine).

For the case $k = 0.01$, the results are shown in Fig. 10.3. Again, it can be seen that the design based on the Euler model gives inferior performance to the design based on the TTS model.

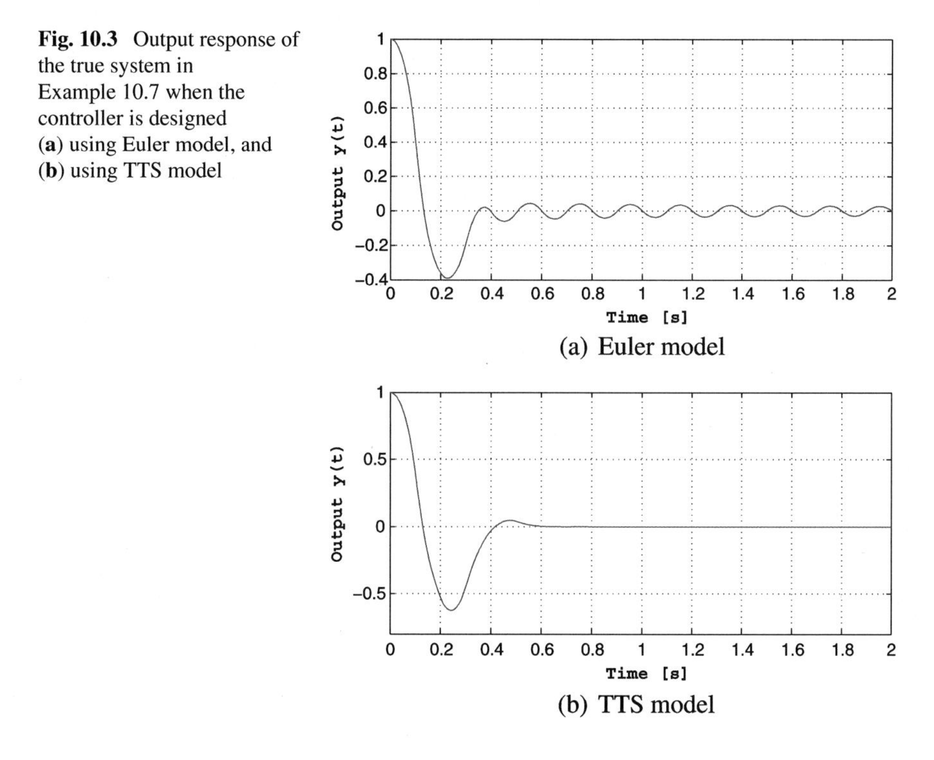

(a) Euler model

(b) TTS model

Fig. 10.3 Output response of the true system in Example 10.7 when the controller is designed (**a**) using Euler model, and (**b**) using TTS model

Finally, the case $k = 1$ is tested. This problem has been studied previously in the literature. It is a demanding case because, with rapid transients, the velocity term in the state-space model, x_2, is large and varies rapidly. This means that the Euler approximation used in both models for the second state is suspect. Indeed, when the discrete-time closed-loop poles are assigned to the origin, the two controllers based on both approximate models lead to instability when used from the initial state $(1, 0)$.

Remark 10.8 The above examples confirm the key claim that TTS models can extend the range of applicability of simple sampled-data controllers. Naturally, there is a limit to how far this can be pushed because, for some cases, the use of the Euler model for the rth state turns out to be inadequate even though the TTS model captures the sampling zero dynamics.

10.6 Continuous-Time System Identification

This section examines a different realm of application for sampled-data models. Here the problem of identification of continuous-time systems using least squares based on equation error is studied. In particular, the impact of incorporating, or not incorporating, sampling zeros in the sampled-data model is explored.

The identification algorithm will use the 'full bandwidth' of available data. Hence, based on the remarks made in Sect. 10.1, we anticipate that sampling zero dynamics should play an important role.

The ideas are illustrated by the following example, which considers a deterministic second order system with known input. The parameters of the system are estimated using different sampled-data model structures.

Example 10.9 Consider the continuous-time linear system

$$\left(\rho^2+\alpha_1\rho+\alpha_0\right)y=\beta_0 u \tag{10.24}$$

where ρ denotes the derivative operator.

For illustration purposes, the continuous-time parameters are chosen to be $\alpha_1=3$, $\alpha_0=2$, $\beta_0=2$. System identification is performed assuming three different model structures:

SDRM: Simple Derivative Replacement Model. This corresponds to the structure given in Sect. 8.1.2, where the approximate sampled-data model is obtained by replacing the continuous-time derivatives by divided differences.

MIFZ: Model Including Fixed Sampling Zero. This model appends the asymptotic zero, assuming a structure as in Sect. 8.1.3.

MIPZ: Model Including Parameterised Zero. This model also includes a sampling zero, whose location is included as part of the estimation problem; i.e., a discrete-time model structure is utilised having a zero whose value is not fixed a priori but is determined via the identification procedure.

The three discrete-time models are represented in terms of the δ-operator as follows:

$$G_\delta(\gamma)=\frac{N_\delta(\gamma)}{\gamma^2+\hat{\alpha}_1\gamma+\hat{\alpha}_0} \tag{10.25}$$

where:

$$N_\delta(\gamma)=\begin{cases}\hat{\beta}_0 & \text{(SDRM)}\\ \hat{\beta}_0(1+\frac{\Delta}{2}\gamma) & \text{(MIFZ)}\\ \hat{\beta}_0+\hat{\beta}_1\gamma & \text{(MIPZ)}\end{cases} \tag{10.26}$$

A sampling period of $\Delta=\pi/100$[s] is used. The input u_k is chosen to be a random Gaussian sequence of unit variance. The output sequence $y_k=y(k\Delta)$ has been obtained by simulating the continuous-time system and sampling its output. Also note that the data is free of any measurement noise.

The parameters are estimated using a quadratic cost function on the equation errors, i.e.,

$$J(\hat{\theta})=\frac{1}{N}\sum_{k=0}^{N-1}e_k(\hat{\theta})^2=\frac{1}{N}\sum_{k=0}^{N-1}\left(\delta^2 y_k-\phi_k^T\theta\right)^2 \tag{10.27}$$

Table 10.1 Parameter estimates for a linear system

	Parameters		Estimates		
	CT	Exact DT	SDRM	MIFZ	MIPZ
α_1	3	2.923	2.8804	2.9471	2.9229
α_0	2	1.908	1.9420	1.9090	1.9083
β_1	–	0.0305	–	$\frac{\beta_0 \Delta}{2} = 0.03$	0.0304
β_0	2	1.908	0.9777	1.9090	1.9083

where:

$$\phi_k = \begin{cases} [-\delta y_k, -y_k, u_k]^T \\ [-\delta y_k, -y_k, (1 + \frac{\Delta}{2}\delta)u_k]^T \\ [-\delta y_k, -y_k, \delta u_k, u_k]^T \end{cases} \quad \text{and} \quad \hat{\theta} = \begin{cases} [\alpha_1, \alpha_0, \beta_0]^T & \text{(SDRM)} \\ [\alpha_1, \alpha_0, \beta_0]^T & \text{(MIFZ)} \\ [\alpha_1, \alpha_0, \beta_1, \beta_0]^T & \text{(MIPZ)} \end{cases} \tag{10.28}$$

Table 10.1 shows the estimation results. Note that the system considered is linear; thus, the exact discrete-time parameters can be computed for the given sampling period. These are also given in Table 10.1 (CT: continuous-time, Exact DT: exact discrete-time).

It can be seen from the table that both the MIFZ and MIPZ, which incorporate a sampling zero, are able to recover the continuous-time parameters to a high degree of accuracy. However, the estimate $\hat{\beta}_0$ obtained using the SDRM is clearly biased.

At first sight, the result in the previous example may seem surprising because, even though the SDRM in (10.29) converges to the continuous-time system as the sampling period goes to zero, the estimate $\hat{\beta}_0$ does not converge to the underlying continuous-time parameter. The estimate is biased. Indeed, β_0 is incorrectly estimated by a factor of 2 by the SDRM. The key issue here is that the 'full data bandwidth' has been used for the purpose of identification. Hence, it is not surprising, based on the comments in Sect. 10.1, that sampling zeros are important.

The following result formally establishes the asymptotic bias that was observed experimentally for the SDRM structure in the previous example. In particular, it is shown that β_0 is indeed underestimated by a factor of 2.

Lemma 10.10 *Consider the general continuous-time second order deterministic system given in* (10.9). *Assume that sampled data is collected from the system using a ZOH input generated from a discrete-time white noise sequence u_k, and the sampled output $y_k = y(k\Delta)$ is measured.*

An equation error identification procedure (*see* (10.31)) *is utilised to estimate the parameters of the SDRM*:

$$\delta^2 y + \hat{\alpha}_1 \delta y + \hat{\alpha}_0 y = \hat{\beta}_0 u \tag{10.29}$$

Then the parameter estimates asymptotically converge, as the sampling period Δ goes to zero, as follows:

$$\hat{\alpha}_1 \to \alpha_1, \quad \hat{\alpha}_0 \to \alpha_0, \quad \textit{and} \quad \hat{\beta}_0 \to \frac{1}{2}\beta_0 \tag{10.30}$$

Proof The parameters of the approximate SDRM (10.29) are obtained by simple least squares, minimising the equation error cost function:

$$J(\hat{\theta}) = \lim_{N\to\infty} \frac{1}{N} \sum_{0}^{N-1} e_k(\hat{\theta})^2 = E\{e_k(\hat{\theta})^2\} \tag{10.31}$$

where $e_k = \delta^2 y + \hat{\alpha}_1 \delta y + \hat{\alpha}_0 y - \hat{\beta}_0 u$. The parameter estimates are given by the solution of $\frac{dJ(\hat{\theta})}{d\hat{\theta}} = 0$. Thus, differentiating the cost function with respect to each of the parameter estimates, the following set of equations is obtained:

$$\begin{bmatrix} E\{(\delta y)^2\} & E\{(\delta y) y\} & -E\{(\delta y) u\} \\ E\{(\delta y) y\} & E\{y^2\} & -E\{yu\} \\ -E\{y^2\} & -E\{yu\} & E\{u^2\} \end{bmatrix} \begin{bmatrix} \hat{\alpha}_1 \\ \hat{\alpha}_0 \\ \hat{\beta}_0 \end{bmatrix} = \begin{bmatrix} -E\{(\delta y)(\delta^2 y)\} \\ -E\{y \delta^2 y\} \\ E\{u \delta^2 y\} \end{bmatrix} \tag{10.32}$$

This equation set can be rewritten in terms of (discrete-time) correlations as:

$$\begin{bmatrix} \frac{2r_y(0)-2r_y(1)}{\Delta^2} & \frac{r_y(1)-r_y(0)}{\Delta} & \frac{r_{yu}(0)-r_{yu}(1)}{\Delta} \\ \frac{r_y(1)-r_y(0)}{\Delta} & r_y(0) & -r_{yu}(0) \\ \frac{r_{yu}(0)-r_{yu}(1)}{\Delta} & -r_{yu}(0) & r_u(0) \end{bmatrix} \begin{bmatrix} \hat{\alpha}_1 \\ \hat{\alpha}_0 \\ \hat{\beta}_0 \end{bmatrix} = \begin{bmatrix} \frac{3r_y(0)-4r_y(1)+r_y(2)}{\Delta^3} \\ \frac{-r_y(0)+2r_y(1)-r_y(2)}{\Delta^2} \\ \frac{r_{yu}(0)-2r_{yu}(1)+r_{yu}(2)}{\Delta^2} \end{bmatrix} \tag{10.33}$$

To continue with the proof, expressions for the correlations involved in the last equation are required. Note that the input sequence is discrete-time white noise having unit variance. Hence,

$$r_u(k) = \delta_K[k] \quad \Longleftrightarrow \quad \Phi_u^q(e^{j\omega\Delta}) = 1 \tag{10.34}$$

The other correlation functions can be obtained from the relations:

$$\begin{aligned} r_{yu}(k) &= \mathcal{F}_d^{-1}\{\Phi_{yu}^q(e^{j\omega\Delta})\} = \mathcal{F}_d^{-1}\{G_q(e^{j\omega\Delta})\Phi_u^q(e^{j\omega\Delta})\} \\ &= \mathcal{F}_d^{-1}\{G_q(e^{j\omega\Delta})\} \end{aligned} \tag{10.35}$$

$$\begin{aligned} r_y(k) &= \mathcal{F}_d^{-1}\{\Phi_y^q(e^{j\omega\Delta})\} = \mathcal{F}_d^{-1}\{G_q(e^{-j\omega\Delta})\Phi_{yu}^q(e^{j\omega\Delta})\} \\ &= \mathcal{F}_d^{-1}\{|G_q(e^{j\omega\Delta})|^2\} \end{aligned} \tag{10.36}$$

where $G_q(e^{j\omega\Delta})$ is the exact sampled-data model corresponding to the continuous-time system (10.9). Given a sampling period Δ, the exact discrete-time model is given by

$$G_q(z) = \frac{\beta_0(c_1 z + c_0)}{(z - e^{\lambda_1\Delta})(z - e^{\lambda_2\Delta})} \tag{10.37}$$

where:

$$c_1 = \frac{(e^{\lambda_1 \Delta} - 1)\lambda_2 - (e^{\lambda_2 \Delta} - 1)\lambda_1}{(\lambda_1 - \lambda_2)\lambda_1 \lambda_2} = \frac{\Delta^2}{2} + \frac{\Delta^3}{6}(\lambda_1 + \lambda_2) + \cdots \tag{10.38}$$

$$c_0 = \frac{e^{\lambda_1 \Delta}(e^{\lambda_2 \Delta} - 1)\lambda_1 - e^{\lambda_2 \Delta}(e^{\lambda_1 \Delta} - 1)\lambda_2}{(\lambda_1 - \lambda_2)\lambda_1 \lambda_2} = \frac{\Delta^2}{2} + \frac{\Delta^3}{3}(\lambda_1 + \lambda_2) + \cdots \tag{10.39}$$

and λ_1 and λ_2 are the continuous-time system (stable) poles of system (10.29), i.e., $\alpha_1 = -(\lambda_1 + \lambda_2)$ and $\alpha_0 = \lambda_1 \lambda_2$.

The exact discrete-time model (10.37) can be rewritten as

$$G_q(z) = \frac{C_1}{z - e^{\lambda_1 \Delta}} + \frac{C_2}{z - e^{\lambda_2 \Delta}} \tag{10.40}$$

where $C_1 = \frac{\beta_0(c_1 e^{\lambda_1 \Delta} + c_0)}{(e^{\lambda_1 \Delta} - e^{\lambda_2 \Delta})}$ and $C_2 = \frac{\beta_0(c_1 e^{\lambda_2 \Delta} + c_0)}{(e^{\lambda_2 \Delta} - e^{\lambda_1 \Delta})}$. Substituting into (10.36) leads to

$$r_{yu}(k) = \mathcal{F}_d^{-1}\{G_q(e^{j\omega\Delta})\} = (C_1 e^{\lambda_1 \Delta(k-1)} + C_2 e^{\lambda_2 \Delta(k-1)})\mu[k-1] \tag{10.41}$$

where $\mu[k]$ is the discrete-time unitary step function. From (10.37), it follows that:

$$\begin{aligned} G_q(z)G_q(z^{-1}) = {} & K_1\left(\frac{e^{\lambda_1 \Delta}}{z - e^{\lambda_1 \Delta}} + \frac{e^{-\lambda_1 \Delta}}{z - e^{-\lambda_1 \Delta}}\right) \\ & + K_2\left(\frac{e^{\lambda_2 \Delta}}{z - e^{\lambda_2 \Delta}} + \frac{e^{-\lambda_2 \Delta}}{z - e^{-\lambda_2 \Delta}}\right) \end{aligned} \tag{10.42}$$

$$K_1 = \frac{\beta_0^2(c_1^2 e^{\lambda_1 \Delta} + c_0 c_1 + c_0 c_1 e^{2\lambda_1 \Delta} + c_0^2 e^{\lambda_1 \Delta})}{(e^{2\lambda_1 \Delta} - 1)(e^{\lambda_1 \Delta} e^{\lambda_2 \Delta} - 1)(e^{\lambda_1 \Delta} - e^{\lambda_2 \Delta})} \tag{10.43}$$

$$K_2 = \frac{\beta_0^2(c_1^2 e^{\lambda_2 \Delta} + c_0 c_1 + c_0 c_1 e^{2\lambda_2 \Delta} + c_0^2 e^{\lambda_2 \Delta})}{(e^{2\lambda_2 \Delta} - 1)(e^{\lambda_2 \Delta} e^{\lambda_1 \Delta} - 1)(e^{\lambda_2 \Delta} - e^{\lambda_1 \Delta})} \tag{10.44}$$

Substituting into (10.36) leads to

$$r_y(k) = \mathcal{F}_d^{-1}\{|G_q(z = e^{j\omega\Delta})|^2\} = K_1 e^{\lambda_1 \Delta |k|} + K_2 e^{\lambda_2 \Delta |k|}; \quad \forall k \in \mathbb{Z} \tag{10.45}$$

The correlations (10.34), (10.41), and (10.45) can be used in the normal equation (10.32) to obtain:

$$\begin{bmatrix} \frac{-\beta_0^2}{2(\lambda_1+\lambda_2)}\Delta & 0 + \mathcal{O}(\Delta^2) & -\frac{\beta_0}{2}\Delta \\ 0 + \mathcal{O}(\Delta^2) & \frac{-\beta_0^2}{2(\lambda_1+\lambda_2)\lambda_1\lambda_2}\Delta & 0 \\ -\frac{\beta_0}{2}\Delta & 0 & 1 \end{bmatrix} \begin{bmatrix} \hat{\alpha}_1 \\ \hat{\alpha}_0 \\ \hat{\beta}_0 \end{bmatrix} = \begin{bmatrix} \Delta \frac{\beta_0^2}{4} \\ \Delta \frac{-\beta_0^2}{2(\lambda_1+\lambda_2)} \\ \frac{\beta_0}{2} + \mathcal{O}(\Delta) \end{bmatrix} \tag{10.46}$$

Finally, if terms of up to order Δ are combined, then:

$$\begin{bmatrix} \hat{\alpha}_1 \\ \hat{\alpha}_0 \\ \hat{\beta}_0 \end{bmatrix} = \begin{bmatrix} \frac{-2(\lambda_1+\lambda_2)}{2+(\lambda_1+\lambda_2)\Delta} \\ \lambda_1\lambda_2 \\ \frac{\beta_0(2-(\lambda_1+\lambda_2)\Delta)}{2(2+(\lambda_1+\lambda_2)\Delta)} \end{bmatrix} \xrightarrow{\Delta\to 0} \begin{bmatrix} -(\lambda_1+\lambda_2) \\ \lambda_1\lambda_2 \\ \beta_0/2 \end{bmatrix} \tag{10.47}$$

which corresponds to the result in (10.30). □

10.7 Predictive Control of an Electrical Machine

Finally, the ideas covered in Part I of the book are applied to the development of a predictive strategy for the speed control of a two-mass system driven by a permanent magnet synchronous motor (PMSM), as shown in Fig. 10.4. The proposed approach allows one to manipulate all the system variables simultaneously, including mechanical and electrical variables in a single control law.

The first step when developing the predictive control strategy is to obtain a suitable model for the system. This model is used to predict the future states of the system for a given actuation sequence. A continuous-time state-space model for the PMSM and the two-mass mechanical system is first obtained. Then an approximate sampled-data model is obtained by applying a discretisation strategy including the sampling zero dynamics.

Using space vector notation, the stator dynamics of the PMSM is described by

$$\vec{v}_s = R_s\vec{i}_s + \frac{d\vec{\psi}_s}{dt} \tag{10.48}$$

where R_s is the winding resistance and the variables $\vec{v}_s$, $\vec{i}_s$, and $\vec{\psi}_s$ are the stator voltage vector, the stator current vector, and the stator flux linkage vector, respectively.

The flux linked by the stator windings $\vec{\psi}_s$ is given by

$$\vec{\psi}_s = L_s\vec{i}_s + \psi_m e^{j\theta_r} \tag{10.49}$$

where L_s is the stator self-inductance, ψ_m is the magnitude of the stator flux linkage due to the rotor magnets, and θ_r is the position of the magnetic axis of the rotor with respect to the stator windings. Substituting (10.49) into (10.48) yields

$$\vec{v}_s = R_s\vec{i}_s + L_s\frac{d\vec{i}_s}{dt} + j\psi_m\omega_r e^{j\theta_r} \tag{10.50}$$

where $\omega_r \triangleq \frac{d\theta_r}{dt}$ is the electrical angular frequency of the motor. Equation (10.50) describes the stator dynamics in the stationary frame and is used in both the predictive model and a state observer. In this equation, the term $j\psi_m\omega_r e^{j\theta_r}$ represents a voltage source that models the *back-emf*. The electrical power absorbed by this voltage source is transferred to the shaft as rotational power. From the power balance

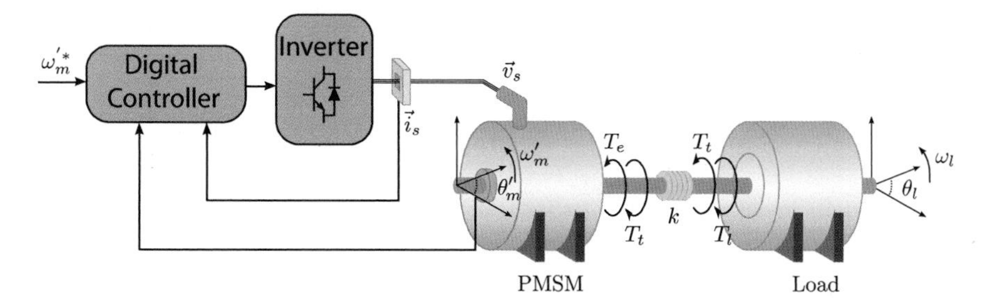

Fig. 10.4 Sampled-data control scheme for a 4 kW permanent magnet synchronous motor coupled with a 4 kW induction machine that acts as a load. The control algorithm is model predictive control (MPC) with a sampling period of 30 μs (that is, the sampling frequency is 33 kHz)

between the electrical and the mechanical systems, the electric torque generated by the motor T_e is given by

$$T_e = \frac{3}{2} p\, \psi_m \big(-\sin(\theta_r) i_\alpha + \cos(\theta_r) i_\beta\big) \tag{10.51}$$

where i_α and i_β are the stator current components, i.e., $\vec{i}_s = i_\alpha + j i_\beta$, and p is the number of poles of the machine.

The position of the magnetic axis of the rotor with respect to the stator windings θ_r is related to the shaft position θ_m through the pair of poles, i.e., $\theta_r = p\theta_m$. Consequently, the rotor frequency ω_r and the shaft angular speed ω_m are related by $\omega_r = p\omega_m$.

The two-mass resonant model is given by:

$$\frac{d\omega_m}{dt} = \frac{1}{J_m}(T_e - T_t) - \frac{B_m}{J_m}\omega_m \tag{10.52}$$

$$\frac{d\omega_l}{dt} = \frac{1}{J_l}(T_t - T_l) - \frac{B_l}{J_l}\omega_l \tag{10.53}$$

$$\frac{d\theta_m}{dt} = \omega_m \tag{10.54}$$

$$\frac{d\theta_l}{dt} = \omega_l \tag{10.55}$$

where θ_l is the shaft position at the load side, T_t is the torsional torque given by

$$T_t = k(\theta_m - \theta_l) \tag{10.56}$$

and T_l is the load torque, which, in general, is an unknown external disturbance. The parameters J_m, B_m, J_l, and B_l are the motor and load inertia and friction coefficients, respectively, and k is the elastic constant of the shaft. If the friction coefficients are assumed to be zero, the resonance frequency of the system is given by

$$\omega_n = \sqrt{k\frac{J_l + J_m}{J_l\, J_m}} \tag{10.57}$$

For simplicity of the model, (10.56) is differentiated to obtain

$$\frac{dT_t}{dt} = k(\omega_m - \omega_l). \tag{10.58}$$

Thus, the torque T_t is considered as a state of the system instead of θ_l.

The system equations are summarised in the following nonlinear state-space representation for the PMSM:

$$\frac{dx(t)}{dt} = f\big(x(t), u(t), d(t)\big) \tag{10.59}$$

$$y(t) = c\big(x(t)\big) \tag{10.60}$$

where $x = [i_\alpha\ i_\beta\ \theta_m\ \omega_m\ \omega_l\ T_t]^T$, $u = [v_\alpha\ v_\beta]^T$, and $d = T_l$. The (nonlinear) functions are given by:

$$f(\cdot) = \begin{pmatrix} -\frac{1}{\tau_s} i_\alpha + \frac{\psi_m}{L_s}\omega_r \sin(\theta_r) + \frac{1}{L_s} v_\alpha \\ -\frac{1}{\tau_s} i_\beta - \frac{\psi_m}{L_s}\omega_r \cos(\theta_r) + \frac{1}{L_s} v_\beta \\ \omega_m \\ \frac{1}{J_m}(T_e - T_t) - \frac{B_m}{J_m}\omega_m \\ \frac{1}{J_l}(T_t - T_l) - \frac{B_l}{J_l}\omega_l \\ k(\omega_m - \omega_l) \end{pmatrix}, \qquad c(\cdot) = \begin{pmatrix} i_\alpha \\ i_\beta \\ \theta_m \end{pmatrix} \tag{10.61}$$

where $\tau_s \triangleq \frac{L_s}{R_s}$ is the stator time constant.

An exact discrete-time model for (10.59)–(10.61) is difficult (or even impossible) to obtain.

In order to effectively evaluate the impact of using a given actuation on every system state, an approximate sampled-data predictive model is required. To achieve this, a sufficiently high order expansion is performed for each state component x_j up to an order N_j such that the input appears in the expansion explicitly. For example, for the stator currents dynamics (10.50), where the voltage actuation has an immediate effect on the state derivative, the order used is $N_j = 1$. On the other hand, for the mechanical variables, higher order expansions are used. For the motor speed ω_m, where the effect of the voltage actuation in the currents is transmitted by the electric torque (10.51) to the speed through the mechanical dynamics (10.52), an expansion of order $N_j = 2$ is used. Consequently, an expansion of order $N_j = 3$ is needed for the motor position θ_m. As a result of the sampling procedure, the obtained approximate discrete-time model can be written in the general form

$$x[k+1] = f_d\big(x[k], u[k], d[k]\big) \tag{10.62}$$

Note that this is exactly the nonlinear model described in Sect. 9.3. The model contains, inter alia, the asymptotic sampling zero dynamics.

The PMSM is assumed to be driven by a conventional two-level voltage source inverter. This converter has eight admissible switching states, generating seven different voltage vectors that can be applied to the load. These vectors constitute the set of available actuations of the power converter and, in the stationary frame, can be calculated as

$$\vec{v}_s = \frac{2}{3} V_{dc}\big(S_a + \vec{a} S_b + \vec{a}^2 S_c\big) \tag{10.63}$$

where $\vec{a} \triangleq e^{j2\pi/3}$, V_{dc} is the *dc link* voltage, and each switching function S_a, S_b, and S_c can take values in the set $\{0, 1\}$.

The future state of the system is predicted by applying each one of the allowable values to (10.62). The performance of each vector is evaluated using a cost function, described below. It is important to note that, in order to take into account the algorithm calculation time, a two-step prediction scheme is used.

The cost function is a non-negative function that evaluates the 'distance' of each predicted system state from the desired values. These desired values are given by different control goals, such as load speed reference tracking, maximisation of the torque per ampere ratio, smooth behaviour of the electrical torque, minimisation of torsional vibrations, and current amplitude limitation. The choice of the cost function simultaneously takes into account all these design criteria. This constitutes a multiobjective optimisation problem, because an inherent trade-off exists between the above-mentioned objectives. For example, improved speed tracking in the load may come at the cost of large stator currents during the transient response. Different weights are introduced to reflect the priorities amongst the different objectives. The chosen cost function structure is given by

$$\begin{aligned} F_c = {} & \underbrace{\lambda_\omega\big(\omega_l^* - \omega_l\big)^2}_{(a)} + \underbrace{\lambda_{id} i_d^2}_{(b)} + \underbrace{\lambda_{iqf} i_{qf}^2 + \lambda'_{iqf}\delta_\omega i'^2_{qf}}_{(c)} \\ & + \underbrace{\big(\lambda_{\Delta\omega} + \lambda'_{\Delta\omega}\delta_\omega\big)(\omega_m - \omega_l)^2}_{(d)} + \underbrace{\hat{f}(i_d, i_q)}_{(e)} \end{aligned} \tag{10.64}$$

Term (a) weights the error between the load speed and its reference ω_l^*. This term favours the voltage vectors that achieve load speed reference tracking.

Term (b) weights the magnitude of the current component in the d axis, penalising the generation of reactive power, and hence, maximising the torque per ampere ratio.

Term (c) weights i_{qf} and i'_{qf} which are high-pass filtered versions of the torque producing current i_q. This penalizes the switching states that generate high frequencies in the stator current. As a consequence, high frequency torque pulsations are mitigated, allowing smooth torque control using only one-step predictions. Both i_{qf}

and i'_{qf} are calculated using biquad filters:

$$i_{qf}[k] = b_0 w[k] + b_1 w[k-1] + b_2 w[k-2] \tag{10.65}$$

$$w[k] = i_q[k] - a_1 w[k-1] - a_2 w[k-2] \tag{10.66}$$

The cutoff frequencies f_c and f'_c (for i_{qf} and i'_{qf}, respectively) do not necessarily have the same value. The filter states w are considered as system states and included in the predictive model for online calculations.

Term (d) weights the error between the motor and the load speed. This error has a peak at the shaft resonance frequency (10.57), and hence, this term penalises the excitation of the shaft resonant mode. In terms (c) and (d) the factor δ_ω is given by

$$\delta_\omega \triangleq \begin{cases} (\omega_l^* - \omega_l)^2 & \text{if } (\omega_l^* - \omega_l)^2 < \hat{\delta}_\omega \\ \hat{\delta}_\omega & \text{if } (\omega_l^* - \omega_l)^2 \geq \hat{\delta}_\omega \end{cases} \tag{10.67}$$

The value of δ_ω grows quadratically with the load speed error up to an arbitrary constant in order to increase the relevance of terms (c) and (d) during speed transients. This generates a smooth torque response and prevents the excitation of the shaft resonant mode during speed transients.

Finally, term (e) is a nonlinear function of the predicted currents given by:

$$\hat{f}(i_d, i_q) \triangleq \begin{cases} K & \text{if } |i_q| > \hat{i}_q \text{ or } |i_d| > \hat{i}_d \\ 0 & \text{if } |i_q| \leq \hat{i}_q \text{ and } |i_d| \leq \hat{i}_d \end{cases} \tag{10.68}$$

where K is a large constant. Hence, the function $\hat{f}(\cdot)$ takes a large value when the predicted currents exceed a given limit, acting in practice as constraints on the current magnitudes.

The λ symbols in the cost function (10.64) are non-negative weighting factors used for tuning purposes.

The cost function (10.64) is evaluated using the predictions for each allowable voltage vector at each sampling instant. The optimal voltage vector is chosen which minimises the cost function. This is then applied to the PMSM.

Simulated and experimental results related to the above controller are available in the publications listed at the end of the chapter. The achieved results are excellent and support the use of the chosen sampled-data model. Again, the importance of sampling zero dynamics is central to the proposed solution.

10.8 Summary

This chapter has presented a number of applications to illustrate the benefit of using more sophisticated models than those obtained by simple Euler integration (i.e., simple derivative replacement). In particular, the chapter includes:

- design of sampled-data linear observers
- design of sampled-data linear state feedback controllers
- design of sampled-data linear output feedback controllers
- design of sampled-data nonlinear state feedback controllers
- deterministic system identification based on equation error minimisation
- predictive control of a two-mass system driven by a permanent magnet synchronous motor

In all cases, we have seen that sampling zeros are crucial to obtaining a clear understanding of the effect of sampling on performance.

Further Reading

The examples in Sect. 10.5 appear in

Grizzle JW, Kokotovic PV (1988) Feedback linearization of sampled-data systems. IEEE Trans Autom Control 33(9):857–859

The problem of predictive control of a two-mass system driven by a permanent magnet synchronous motor is described in more detail in

Fuentes E, Silva CA, Yuz JI (2012) Predictive speed control of a two-mass system driven by a permanent magnet synchronous motor. IEEE Trans Ind Electron 59(7):2840–2848

Silva C, Yuz JI (2010) On sampled-data models for model predictive control. In: 36th annual conference of the IEEE industrial electronics society, IECON 2010, Phoenix, AZ, USA

Further details of the examples in Sect. 10.5 can be found in

Carrasco DS, Goodwin GC (2013) The role of asymptotic sampling zero dynamics in the sampled-data control of continuous nonlinear systems. In: 52nd IEEE conference on decision and control, Firenze, Italy

Part II
Stochastic Systems

Chapter 11
Background on Sampling of Stochastic Signals

This chapter extends the review of deterministic signals presented in Chap. 2 to cover sampling and Fourier analysis of stochastic signals. As in Chap. 2, only a brief summary of key concepts is provided so as to establish notation and core concepts. Some readers will be familiar with this background. In this case, they can proceed immediately to Chap. 12. Other readers may wish to quickly read this chapter before proceeding.

11.1 Continuous-Time Stochastic Processes

A cornerstone concept is that of a continuous-time noise process. Attention is restricted to *wide-sense* stationary processes, i.e., processes where the first and second order properties are time invariant. Thus, consider a wide-sense stationary continuous-time stochastic process $z(t)$ having zero mean, i.e.,

$$E\{z(t)\} = 0 \tag{11.1}$$

and auto-correlation function given by

$$r_z(\tau) = E\{z(t+\tau)z(t)\}. \tag{11.2}$$

The associated power spectral density (PSD, sometimes called the *spectrum*) of this process is defined to be the Fourier transform of the auto-correlation function (11.2), i.e.,

$$\Phi_z(\omega) = \mathcal{F}\{r_z(\tau)\} = \int_{-\infty}^{\infty} r_z(\tau) e^{-j\omega\tau}\, d\tau \tag{11.3}$$

where $\omega \in (-\infty, \infty)$.

J.I. Yuz, G.C. Goodwin, *Sampled-Data Models for Linear and Nonlinear Systems*, Communications and Control Engineering, DOI 10.1007/978-1-4471-5562-1_11,

Lemma 11.1 *The variance of the zero-mean process $z(t)$ is equal to the integral (over all frequencies) of its power spectral density $\Phi_z(\omega)$, i.e.,*

$$E\{z(t)^2\} = \frac{1}{2\pi}\int_{-\infty}^{\infty} \Phi_z(\omega)\, d\omega \tag{11.4}$$

Proof The result follows by applying the inverse Fourier transform to the definition of the PSD in (11.3), i.e.,

$$\begin{aligned} E\{z(t+\tau)z(t)\} &= \mathcal{F}^{-1}\{\Phi_z(\omega)\} \\ &= \frac{1}{2\pi}\int_{-\infty}^{\infty} \Phi_z(\omega)e^{j\omega\tau}\, d\omega \end{aligned} \tag{11.5}$$

The result in (11.4) follows on setting $\tau = 0$. □

An important building block in signal analysis is that of *white noise*. A simple formulation is that a continuous-time process $\dot{v}(t)$ is *white noise* if it satisfies the following three conditions:

(i) $E\{\dot{v}(t)\} = 0$,
(ii) $\dot{v}(t)$ is independent of $\dot{v}(s)$, i.e., $E\{\dot{v}(t)\dot{v}(s)\} = 0$, if $t \neq s$, and
(iii) $\dot{v}(t)$ has continuous sample paths.

However, a stochastic process satisfying the above three conditions happens to be equal to zero in the mean square sense, i.e., $E\{\dot{v}(t)^2\} = 0$, for all t. This suggests that difficulties will arise, because the process $\dot{v}(t)$ does not exist in any meaningful sense. However, these difficulties can be circumvented by expressing the noise in its integrated form, $v(t)$.

The process $v(t)$ is a *Wiener* process and has the following properties:

(i) It has zero mean, i.e., $E\{v(t)\} = 0$, for all t;
(ii) Its increments are independent, i.e., $E\{(v(t_1) - v(t_2))(v(s_1) - v(s_2))\} = 0$, for all $t_1 > t_2 > s_1 > s_2 \geq 0$; and
(iii) For every s and t, $s \leq t$, the increments $v(t) - v(s)$ have a Gaussian distribution with zero mean and variance $E\{(v(t) - v(s))^2\} = \sigma_v^2|t - s|$

This process is not differentiable *almost everywhere*. However, the *continuous-time white noise* (CTWN) process, $\dot{v}(t)$, formally defined as the derivative of $v(t)$, is often a useful heuristic device. CTWN is thus introduced here for conceptual completeness. Note that the third condition above implies that the CTWN will have infinite variance:

$$E\{dv\,dv\} = E\{(v(t+dt) - v(t))^2\} = \sigma_v^2\,dt$$
$$\xrightarrow{dt\to 0} E\{\dot{v}^2\} = \infty \tag{11.6}$$

Remark 11.2 CTWN is a mathematical abstraction, but it can be *approximated* to any desired degree of accuracy by conventional stochastic processes with broad-band spectra. Conversely, broad-band noise can be modelled as white noise having the same spectral density over the frequency range of interest.

Three alternative interpretations of σ_v^2 are given below:

(i) The quantity σ_v^2 can be interpreted as the *incremental variance* of the Wiener process $v(t)$:

$$E\{(dv)^2\} = \sigma_v^2\,dt \tag{11.7}$$

(ii) If $\dot{v}(t)$ is considered as a *generalised* process, and a Dirac delta function is used to define its auto-correlation structure, then

$$r_{\dot{v}}(t-s) = E\{\dot{v}(t)\,\dot{v}(s)\} = \sigma_v^2\delta(t-s) \tag{11.8}$$

(iii) In the frequency domain, σ_v^2 can be interpreted as the *power spectral density* of $\dot{v}(t)$. Indeed, from (11.8), it follows that the spectral density satisfies

$$\Phi_{\dot{v}}(\omega) = \int_{-\infty}^{\infty} r_{\dot{v}}(\tau)e^{-j\omega\tau}\,d\tau = \sigma_v^2 \tag{11.9}$$

for $\omega \in (-\infty, \infty)$. The spectral density of $\dot{v}(t)$ is thus seen to be constant for all ω, which corresponds to the usual heuristic notion of *white* noise.

Under suitable conditions, it is also possible to express any PSD $\Phi_y(\omega)$ of a process $y(t)$ in the form

$$\Phi_y(\omega) = |H(j\omega)|^2\sigma_v^2 \tag{11.10}$$

where $H(s)$ is a rational function such that $\lim_{s\to\infty} H(s) = 1$. A systems' interpretation of the above result will be provided later.

11.2 Power Spectral Density of a Sampled Process

Next consider the process obtained by sampling a stationary continuous-time stochastic process $y(t)$.

If instantaneous sampling is used, with sampling period Δ, then the sampled sequence $y_k = y(k\Delta)$ is obtained. It follows that the covariance $r_y^d[\ell]$ of the sequence is equal to the continuous-time signal covariance *at the sampling instants*, i.e.,

$$r_y^d[\ell] = E\{y_{k+\ell} y_k\} = E\big\{y(k\Delta + \ell\Delta) y(k\Delta)\big\} = r_y(\ell\Delta) \tag{11.11}$$

Next the PSD of the sampled signal is defined. The PSD is given by the discrete-time Fourier transform (DTFT) of the covariance function, namely:

$$\Phi_y^d(\omega) = \Delta \sum_{k=-\infty}^{\infty} r_y^d[k] e^{-j\omega k\Delta}; \quad \omega \in \left[\frac{-\pi}{\Delta}, \frac{-\pi}{\Delta}\right] \tag{11.12}$$

Remark 11.3 Note that the DTFT as defined in (2.6) includes the sampling period Δ as a scaling factor. As a consequence, the DTFT defined in this fashion converges to the continuous-time Fourier transform as the sampling period Δ goes to zero.

Remark 11.4 The continuous-time and discrete-time PSD expressions in (11.3) and (11.12), respectively, are *real* functions of the frequency ω. However, to make the connections to the deterministic case more apparent, it is often convenient to express the continuous-time PSD in terms of the complex variable $s = j\omega$, and the discrete-time PSD in terms of $z = e^{j\omega\Delta}$, respectively, i.e.,

$$\text{(Continuous-time PSD)} \quad \Phi_y(\omega) = \bar{\Phi}_y(j\omega) = \bar{\Phi}_y(s)|_{s=j\omega} \tag{11.13}$$

$$\text{(Discrete-time PSD)} \quad \Phi_y^d(\omega) = \bar{\Phi}_y^d\left(e^{j\omega\Delta}\right) = \bar{\Phi}_y^d(z)\big|_{z=e^{j\omega\Delta}} \tag{11.14}$$

The following lemma then relates the PSD of a sampled sequence to the PSD of the original continuous-time process.

Lemma 11.5 *Consider a stochastic process $y(t)$, with the PSD given by* (11.3), *together with an associated sequence of samples $y_k = y(k\Delta)$, with the discrete-time PSD given by* (11.12). *Then the following relationship holds:*

$$\Phi_y^d(\omega) = \sum_{\ell=-\infty}^{\infty} \Phi_y\left(\omega + \frac{2\pi}{\Delta}\ell\right) \tag{11.15}$$

Proof The discrete-time PSD (11.12) can be rewritten in terms of the inverse (continuous-time) Fourier transform of the covariance function:

$$\begin{aligned}\Phi_y^d(\omega) &= \Delta \sum_{k=-\infty}^{\infty} r_y(k\Delta)e^{-j\omega k\Delta} = \Delta \sum_{k=-\infty}^{\infty}\left[\frac{1}{2\pi}\int_{-\infty}^{\infty}\Phi_y(\eta)e^{j\eta k\Delta}\,d\eta\right]e^{-j\omega k\Delta} \\ &= \frac{\Delta}{2\pi}\int_{-\infty}^{\infty}\Phi_y(\eta)\left[\sum_{k=-\infty}^{\infty} e^{j(\eta-\omega)k\Delta}\right]d\eta \end{aligned} \tag{11.16}$$

As seen in the proof of Lemma 2.3, a sum of complex exponentials can be rewritten as an infinite sum of Dirac impulses spread every $\frac{2\pi}{\Delta}$, i.e.,

$$\sum_{k=-\infty}^{\infty} e^{j(\eta-\omega)k\Delta} = \frac{2\pi}{\Delta}\sum_{\ell=-\infty}^{\infty}\delta\left(\eta-\omega-\frac{2\pi}{\Delta}\ell\right) \tag{11.17}$$

Using (11.17) in (20.8), the following result is obtained:

$$\begin{aligned}\Phi_y^d(\omega) &= \int_{-\infty}^{\infty}\Phi_y(\eta)\sum_{\ell=-\infty}^{\infty}\delta\left(\eta-\omega-\frac{2\pi}{\Delta}\ell\right)d\eta \\ &= \sum_{\ell=-\infty}^{\infty}\int_{-\infty}^{\infty}\Phi_y(\eta)\delta\left(\eta-\omega-\frac{2\pi}{\Delta}\ell\right)d\eta = \sum_{\ell=-\infty}^{\infty}\Phi_y\left(\omega+\frac{2\pi}{\Delta}\ell\right)\end{aligned} \tag{11.18}$$

This completes the proof. □

Equation (11.15) again reflects the following fact, noted in Chap. 2.

Sampling in the time domain produces folding in the frequency domain.

Remark 11.6 As an aside, note that there are alternative forms for the expression (11.15). For example (see the references at the end of the chapter), the sum in (11.15) can also be calculated by using the Hurwitz zeta function:

$$\zeta(r,a) = \sum_{k=0}^{\infty} \frac{1}{(k+a)^r}, \quad \text{Re}\{r\} > 1,\ a \notin \mathcal{Z}_0^- \tag{11.19}$$

This can be helpful in some derivations but will not be pursued further here.

The result in (11.15) is usually referred to as the *aliasing* effect. For deterministic systems, an analogous result was obtained in (2.19). For the stochastic case considered here, the discrete-time PSD is obtained by folding high frequency components of the continuous-time PSD.

Most of this book deals with univariate systems. However, it is straightforward to show that the result in Lemma 11.5 also applies (component-wise) to multivariate signals.

Another general result linking continuous-time and discrete-time PSDs is the following (see the references at the end of the chapter).

Lemma 11.7 *If the continuous-time PSD is given by*

$$\Phi_z^c(j\omega) = C(sI - A)^{-1}BB^T\left(-sI - A^T\right)^{-1}C^T, \quad s = j\omega \tag{11.20}$$

then the discrete-time PSD is given by

$$\Phi_z^d\left(e^{j\omega t}\right) = C\left(zI - e^{A\Delta}\right)^{-1}Q\left(z^{-1}I - e^{A^T\Delta}\right)^{-1}C^T, \quad z = e^{j\omega t} \tag{11.21}$$

where Q is

$$Q = \int_0^{\Delta} e^{A\tau}BB^Te^{A^T\tau}\,d\tau \tag{11.22}$$

Remark 11.8 The quantities $\Phi_z^d(e^{j\omega t})$ in (11.21) and (11.15) are equivalent. This can be readily verified—see, for example, the references at the end of the chapter.

11.3 Anti-aliasing Filtering

Equations (2.19) and (11.15) have important implications in sampling. In fact, they provide further motivation to include an *anti-aliasing filter* (AAF) prior to sampling. For example, say that the continuous-time process of interest contains a *signal* component, $s(t)$, whose PSD is (essentially) limited to the range $\omega \in (\frac{-\pi}{\Delta}, \frac{\pi}{\Delta})$, and another *noise* component whose PSD covers a broad frequency range. Then, if the composite process is directly sampled, the sampled PSD will be dominated by noise, because all of the noise PSD will be brought to the range $(\frac{-\pi}{\Delta}, \frac{\pi}{\Delta})$ by the folding process (11.15). However, if an ideal low-pass filter (with cutoff frequency equal to $\frac{\pi}{\Delta}$) is applied prior to sampling, the signal will be passed unaltered, and folding of the noise will be prevented.

Broad-band measurement noise frequently appears in practice. Thus, it is usually highly desirable to include an AAF prior to taking samples. Two AAF's used in practice are:

(i) A simple analogue low-pass filter,
(ii) An averaging AAF (also called an integrate and reset filter) where

$$\bar{z}_k = \bar{z}(k\Delta) = \frac{1}{\Delta}\int_{(k-1)\Delta}^{k\Delta} z(t)\,dt \tag{11.23}$$

An interesting property of an averaging AAF is that it converts continuous-time *white noise* of power spectral density R into discrete-time white noise of power spectral density R.

This fact can be seen as follows. The auto-correlation of the sampled output satisfies

$$r_{\bar{z}}^d[\ell] = E\{\bar{z}_k\bar{z}_{k+\ell}\} = \frac{R}{\Delta}\delta_K[\ell] \tag{11.24}$$

where $\delta_K[\ell]$ is the Kronecker delta. Applying (11.12), it follows that the discrete-time PSD of $\{\bar{z}_k\}$ is

$$\Phi_{\bar{z}}^d(\omega) = \Delta\sum_{k=-\infty}^{\infty} r_{\bar{z}}^d[k]e^{-j\omega k\Delta} = R \tag{11.25}$$

for all $\omega \in [\frac{-\pi}{\Delta}, \frac{\pi}{\Delta}]$. Hence, the sampled noise sequence has the same *PSD* as does the underlying continuous-time signal. This is an important property, as it implies that sampling (with an averaging AAF) preserves the spectral properties of white noise.

Remark 11.9 The previous result is quite elegant. This again highlights the importance of the proper use of the scaling factor Δ in all appropriate definitions to achieve consistency between continuous-time and discrete-time results.

Equation (11.24) is also interesting, because it implies that, as $\Delta \to 0$, the variance of the sampled white noise sequence goes to ∞! This is consistent with the fact that the variance of the underlying continuous-time signal is also (theoretically) infinite (see (11.8)).

Indeed, the above observation will be a recurring theme in this part of the book: discrete-time variances can be poorly scaled and either tend to infinity or zero as $\Delta \to 0$. Thus, it is usually preferable to describe both continuous-time and discrete-time noise sequences via their *power spectral density properties*, or equivalently their incremental covariances.

11.4 Summary

The key points covered in this chapter are:

- The definition of power spectral density of a continuous-time stochastic process (11.3).
- The definition of power spectral density of a discrete-time stochastic process (11.12).
- The folding (or *aliasing*) formula for stochastic processes, i.e.,

$$\Phi_y^d(\omega) = \sum_{\ell=-\infty}^{\infty} \Phi_y\left(\omega + \frac{2\pi}{\Delta}\ell\right) \tag{11.26}$$

- The introduction of the need for anti-aliasing filters prior to sampling. Such filters are used to prevent wide-band spectral components of the noise being folded back to lower frequencies by the sampling process. If an anti-aliasing filter is not included in the sampling scheme, the sampled signal may be overwhelmed by folded high frequency noise.
- The definition of the averaging anti-aliasing filter

$$\bar{z}_k = \bar{z}(k\Delta) = \frac{1}{\Delta}\int_{(k-1)\Delta}^{k\Delta} z(t)\,dt \tag{11.27}$$

Further Reading

Further background on stochastic processes can be found in

Åström KJ (1970) Introduction to stochastic control theory. Academic Press, New York
Brillinger DR (1974) Fourier analysis of stationary processes. Proc IEEE 62(12):1628–1643
Jazwinski AH (1970) Stochastic processes and filtering theory. Academic Press, San Diego
Oppenheim AV, Schafer RW (1999) Discrete-time signal processing, 2nd edn. Prentice Hall, New York
Papoulis A, Pillai SU (2002) Probability, random variables, and stochastic processes, 4th edn. McGraw-Hill, New York
Söderström T (2002) Discrete-time stochastic systems—estimation and control, 2nd. edn. Springer, London

The proof of a result closely related to the idea described in Remark 11.8 is given in

Feuer A, Goodwin GC (1996) Sampling in digital signal processing and control Birkhäuser, Boston, p 180 (Lemma 4.6.1)

Additional background on the use of the Hurwitz zeta function to derive Eq. (11.19) can be found in

Adamchik VS (2007) On the Hurwitz function for rational arguments. Appl Math Comput 187(1):3–12
Apostol TM (1976) Introduction to analytic number theory. Springer, Berlin

Chapter 12
Sampled-Data Models for Linear Stochastic Systems

Chapter 11 briefly reviewed sampling issues for stochastic signals. Here these ideas are used to formulate models for stochastic linear systems. As in Chap. 3, we begin with a brief introduction to continuous-time (stochastic) models. The impact of sampling is then added.

12.1 Continuous-Time Stochastic Linear Systems

The stochastic counterpart of the model (3.1)–(3.2) given in Sect. 3.1 on p. 21 is a linear continuous-time stochastic differential equation:

A linear time-invariant continuous-time stochastic differential equation can be written in incremental form as

$$dx = Ax\,dt + dw \tag{12.1}$$

$$dy = Cx\,dt + dv \tag{12.2}$$

where dw, dv are independent vector Wiener processes having incremental covariance $Q\,dt$ and $R\,dt$, respectively, i.e.,

$$E\left\{\begin{bmatrix} dw \\ dv \end{bmatrix}\begin{bmatrix} dw \\ dv \end{bmatrix}^T\right\} = \begin{bmatrix} Q & 0 \\ 0 & R \end{bmatrix} dt \tag{12.3}$$

Note the link to the deterministic state-space model (3.3)–(3.4) on p. 22.

Actually the model (12.1), (12.2) is a special case of a nonlinear stochastic differential equation. A detailed technical treatment of such equations will be presented in Chap. 17. For the moment, a simpler approach is used, because many of the technicalities can be ignored in the linear case.

J.I. Yuz, G.C. Goodwin, *Sampled-Data Models for Linear and Nonlinear Systems*, Communications and Control Engineering, DOI 10.1007/978-1-4471-5562-1_12,

It is also possible to describe the above model by the following formal state-space description with derivatives:

$$\frac{dx}{dt} = Ax + \dot{w} \tag{12.4}$$

$$z = \frac{dy}{dt} = Cx + \dot{v} \tag{12.5}$$

where $\dot{w}$ and $\dot{v}$ are independent continuous-time 'white noise' processes:

$$E\{\dot{w}(t)\dot{w}(s)\} = Q\delta(t-s) \tag{12.6}$$

$$E\{\dot{v}(t)\dot{v}(s)\} = R\delta(t-s) \tag{12.7}$$

The corresponding continuous-time (output) power spectral density for z can be written as

$$\Phi_z(\omega) = C(j\omega I - A)^{-1}Q(-j\omega I - A)^{-T}C^T + R \tag{12.8}$$

Remark 12.1 A simpler model driven by one noise source can also be obtained. In particular, by continuous-time spectral factorisation, the power spectral density (12.8) can be written in the form

$$\Phi_z(\omega) = |H(j\omega)|^2 \sigma_n^2 \tag{12.9}$$

Using this idea, the power spectral density (12.8) can be described by a simpler single-input single-output model (the innovations model):

$$dx = Ax\,dt + B\,dn \tag{12.10}$$

$$z\,dt = dy = Cx\,dt + D\,dn \tag{12.11}$$

where dn is a *scalar* Wiener process with incremental variance $\sigma_n^2\,dt$ and where

$$H(s) = C(sI - A)^{-1}B + D \tag{12.12}$$

In the next two sections, sampled-data models are obtained for stochastic systems. As discussed in Sect. 11.3, a key issue is that an anti-aliasing filter (AAF) is generally needed prior to sampling, especially if the output is contaminated by broad-band noise. For example, if the output were to be contaminated by 'white noise' and an AAF were not used, then the resulting sequence of output samples would have infinite variance and, thus, a sampled-data model would not be meaningful.

12.2 Sampled-Data Model for Systems Having Relative Degree Greater than Zero

First consider continuous-time systems where the output is not directly contaminated by white (broad-band) noise. For example, this could arise when a finite dimensional AAF is used. If an AAF is present, the filter can be directly incorporated into the continuous-time model. In this case, one can sample the (filtered) output without difficulty. This simplification will be adopted in the remainder of this section.

Consider the model

$$dx = Ax\,dt + dw \tag{12.13}$$

$$z = Cx \tag{12.14}$$

where w is a vector Wiener process which has incremental covariance matrix $Q\,dt$. Assume that z is a scalar for ease of presentation. As in (12.4), the model (12.13) can also be rewritten using the formal derivative of the Wiener process, i.e.,

$$\frac{dx}{dt} = Ax + \dot{w} \tag{12.15}$$

$$z = Cx \tag{12.16}$$

where the input $\dot{w}$ is a vector continuous-time white noise (CTWN) process with (constant) power spectral density Q.

Since the above model does not have white noise directly appearing on the output (e.g., the model may already incorporate an AAF), instantaneous sampling can be used without difficulty.

The following result gives the sampled-data model when considering instantaneous sampling of the output (12.14) or equivalently (12.16).

Lemma 12.2 *Consider the stochastic system defined in state-space form (12.15)–(12.16), where the input w is a Wiener process of incremental variance $Q\,dt$ (i.e., $\dot{w}$ is a CTWN process with constant power spectral density Q). Assume that the output z is instantaneously sampled, with sampling period Δ. Then a discrete-time model describing the evolution of the samples is given by:*

$$x_{k+1} = A_q x_k + w_k \tag{12.17}$$

$$z_k = C_q x_k \tag{12.18}$$

where $C_q = C$, $A_q = e^{A\Delta}$, and the sequence w_k is a discrete-time white noise (DTWN) process, with zero mean and covariance structure given by

$$E\{w_k w_\ell^T\} = Q_q \delta_K[k-\ell] \tag{12.19}$$

where

$$Q_q = \int_0^{\Delta} e^{A\eta} Q e^{A^T\eta}\, d\eta \tag{12.20}$$

Proof Arguing as in (3.24), it follows that

$$x_{k+1} = e^{A\Delta} x_k + \underbrace{\int_{k\Delta}^{k\Delta+\Delta} e^{A(k\Delta+\Delta-\eta)} \dot{w}(\eta)\, d\eta}_{w_k} \tag{12.21}$$

The covariance of the noise sequence w_k in (12.21) is given by:

$$\begin{aligned} E\{w_k\, w_\ell^T\} &= E\left\{\left(\int_{k\Delta}^{k\Delta+\Delta} e^{A(k\Delta+\Delta-\eta)} \dot{w}(\eta)\, d\eta\right)\left(\int_{\ell\Delta}^{\ell\Delta+\Delta} e^{A(\ell\Delta+\Delta-\xi)} \dot{w}(\xi)\, d\xi\right)^T\right\} \\ &= \int_{k\Delta}^{k\Delta+\Delta} \int_{\ell\Delta}^{\ell\Delta+\Delta} e^{A(k\Delta+\Delta-\eta)} \underbrace{E\{\dot{w}(\eta)\dot{w}(\xi)\}}_{Q\delta(\eta-\xi)} e^{A^T(\ell\Delta+\Delta-\xi)}\, d\xi\, d\eta \end{aligned} \tag{12.22}$$

The double integral above is non-zero only when $k = \ell$ and $\eta = \xi$. Thus,

$$E\{w_k\, w_\ell^T\} = \int_{k\Delta}^{k\Delta+\Delta} e^{A(k\Delta+\Delta-\eta)} Q e^{A^T(k\Delta+\Delta-\eta)}\, d\eta\, \delta_K[k-\ell] \tag{12.23}$$

Changing variables in the integral, the above expression can be shown to be equivalent to (12.19)–(12.20). □

12.3 Sampled-Data Models for Systems Having Averaging Filter

Next consider systems where the output is contaminated by white (broad-band) noise. In this case, some form of AAF is essential prior to sampling. By way of illustration, assume that an averaging AAF (as described in Sect. 11.3) is used. An advantage of using this AAF is that (as seen in Sect. 11.3) such a filter converts continuous-time white noise into discrete-time white noise of the same power spectral density.

Including the averaging AAF before taking the sample leads to a sampled output satisfying:

$$\bar{y}_k = \frac{1}{\Delta}\int_{(k-1)\Delta}^{k\Delta} dy = \frac{1}{\Delta}[y_k - y_{k-1}] \tag{12.24}$$

The associated sampled-data model can then be obtained by simply 'integrating' the continuous-time model to obtain a discrete-time model having the same second order statistics at the sampling instants.

Lemma 12.3 *Consider the continuous-time stochastic model* (12.4)–(12.5) *and the output averaging AAF* (12.24). *The following discrete-time model has the same second order stochastic properties as the resultant continuous-time system* at the sampling instants:

$$x_{k+1} = A_q x_k + w_k \tag{12.25}$$

$$\bar{y}_{k+1} = C_q x_k + v_k \tag{12.26}$$

where $k \in \mathcal{Z}$ is the discrete-time index, $\{x_k\} \in \mathbb{R}^n$ is the state of the discrete-time system, and $\{y_k\} \in \mathbb{R}^p$ is the output.

In (12.25), (12.26), the system matrices $A_q \in \mathbb{R}^{n\times n}$ and $C_q \in \mathbb{R}^{p\times n}$ are given by:

$$A_q = e^{A\Delta} \tag{12.27}$$

$$C_q = \frac{1}{\Delta} C \int_0^{\Delta} e^{A\tau}\, d\tau = \frac{1}{\Delta} C A^{-1}\left(e^{A\Delta} - I\right) \tag{12.28}$$

The process noise $w_k \in \mathbb{R}^n$ and the output measurement noise $v_k \in \mathbb{R}^p$ are white with zero mean, and joint covariance

$$\Sigma_q = E\left\{\begin{bmatrix} w_k \\ v_k \end{bmatrix}\begin{bmatrix} w_k \\ v_k \end{bmatrix}^T\right\} = \begin{bmatrix} Q_q & S_q \\ S_q^T & R_q \end{bmatrix} \tag{12.29}$$

where the matrices $\Sigma_q \in \mathbb{R}^{(n+p)\times(n+p)}$, $Q_q \in \mathbb{R}^{n\times n}$, and $R_q \in \mathbb{R}^{p\times p}$ are positive semi-definite matrices. The covariance matrix, Σ_q, of the noise process is given by

$$\Sigma_q = \begin{bmatrix} I & 0 \\ 0 & \frac{1}{\Delta} \end{bmatrix} \int_0^{\Delta} e^{\bar{A}\tau} \begin{bmatrix} Q & 0 \\ 0 & R \end{bmatrix} e^{\bar{A}^T\tau}\, d\tau \begin{bmatrix} I & 0 \\ 0 & \frac{1}{\Delta} \end{bmatrix} \tag{12.30}$$

where

$$\bar{A} = \begin{bmatrix} A & 0 \\ C & 0 \end{bmatrix} \Rightarrow e^{\bar{A}\tau} = \begin{bmatrix} e^{A\tau} & 0 \\ C\int_0^\tau e^{A\sigma}\,d\sigma & I \end{bmatrix} \tag{12.31}$$

Proof Arguing as in (3.24), it follows that the exact state and output trajectories satisfy

$$\begin{bmatrix} x_{k+1} \\ y_{k+1} \end{bmatrix} = e^{\bar{A}\Delta} \begin{bmatrix} x_k \\ y_k \end{bmatrix} + \underbrace{\int_{k\Delta}^{k\Delta+\Delta} e^{\bar{A}(k\Delta+\Delta-\eta)} \begin{bmatrix} \dot{w}(\eta) \\ \dot{v}(\eta) \end{bmatrix} d\eta}_{n_k} \tag{12.32}$$

where $\bar{A}$ is given in (12.31). The covariance of the discrete-time noise process n_k is given by

$$\begin{aligned} &E\{n_k n_\ell^T\} \\ &\quad = E\left\{\left(\int_{k\Delta}^{k\Delta+\Delta} e^{\bar{A}(k\Delta+\Delta-\eta)} \begin{bmatrix} \dot{w}(\eta) \\ \dot{v}(\eta) \end{bmatrix} d\eta\right)\left(\int_{\ell\Delta}^{\ell\Delta+\Delta} e^{\bar{A}(\ell\Delta+\Delta-\xi)} \begin{bmatrix} \dot{w}(\xi) \\ \dot{v}(\xi) \end{bmatrix} d\xi\right)^T\right\} \\ &\quad = \int_{k\Delta}^{k\Delta+\Delta} \int_{\ell\Delta}^{\ell\Delta+\Delta} e^{\bar{A}(k\Delta+\Delta-\eta)} E\left\{\begin{bmatrix} \dot{w}(\eta) \\ \dot{v}(\eta) \end{bmatrix} \begin{bmatrix} \dot{w}(\xi) \\ \dot{v}(\xi) \end{bmatrix}^T\right\} e^{\bar{A}^T(\ell\Delta+\Delta-\xi)}\,d\xi\,d\eta \end{aligned} \tag{12.33}$$

Note that

$$E\left\{\begin{bmatrix} \dot{w}(\eta) \\ \dot{v}(\eta) \end{bmatrix} \begin{bmatrix} \dot{w}(\xi) \\ \dot{v}(\xi) \end{bmatrix}^T\right\} = \begin{bmatrix} Q & 0 \\ 0 & R \end{bmatrix} \delta(\eta-\xi) \tag{12.34}$$

Thus, the double integral in (12.33) is non-zero only when $k=\ell$ and $\eta=\xi$. This fact implies that n_k is a discrete-time white noise vector sequence. Thus,

$$\begin{aligned} E\{n_k n_\ell^T\} &= \int_{k\Delta}^{k\Delta+\Delta} e^{\bar{A}(k\Delta+\Delta-\eta)} \begin{bmatrix} Q & 0 \\ 0 & R \end{bmatrix} e^{\bar{A}^T(k\Delta+\Delta-\eta)}\,d\eta\,\delta_K[k-\ell] \\ &= \int_0^\Delta e^{\bar{A}\eta} \begin{bmatrix} Q & 0 \\ 0 & R \end{bmatrix} e^{\bar{A}^T\eta}\,d\eta\,\delta_K[k-\ell] \end{aligned} \tag{12.35}$$

where δ_K is a Kronecker delta.

The exponential matrix (12.31) can be obtained from a series expansion, i.e.,

$$\begin{aligned} e^{\bar{A}\tau} &= I + \bar{A}\tau + \bar{A}^2\frac{\tau^2}{2!} + \cdots + \bar{A}^n\frac{\tau^n}{n!} + \cdots \\ &= \begin{bmatrix} I + A\tau + A^2\frac{\tau^2}{2!} + \cdots & 0 \\ C\tau + CA\frac{\tau^2}{2!} + CA^2\frac{\tau^3}{3!} + \cdots & I \end{bmatrix} \\ &= \begin{bmatrix} e^{A\tau} & 0 \\ C\int_0^\tau e^{A\sigma}\,d\sigma & I \end{bmatrix} \end{aligned} \tag{12.36}$$

Thus, substituting (12.36) into (12.32), it follows that

$$\begin{bmatrix} x_{k+1} \\ y_{k+1} - y_k \end{bmatrix} = \begin{bmatrix} e^{A\Delta} \\ C \int_0^\Delta e^{A\sigma}\, d\sigma \end{bmatrix} x_k + n_k \tag{12.37}$$

Finally, from the AAF relationship (12.24), it is necessary to scale the output equation. Hence, the noise process is scaled as follows:

$$\begin{bmatrix} w_k \\ v_k \end{bmatrix} = \begin{bmatrix} I & 0 \\ 0 & \frac{1}{\Delta} I \end{bmatrix} n_k \tag{12.38}$$

from which (12.30) follows. □

Lemma 12.3 presents a sampled-data model for the linear continuous-time stochastic system described by (12.1)–(12.2) with averaging AAF. This model is exact in the sense that all of the properties of the state and output sequence are the same as the sampled output of the continuous-time system at the sampling instants, i.e.,

$$r^d_{\bar{y},k} = E\{\bar{y}_\ell \bar{y}^T_{\ell+k}\} = E\{\bar{y}(\ell\Delta) y^T((\ell+k)\Delta)\} = r_{\bar{y}}(k\Delta) \tag{12.39}$$

(See also Sect. 11.2 on p. 142.)

Remark 12.4 An alternative way of describing the relationship between (12.25), (12.26) and (12.13), (12.14), (12.24) is that they produce the same sample paths when driven by the same underlying continuous-time Wiener noise processes. This alternative connection will be used later in the book.

An interesting fact about the model in (12.25)–(12.31) is that the model describing the sampled output has correlation between the process noise $\{w_k\}$ and the measurement noise $\{v_k\}$. This occurs even though the underlying continuous-time system has independent process and measurement noise.

12.4 Summary

The key ideas covered in this chapter are:

- Continuous-time stochastic linear models in state-space incremental form (Eqs. (12.1) to (19.32)).
- Sampling of continuous-time stochastic systems where no white noise appears directly on the measured output.
- Sampling of continuous-time linear stochastic systems where white noise appears directly on the measured output and an averaging anti-aliasing filter is used prior to taking samples.

Further Reading

Further information on continuous-time and sampled-data models for linear stochastic systems can be found in

Åström KJ (1970) Introduction to stochastic control theory. Academic Press, New York
Jazwinski AH (1970) Stochastic processes and filtering theory. Academic Press, San Diego
Farrell J, Livstone M (1996) Calculation of discrete-time process noise statistics for hybrid continuous/discrete-time applications. Optim Control Appl Methods 17:151–155
Feuer A, Goodwin GC (1996) Sampling in digital signal processing and control. Birkhäuser, Boston
Ljung L, Wills A (2010) Issues in sampling and estimating continuous-time models with stochastic disturbances. Automatica 46(5):925–931

Chapter 13
Incremental Stochastic Sampled-Data Models

This chapter further embellishes the ideas described previously. Here the extension of incremental models to the stochastic case is studied. In particular, Chap. 4 discussed linear deterministic sampled-data models based on the shift operator q. It was shown that the limit (as the sample period, Δ, approaches zero) is not well defined for shift operator models. The remedy used in Chap. 4 was to make a simple change in the operator by differencing and scaling the variables by the sample period. This led to deterministic incremental sampled-data models. Here this circle of ideas is extended to the stochastic case.

13.1 Incremental Model

The shift operator models described in Chap. 12 lead to difficulties if we take the limit as the sampling frequency is increased. For example, consider the model obtained in Sect. 12.3. Then, as $\Delta \to 0$,

$$A_q = I + A\Delta + \frac{1}{2}A^2\Delta^2 + \cdots \to I \tag{13.1}$$

Also, the noise covariance structure (12.30) tends to the following values as $\Delta \to 0$:

$$Q_q = \Delta Q + \frac{\Delta^2}{2}\left(AQ + QA^T\right) + \cdots \to 0 \tag{13.2}$$

$$S_q = \frac{\Delta}{2}QC^T + \frac{\Delta^2}{3!}\left(2AQ + QA^T\right)C^T + \cdots \to 0 \tag{13.3}$$

$$R_q = \frac{1}{\Delta}R + \frac{\Delta}{3}CQC^T + \frac{\Delta^2}{4}C\left(AQ + QA^T\right)C^T + \cdots \to \infty \tag{13.4}$$

The convergence results presented above indicate that traditional discrete-time models (expressed in the shift operator domain) will have conceptual and numerical difficulties for small sampling intervals. Indeed, at fast sampling rates, and for

J.I. Yuz, G.C. Goodwin, *Sampled-Data Models for Linear and Nonlinear Systems*, Communications and Control Engineering, DOI 10.1007/978-1-4471-5562-1_13,

all systems, the A_q matrix tends to the identity matrix. Also the noise covariance matrix Σ_q in (12.29) tends to the troublesome values given in (13.2)–(13.4). As in Chap. 4, the source of the difficulty is that the system equation describes the next value of x_k rather than the change (or increment) in the value. This motivates proceeding as in Chap. 4 to develop the corresponding incremental models for the stochastic case.

13.2 Incremental Model for Instantaneously Sampled Systems

Consider the case of a continuous-time system having relative degree greater than zero. The corresponding discrete-time model (in shift operator form) was described in Sect. 12.2. The following result gives the equivalent incremental model.

Lemma 13.1 *Consider a linear continuous-time stochastic system having relative degree greater than* 1. *Assume that the output is instantaneously sampled. Then the corresponding sampled-data model can be written in incremental form as follows*:

$$dx_k^+ = x_{k+1} - x_k = A_i x_k \Delta + \bar{w}_k \tag{13.5}$$

$$z_k = Cx_k \tag{13.6}$$

where

$$A_i = \frac{e^{A\Delta} - I}{\Delta} \tag{13.7}$$

and the sequence $\bar{w}_k$ is a discrete-time white noise (DTWN) process, with zero mean and incremental covariance structure given by

$$E\{\bar{w}_k \bar{w}_\ell^T\} = Q_i \Delta \delta_K[k-\ell] \tag{13.8}$$

where

$$Q_i = \frac{1}{\Delta}\int_0^\Delta e^{A\eta} Q e^{A^T\eta}\, d\eta \tag{13.9}$$

Proof The proof follows by rewriting the model (12.17) in terms of differences, i.e.,

$$dx_k^+ = x_{k+1} - x_k = (A_q - I)x_k + w_k \tag{13.10}$$

from which (13.5) follows and $\bar{w}_k = w_k$. As a consequence, $E\{\bar{w}_k \bar{w}_\ell^T\} = E\{w_k w_\ell^T\}$, and thus $Q_i \Delta = Q_q$, from which (13.9) readily follows using (12.20). □

It is important to note that the matrix Q_i in Lemma 13.1 corresponds to the (constant) power spectral density (PSD) of the noise vector $\bar{w}_k$, as can be seen by applying the discrete-time Fourier transform to (12.19):

$$\mathcal{F}_d\left\{Q_i\frac{\delta_K[k]}{\Delta}\right\} = \Delta\sum_{k=-\infty}^{\infty} Q_i\frac{\delta_K[k]}{\Delta}e^{-j\omega k\Delta} = Q_i; \quad \omega\in\left[-\frac{\pi}{\Delta},\frac{\pi}{\Delta}\right] \tag{13.11}$$

Additionally, Lemma 13.1 shows that the continuous-time stochastic description (12.15) is recovered as the sampling period Δ goes to zero; namely, from (13.5),

$$\frac{dx_k^+}{\Delta} = A_i x_k + \frac{\bar{w}_k}{\Delta} \quad \xrightarrow{\Delta\to 0} \quad \frac{dx}{dt} = Ax + \dot{w} \tag{13.12}$$

In particular, the covariance of $\bar{w}_k/\Delta$ converges to the continuous-time covariance of the vector continuous-time white noise (CTWN) process $\dot{w}$:

$$E\left\{\frac{\bar{w}_k}{\Delta}\frac{\bar{w}_\ell^T}{\Delta}\right\} = Q_i\frac{\delta_K[k-\ell]}{\Delta} \quad \xrightarrow{\Delta\to 0} \quad Q\delta(t_k-t_\ell) = E\left\{\dot{w}\dot{w}^T\right\} \tag{13.13}$$

Finally the matrix Q_q in (12.20) can be computed by solving the discrete-time Lyapunov equation

$$Q_q = P - A_q P A_q^T \tag{13.14}$$

or, equivalently, using the incremental (or δ-domain) version:

$$Q_i = A_i P + P A_i^T + \Delta A_i P A_i^T \tag{13.15}$$

where P satisfies the continuous-time Lyapunov equation $AP + PA^T + Q = 0$, for stable systems, or $AP + PA^T - Q = 0$, for anti-stable systems.

13.3 Incremental Model for Sampled Systems with Averaging AAF

Next consider the shift operator model of Sect. 12.3. This model can be expressed in incremental form as described below:

Lemma 13.2 *Consider the continuous-time stochastic model* (12.4)–(12.5) *and the output averaging anti-aliasing filter (AAF)* (12.24). *The following incremental discrete-time model is a reparameterisation of the model in Lemma* 12.3. *It has the same (state and output) stochastic properties as the continuous-time system (with AAF)* at the sampling instants:

$$dx_k^+ = x_{k+1} - x_k = A_i x_k \Delta + \bar{w}_k \tag{13.16}$$

$$dy^+ = y_{k+1} - y_k = C_i x_k \Delta + \bar{v}_k = \Delta \bar{y}_{k+1} \tag{13.17}$$

where

$$A_i = \frac{A_q - I}{\Delta}, \qquad C_i = C_q \tag{13.18}$$

and where the noise sequences are

$$\bar{w}_k = w_k \quad \Rightarrow \quad E\{w_k w_k^T\} = Q_q = Q_i \Delta \tag{13.19}$$

$$\bar{v}_k = v_k \Delta \quad \Rightarrow \quad E\{\bar{v}_k \bar{v}_k^T\} = R_q \Delta^2 = R_i \Delta \tag{13.20}$$

The joint noise covariance satisfies

$$E\left\{\begin{bmatrix} \bar{w}_k \\ \bar{v}_k \end{bmatrix} \begin{bmatrix} \bar{w}_{k-\ell} \\ \bar{v}_{k-\ell} \end{bmatrix}^T\right\} = \Sigma_i \Delta \delta_K(\ell) \tag{13.21}$$

where

$$\Sigma_i = \frac{1}{\Delta} \int_0^\Delta e^{\bar{A}\tau} \begin{bmatrix} Q & 0 \\ 0 & R \end{bmatrix} e^{\bar{A}^T \tau} d\tau = \begin{bmatrix} Q_i & S_i \\ S_i^T & R_i \end{bmatrix} \tag{13.22}$$

Proof The result follows immediately from the results in Lemma 12.3. In particular, (13.18)–(13.20) follow by rewriting (12.25)–(12.26), in incremental form (13.16)–(13.17). Then, (13.22) follows by introducing the reparameterisation of the noise sequences in (13.19)–(13.20) and the Σ_q matrix found earlier in (12.30). □

The reparameterisation of the system matrices presented above again ensures a smooth convergence to the continuous-time matrices as the sampling interval goes to zero. In fact, it follows that, as $\Delta \to 0$,

$$A_i = A + \frac{1}{2} A^2 \Delta + \cdots \to A \tag{13.23}$$

$$C_i = C\left(I + A\frac{\Delta}{2}\right) + \cdots \to C \tag{13.24}$$

$$Q_i = \frac{1}{\Delta}Q_q = Q + \frac{\Delta}{2}\left(AQ + QA^T\right) + \cdots \to Q \tag{13.25}$$

$$S_i = S_q = \frac{\Delta}{2}QC^T + \cdots \to 0 \tag{13.26}$$

$$R_i = R_q\Delta = R + \frac{\Delta}{3}CQC^T + \cdots \to R \tag{13.27}$$

In particular, the matrix Σ_i converges to its continuous-time counterpart, i.e.,

$$\Sigma_i = \begin{bmatrix} Q_i & S_i \\ S_i^T & R_i \end{bmatrix} \xrightarrow{\Delta \to 0} \begin{bmatrix} Q & 0 \\ 0 & R \end{bmatrix} \tag{13.28}$$

13.4 Output Power Spectral Density

This chapter is concluded by the observation that all of the sampled-data models described in Chaps. 11 and 12 will have outputs which are discrete-time stationary stochastic processes. The corresponding PSD properties can be readily obtained from the corresponding state-space model. The details are given below.

Lemma 13.3 *Consider a continuous-time linear stochastic system having relative degree greater than zero and assume that the output is sampled instantaneously. Then the output PSD is given by:*

(i) *For the incremental form of the model*

$$\Phi_z^\delta(\gamma_\omega) = C_i(\gamma_\omega I_n - A_i)^{-1}Q_i\left(\gamma_\omega^* I_n - A_i^T\right)^{-1}C_i^T \tag{13.29}$$

where $\gamma_\omega = \frac{1}{\Delta}(e^{j\omega\Delta} - 1)$ *and* $*$ *denotes complex conjugation.*

(ii) *For the shift operator form of the model*

$$\Phi_z^q\left(e^{j\omega\Delta}\right) = \Delta C_q\left(e^{j\omega\Delta}I_n - A_q\right)^{-1}Q_q\left(e^{-j\omega\Delta}I_n - A_q^T\right)^{-1}C_q^T \tag{13.30}$$

Proof The result for the incremental form of the model follows immediately from Lemma 13.1. The model equations define a discrete-time linear system with a vector input $\bar{\omega}_k$ and output z_k. The output PSD is then given by

$$\Phi_z^d(\omega) = H_i(\gamma_\omega)\Phi_\omega^d(\omega)H_i(\gamma_\omega^*)^T \tag{13.31}$$

where $H_i(\gamma_\omega) = C_i(\gamma_\omega I_n - A_i)^{-1}$, and the PSD of the input noise is $\Phi_\omega^d(\omega) = Q_i$ Equation (13.30) follows from the relations $\gamma_\omega = \frac{e^{j\omega\Delta}-1}{\Delta}$, $A_i = \frac{A_q - 1}{\Delta}$, and $Q_i = \frac{Q_q}{\Delta}$. □

Lemma 13.4 *Consider a continuous-time linear stochastic system and assume that the output is passed through an averaging AAF prior to sampling. Then the sampled output PSD is given by*:

(i) *Incremental model*

$$\Phi_z^\delta(\gamma_\omega) = \begin{bmatrix} C_i(\gamma_\omega I_n - A_i)^{-1} & 1 \end{bmatrix} \begin{bmatrix} Q_i & S_i \\ S_i^T & R_i \end{bmatrix} \begin{bmatrix} (\gamma_\omega^* I_n - A_i^T)^{-1} C_i^T \\ 1 \end{bmatrix} \tag{13.32}$$

where C_i, Q_i, S_i, R_i, A_i are as in Lemma 13.2, *and where* $\gamma_\omega = \frac{1}{\Delta}(e^{j\omega\Delta} - 1)$. *Here* * *denotes complex conjugation.*

(ii) *Shift operator model*

$$\Phi_{\bar{y}}^q(e^{j\omega\Delta}) = \Delta \begin{bmatrix} C_q(e^{j\omega\Delta} I_n - A_q)^{-1} & 1 \end{bmatrix} \times \begin{bmatrix} Q_q & S_q \\ S_q^T & R_q \end{bmatrix} \begin{bmatrix} (e^{-j\omega\Delta} I_n - A_q^T)^{-1} C_q^T \\ 1 \end{bmatrix} \tag{13.33}$$

where C_q, Q_q, S_q, R_q, A_q are as in Lemma 12.3.

Proof The proof follows the same lines as that of Lemma 13.3. Equations (13.16)–(13.22) define a discrete-time linear system with vector inputs v_k and w_k, and output y_k. The output PSD is then given by

$$\Phi_{\bar{y}}^d(\omega) = H_\delta(\gamma_\omega) \begin{bmatrix} \Phi_v^d(\omega) & \Phi_{vw}^d(\omega) \\ (\Phi_{vw}^d(\omega))^T & \Phi_w^d(\omega) \end{bmatrix} H_\delta(\gamma_\omega^*)^T \tag{13.34}$$

where

$$H_\delta(\gamma_\omega) = \begin{bmatrix} C_i(\gamma_\omega I_n - A_i)^{-1} \\ 1 \end{bmatrix} \tag{13.35}$$

and the PSD of the input noise is obtained from (13.22), i.e.,

$$\begin{bmatrix} \Phi_v^d(\omega) & \Phi_{vw}^d(\omega) \\ (\Phi_{vw}^d(\omega))^T & \Phi_w^d(\omega) \end{bmatrix} = \begin{bmatrix} Q_i & S_i \\ S_i^T & R_i \end{bmatrix} \tag{13.36}$$

Equation (13.33) follows using the relationships $\gamma_\omega = \frac{e^{j\omega\Delta}-1}{\Delta}$, $A_i = \frac{A_q - 1}{\Delta}$. □

Next, simple connections between the sampled-data PSDs and those of the underlying continuous-time system will be explored.

13.5 Simple Connections Between Continuous-Time and Discrete-Time PSDs

Here, simple connections between the underlying continuous-time PSD and the discrete-time PSD, resulting from sampling with small sample period Δ, are explored.

To study this question, recall the *folding* relationship given in Lemma 11.5. An immediate consequence of the folding relationship is the following result, which shows that, provided the continuous-time PSD satisfies a mild decay condition, the corresponding discrete-time PSD is close in a well-defined sense. Indeed, the error is of order Δ^2.

Lemma 13.5 *Assuming that there exists an ω_N such that for $\omega \geq \omega_N$,*

$$\left|\Phi^c(j\omega)\right| \leq \frac{\beta}{\omega^2} \tag{13.37}$$

then

$$\left|\Phi^d(e^{j\omega t}) - \Phi^c(j\omega)\right| = \mathcal{O}(\Delta^2), \quad 0 \leq \omega \leq \frac{\pi}{\Delta} \tag{13.38}$$

Proof Using the triangle inequality and Lemma 11.5, it follows that:

$$E = \left|\sum_{k=-\infty}^{\infty} \Phi^c\left(j\omega + j\frac{2\pi}{\Delta}k\right) - \Phi^c(j\omega)\right| \tag{13.39}$$

$$= \left|\sum_{k=-\infty, k\neq 0}^{\infty} \Phi^c\left(j\omega + j\frac{2\pi}{\Delta}k\right)\right| \tag{13.40}$$

$$\leq \sum_{k=-\infty, k\neq 0}^{\infty} \frac{\beta}{(\omega + \frac{2\pi}{\Delta}k)^2} \tag{13.41}$$

$$= \sum_{k=-\infty}^{\infty} \frac{\beta}{(\omega + \frac{2\pi}{\Delta}k)^2} - \frac{\beta}{\omega^2} \tag{13.42}$$

$$= \frac{\beta}{\omega^2}\left[\frac{\frac{\omega\Delta}{2}}{\sin(\frac{\omega\Delta}{2})}\right]^2 - \frac{\beta}{\omega^2} \tag{13.43}$$

$$= \frac{\beta}{\omega^2}\left[\left[\frac{\frac{\omega\Delta}{2}}{\sin(\frac{\omega\Delta}{2})}\right]^2 - 1\right] \tag{13.44}$$

Using a Taylor series expansion for $\sin^2(\frac{\omega\Delta}{2})$,

$$E \leq \frac{\beta}{\omega^2}\left[\frac{(\frac{\omega\Delta}{2})^2}{(\frac{\omega\Delta}{2})^2 - \frac{1}{3}(\frac{\omega\Delta}{2})^4 + \frac{2}{45}(\frac{\omega\Delta}{2})^6 \cdots} - 1\right] \tag{13.45}$$

$$= \frac{\beta}{\omega^2}\left[\frac{\frac{1}{3}(\frac{\omega\Delta}{2})^4 - \frac{2}{45}(\frac{\omega\Delta}{2})^6 \cdots}{(\frac{\omega\Delta}{2})^2 - \frac{1}{3}(\frac{\omega\Delta}{2})^4 + \frac{2}{45}(\frac{\omega\Delta}{2})^6 + \cdots}\right] \tag{13.46}$$

$$= \frac{\beta}{\omega^2}\left[\frac{\frac{1}{3}(\frac{\omega\Delta}{2})^2 - \frac{2}{45}(\frac{\omega\Delta}{2})^4 \cdots}{1 - \frac{1}{3}(\frac{\omega\Delta}{2})^2 + \cdots}\right] \tag{13.47}$$

$$= \frac{\beta\Delta^2}{\omega^2}\left[\frac{\frac{1}{3}(\frac{\omega}{2})^2 - \frac{2}{45}(\frac{\omega}{2})^4\Delta^2 \cdots}{1 - \frac{1}{3}(\frac{\omega\Delta}{2})^2 + \cdots}\right] \tag{13.48}$$

Finally it follows that

$$\lim_{\Delta\to 0} \frac{E}{\Delta^2} \leq \frac{\beta}{12} \tag{13.49}$$

which completes the proof. □

The result in Lemma 13.5 is valid for general systems. It shows that, subject to (13.37), the error between the discrete-time and continuous-time PSDs is of order Δ^2. Hence, it seems that the sampled-data PSD could be simply replaced by the continuous-time PSD when the sample period, Δ, is small. This seems to resolve the question and would suggest that it is not necessary to explore the matter further. However, there is a subtle hidden caveat. Specifically, for high frequencies, the continuous-time PSD also satisfies $\Phi^c(j\omega) = \mathcal{O}(\Delta^2)$. This means that the continuous-time and discrete-time PSDs are not necessarily close to each other at high frequencies, in the sense of the *relative error* being small. The errors are of the same order of magnitude as the function itself. To illustrate the difficulty, Fig. 13.1 shows the relative error defined by

$$R(r,\omega) = \left|\frac{\Phi^d(e^{j\omega t}) - \Phi^c(j\omega)}{\Phi^c(j\omega)}\right| \tag{13.50}$$

for $\Phi^c(s) = \Psi_r^c(s) = 1/s^r$ and different values of r. It can be seen that the relative error between the continuous-time PSD and discrete-time PSD is certainly not small for frequencies close to $\omega = \pi/\Delta$.

The reader should not be surprised by the above results, given the discussion of relative errors in Chap. 16 for the deterministic case.

Thus, Lemma 13.5 does not capture key aspects of the connection between continuous-time and discrete-time PSDs. This raises a follow-up question as to whether $\Phi^c(j\omega)$ is close to $\Phi^d(e^{j\omega t})$ in the sense of small *relative errors* over a restricted bandwidth. The following result is immediate.

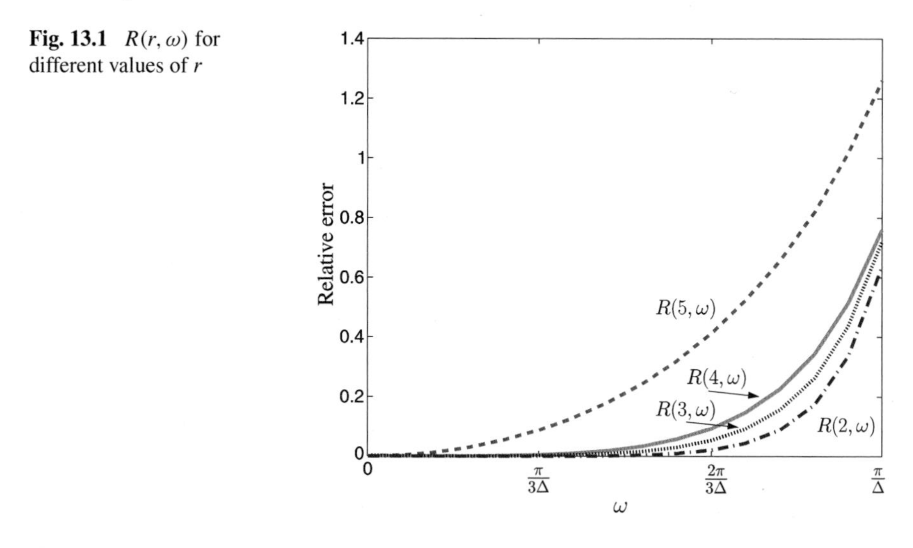

Fig. 13.1 $R(r,\omega)$ for different values of r

Lemma 13.6 *Assume that for* $\omega \geq \omega_N$ $\frac{\alpha}{\omega^p} \leq |\Phi^c(j\omega)| \leq \frac{\beta}{\omega^2}$ $(p \geq 2)$*. Then, for finite bandwidth* ($\omega \leq \omega_B < \pi/\Delta$)*, it follows that*

$$\left|\frac{\Phi^d(e^{j\omega t}) - \Phi^c(j\omega)}{\Phi^c(j\omega)}\right| = \mathcal{O}(\Delta^2) \tag{13.51}$$

Proof Using Eq. (13.48),

$$R = \left|\frac{\sum_{k=-\infty}^{\infty} \Phi^c(j\omega + j\frac{2\pi}{\Delta}k) - \Phi^c(j\omega)}{\Phi^c(j\omega)}\right| \tag{13.52}$$

$$\leq \frac{1}{|\Phi^c(j\omega)|}\frac{\beta\Delta^2}{\omega^2}\left[\frac{\frac{1}{3}(\frac{\omega}{2})^2 - \frac{2}{45}(\frac{\omega}{2})^4\Delta^2\cdots}{1 - \frac{1}{3}(\frac{\omega\Delta}{2})^2 + \cdots}\right] \tag{13.53}$$

$$\leq \frac{\omega^p}{\alpha}\frac{\beta\Delta^2}{\omega^2}\left[\frac{\frac{1}{3}(\frac{\omega}{2})^2 - \frac{2}{45}(\frac{\omega}{2})^4\Delta^2\cdots}{1 - \frac{1}{3}(\frac{\omega\Delta}{2})^2 + \cdots}\right] \tag{13.54}$$

Finally, it follows that

$$\lim_{\Delta\to 0}\frac{R}{\Delta^2} \leq \frac{\omega^p}{12}\frac{\beta}{\alpha} \leq \frac{\omega_B^p}{12}\frac{\beta}{\alpha} \tag{13.55}$$

which completes the proof. □

The above result shows that, *provided frequencies in the near vicinity of the Nyquist frequency are not considered*, the relative errors between the continuous-time and discrete-time PSDs will be small. This raises the obvious next question regarding behaviour near the Nyquist frequency. This question is more difficult and leads to the issue of stochastic sampling zeros. This latter question requires attention to details of the folding process and is explored in detail in the next chapter.

13.6 Summary

The key points covered in this chapter are:

- A convergence result for the incremental model, as the sampling period $\Delta \to 0$, i.e.,

$$A_i \to A, \qquad C_i \to C, \qquad Q_i \to Q, \qquad S_i \to 0, \qquad R_i \to R \tag{13.56}$$

- A sampled-data model for the noise-free output case, i.e.,

$$dx = Ax\,dt + dw \tag{13.57}$$

$$z = Cx \tag{13.58}$$

where $E\{d\omega\, d\omega^T\} = Q\,dt$.

The output of this model can be sampled instantaneously. The resulting sampled model, in shift operator form, is given by

$$x_{k+1} = A_q\, x_k + w_k \tag{13.59}$$

$$z_k = Cx_k \tag{13.60}$$

where $A_q = e^{A\Delta}$ and $E\{w_k\, w_\ell^T\} = Q_q \delta_K[k-\ell]$, where

$$Q_q = \int_0^\Delta e^{A\eta} Q e^{A^T\eta}\, d\eta \tag{13.61}$$

The sampled model can be expressed in incremental form as

$$dx_k^+ = A_i x_k \Delta + \bar{w}_k \tag{13.62}$$

$$z_k = Cx_k \tag{13.63}$$

where $A_i = \frac{e^{A\Delta}-I}{\Delta}$ and $E\{\bar{w}_k \bar{w}_\ell^T\} = Q_i \Delta \delta_K[k-\ell]$ and where

$$Q_i = \frac{1}{\Delta}\int_0^\Delta e^{A\eta} Q e^{A^T\eta}\, d\eta \tag{13.64}$$

As the sampling period goes to 0, it follows that:

$$A_i \xrightarrow{\Delta\to 0} A, \qquad Q_i \xrightarrow{\Delta\to 0} Q \tag{13.65}$$

- The corresponding output power spectral density can be expressed as

$$\Phi_z^\delta(\gamma_\omega) = C_i(\gamma_\omega I_n - A_i)^{-1} Q_i \left(\gamma_\omega^* I_n - A_i^T\right)^{-1} C_i^T \tag{13.66}$$

- Similar results hold for continuous-time systems having white noise on the output when an averaging anti-aliasing filter is deployed.
- Continuous-time and sampled-data power spectral densities approach each other as $\Delta \to 0$.
- The relative error between continuous-time and discrete-time power spectral densities does not go to zero in the vicinity of the Nyquist frequency.

Further Reading

The following two books give a complete background to incremental models for stochastic linear systems

Feuer A, Goodwin GC (1996) Sampling in digital signal processing and control. Birkhäuser, Boston

Middleton RH, Goodwin GC (1990) Digital control and estimation. A unified approach. Prentice Hall, Englewood Cliffs

Chapter 14
Asymptotic Sampling Zeros for Linear Stochastic Systems

This chapter further explores the connection between continuous-time power spectral densities (PSDs) and their sampled-data equivalents. A key issue of importance will turn out to be the existence of sampling zeros in the sampled-data PSD. It will be shown that these stochastic sampling zeros can be asymptotically characterised as the sampling period goes to zero. These results are the stochastic extension of the results presented in Chap. 5 for the deterministic case.

14.1 Discrete-Time PSDs Corresponding to Simple Continuous-Time Systems

Recall the results presented in Sect. 13.5 of the previous chapter, namely that the continuous-time and discrete-time PSDs approach each other as $\Delta \to 0$. However, the relative error in the vicinity of the Nyquist frequency does not go to zero. To better understand this phenomenon we require a deeper understanding of the folding process.

14.1.1 First Order System Having Relative Degree 1

Example 14.1 Consider a first order continuous-time auto-regressive (CAR) system:

$$\frac{dy(t)}{dt} - a_0 y(t) = b_0 \dot{v}(t) \tag{14.1}$$

where $a_0 < 0$ and $\dot{v}(t)$ is a continuous-time white noise (CTWN) process of unitary PSD, i.e., $\sigma_v^2 = 1$. An equivalent state-space model can readily be obtained as:

$$\frac{dx(t)}{dt} = a_0 x(t) + b_0 \dot{v}(t) \tag{14.2}$$

J.I. Yuz, G.C. Goodwin, *Sampled-Data Models for Linear and Nonlinear Systems*, Communications and Control Engineering, DOI 10.1007/978-1-4471-5562-1_14,

$$y(t) = x(t) \tag{14.3}$$

Note that this system has relative degree 1, so it is feasible to instantaneously sample the output.

The corresponding sampled-data model can be expressed in either shift operator or incremental operator form as follows:

$$qx_k = e^{a_0 \Delta} x_k + \tilde{v}_k, \qquad \delta x_k = \left(\frac{e^{a_0 \Delta} - 1}{\Delta} \right) x_k + v_k \tag{14.4}$$

$$y_k = x_k, \qquad y_k = x_k \tag{14.5}$$

where $\tilde{v}_k$ and v_k are discrete-time white noise (DTWN) processes with variance Q_q and $\frac{Q_i}{\Delta}$, respectively. Also, it follows that:

$$Q_q = \Delta Q_i = b_0^2 \frac{(e^{2a_0 \Delta} - 1)}{2a_0} \tag{14.6}$$

$$Q_i = b_0^2 \frac{(e^{2a_0 \Delta} - 1)}{2a_0 \Delta} \tag{14.7}$$

As usual, the incremental model converges to the underlying continuous-time system as $\Delta \to 0$:

$$Q_i = b_0^2 \frac{(e^{2a_0 \Delta} - 1)}{2a_0 \Delta} \xrightarrow{\Delta \to 0} b_0^2 \tag{14.8}$$

$$a_i = \frac{(e^{2a_0 \Delta} - 1)}{\Delta} \xrightarrow{\Delta \to 0} a_0 \tag{14.9}$$

The corresponding PSD is

$$\Phi_y^q(z) = \left(z - e^{a_0 \Delta}\right)\left(1 - e^{a_0 \Delta} z\right) Q_q \tag{14.10}$$

The above result is unremarkable. However, more interesting things happen if a continuous-time system having relative degree 2 is considered. The latter case is explored in the next section.

14.1.2 Second Order System Having Relative Degree 2

Example 14.2 Consider the second order CAR system

$$\frac{d^2 y(t)}{dt} + a_1 \frac{dy(t)}{dt} + a_0 y(t) = \dot{v}(t) \tag{14.11}$$

where $\dot{v}(t)$ is a CTWN process of unitary spectral density, i.e., $\sigma_v^2 = 1$. A suitable state-space model is given by:

$$\frac{dx(t)}{dt} = \begin{bmatrix} 0 & 1 \\ -a_0 & -a_1 \end{bmatrix} x(t) + \begin{bmatrix} 0 \\ 1 \end{bmatrix} \dot{v}(t) \tag{14.12}$$

$$y(t) = \begin{bmatrix} 1 & 0 \end{bmatrix} x(t) \tag{14.13}$$

Using the results in Sects. 12.3 and 13.4, it follows, after some lengthy calculations, that the discrete-time output PSD has the form

$$\Phi_y^q(z) = K \frac{z(b_2 z^2 + b_1 z + b_0)}{(z - e^{\lambda_1 \Delta})(z - e^{\lambda_2 \Delta})(1 - e^{\lambda_1 \Delta} z)(1 - e^{\lambda_2 \Delta} z)} \tag{14.14}$$

where λ_1 and λ_2 are the continuous-time system poles, and:

$$b_2 = (\lambda_1 - \lambda_2)\left[e^{(\lambda_1+\lambda_2)\Delta}\left(\lambda_2 e^{\lambda_1 \Delta} - \lambda_1 e^{\lambda_2 \Delta}\right) + \lambda_1 e^{\lambda_1 \Delta} - \lambda_2 e^{\lambda_2 \Delta}\right] \tag{14.15}$$

$$b_1 = \left[(\lambda_1 + \lambda_2)\left(e^{2\lambda_1 \Delta} - e^{2\lambda_2 \Delta}\right) + (\lambda_1 - \lambda_2)\left(e^{2(\lambda_1+\lambda_2)\Delta} - 1\right)\right] \tag{14.16}$$

$$b_0 = b_2 \tag{14.17}$$

$$K = \frac{\Delta}{2\lambda_1 \lambda_2 (\lambda_1 - \lambda_2)^2 (\lambda_1 + \lambda_2)} \tag{14.18}$$

Performing spectral factorisation on the sampled PSD (14.14), the following sampled-data model is obtained. Note that this model is expressed in terms of only one noise source, i.e.,

$$\Phi_y^q(z) = H_q(z) H_q\left(z^{-1}\right) \tag{14.19}$$

where

$$H_q(z) = \frac{\sqrt{K}(c_1 z + c_0)}{(z - e^{\lambda_1 \Delta})(z - e^{\lambda_2 \Delta})} \tag{14.20}$$

The poles of the above model are simply related to the poles of the continuous-time system. However, a new ingredient arises in the model (14.20) that was not present in the continuous-time model: a sampling zero has appeared at $-\frac{c_0}{c_1}$. This zero is called a 'stochastic sampling zero' by analogy with the deterministic case discussed in Chap. 5.

This example motivates a more in-depth study of sampled-data models arising from continuous-time systems having relative degree greater than 1. This topic will be examined in the next section.

14.2 Asymptotic Sampling Zeros of the Output Power Spectral Density

In the previous section, it was shown that the output PSD of a sampled-data model for a continuous-time system of relative degree 2 contains *sampling zeros* which have no counterpart in the underlying continuous-time system. Similar to the deterministic case, these zeros can be asymptotically characterised. The following two results characterise the asymptotic stochastic sampling zeros of the output PSD for two cases: instantaneous sampling, and use of an integrating pre-filter.

Lemma 14.3 *Consider the instantaneous sampling of the continuous-time process* (12.13), (12.14). *Then,*

$$\Phi_y^d(\omega) \xrightarrow{\Delta\to 0} \Phi_y(\omega) \tag{14.21}$$

uniformly in s, on compact subsets. Moreover, let $\pm z_i$, $i = 1, \dots, m$ be the $2m$ zeros of $\Phi_y(s)$, and $\pm p_i$, $i = 1, \dots, n$ its $2n$ poles. Then:

- *$2m$ zeros of $\Phi_y^d(z)$ will converge to 1 as $e^{\pm z_i \Delta}$;*
- *The remaining $2(n-m)-1$ converge to the zeros of $zB_{2(n-m)-1}(z)$ as Δ goes to zero; and*
- *The $2n$ poles of $\Phi_y^d(z)$ equal $e^{\pm p_i \Delta}$, and will hence go to 1 as Δ goes to zero.*

Proof The proof follows from the fact that, for large $|s|$, the continuous-time PSD $\Phi_y(s)$ can be approximated by a $2(n-m)$-th order integrator. Then the sampled-data PSD can be obtained from the infinite sum (11.15). Using Lemma 5.4, it follows that

$$\sum_{k=-\infty}^{\infty} \frac{1}{(\log z + j2\pi k)^{2(n-m)}} = \frac{zB_{2(n-m)-1}(z)}{(2(n-m)-1)!(z-1)^{2(n-m)}} \tag{14.22}$$

□

When an averaging anti-aliasing filter (AAF) is used, this can be thought of as adding an extra integrator to the continuous-time model. Hence the relative degree is increased by 1. Thus, we have the following.

Lemma 14.4 *Consider the sampling of the continuous-time process* (12.13) *with averaging AAF,* (12.14). *Then the results of Lemma* 14.3 *essentially apply, save that there are now $2(n-m)$ sampling zeros of $\Phi_y^d(z)$. These zeros converge to the zeros of $B_{2(n-m)}(z)$ as Δ goes to zero.*

An alternative viewpoint of these asymptotic stochastic sampling zeros will be presented in Chap. 18.

Example 14.5 Consider again the second order CAR system in Example 14.2. The discrete-time PSD (14.14) was obtained for the case of instantaneous sampling of the output $y(t)$. Exact expressions for the sampling zeros of this PSD were obtained and are quite involved. To gain more insight, a Taylor series expansion is carried out on the numerator. It follows that

$$Kz\left(b_2z^2 + b_1z + b_0\right) = \frac{\Delta^4}{3!}z\left(z^2 + 4z + 1\right) + \mathcal{O}\left(\Delta^5\right) \tag{14.23}$$

which, asymptotically (as Δ goes to zero), is consistent with Lemma 14.3, noting that $B_3(z) = z^2 + 4z + 1$ as in (5.21).

The asymptotic sampled PSD can be obtained as

$$\begin{aligned}\Phi_y^q(z) &= \frac{\Delta^4}{6}\frac{(z+4+z^{-1})}{(z-e^{\lambda_1\Delta})(z-e^{\lambda_2\Delta})(z^{-1}-e^{\lambda_1\Delta})(z^{-1}-e^{\lambda_2\Delta})}\\ &= \frac{\Delta^4}{6(2-\sqrt{3})}\frac{(z+2-\sqrt{3})}{(z-e^{\lambda_1\Delta})(z-e^{\lambda_2\Delta})}\frac{(z^{-1}+2-\sqrt{3})}{(z^{-1}-e^{\lambda_1\Delta})(z^{-1}-e^{\lambda_2\Delta})}\end{aligned} \tag{14.24}$$

This PSD can be written as $\Phi_y^q(z) = H_q(z)H_q(z^{-1})$, where

$$H_q(z) = \frac{\Delta^2}{3-\sqrt{3}}\frac{(z+2-\sqrt{3})}{(z-e^{\lambda_1\Delta})(z-e^{\lambda_2\Delta})} \tag{14.25}$$

The corresponding δ-operator form can be obtained by a simple change of variables using $z = 1 + \gamma\Delta$. This yields the following incremental discrete-time model:

$$H_\delta(\gamma) = \frac{1 + \frac{1}{3-\sqrt{3}}\Delta\gamma}{(\gamma - \frac{e^{\lambda_1\Delta}-1}{\Delta})(\gamma - \frac{e^{\lambda_2\Delta}-1}{\Delta})} \tag{14.26}$$

In summary, the general result can be expressed as follows.

Theorem 14.6 *Consider the continuous-time model*

$$y = \frac{B(s)}{A(s)}\dot{\omega} \tag{14.27}$$

where $B(s)$, $A(s)$ are stable polynomials of order m, n, respectively, and $\dot{\omega}$ is continuous-time 'white noise' with intensity σ^2. The corresponding (asymptotic as $\Delta \to 0$) model for the sampled PSD can be modelled by

$$y_k = \frac{B(\frac{q-1}{\Delta})N^{\delta}_{r-1}(\frac{q-1}{\Delta})}{A(\frac{q-1}{\Delta})}\bar{\omega}_k \tag{14.28}$$

where $r = n - m$ and $\bar{\omega}_k$ is a discrete white noise sequence. Also, $N^{\delta}_{r-1}(\delta) = N^{q}_{r-1}(\Delta\delta + 1)$, where $N^{\delta}_{r-1}(z)$ is the minimum phase spectral factor of $z^{1-r}B_{2r-1}(z)$. ($B_i(z)$ denotes the Euler-Frobenius polynomials.)

In the sequel, the polynomials $N^{q}_{r-1}(z)$ and $N^{\delta}_{r-1}(\delta)$ are said to be the stochastic sampling zero dynamics corresponding to a continuous-time stochastic system of relative degree r.

14.3 Relating Deterministic and Stochastic Sampling Zeros

Exactly the same mechanism applies to both the deterministic and stochastic cases, namely, sampling zeros are a consequence of folding. To elucidate the equivalence between deterministic and stochastic cases, it is important that the effect of the hold (in the deterministic case) and the AAF (in the stochastic case) be considered. For example, in the deterministic case, a zero order hold (ZOH) on the input adds an extra integrator to the folded function. This is clear in the folding expressions given in Lemma 3.5. (See also Eqs. (3.29) and (3.40).) Similarly, in the stochastic case, an averaging AAF adds an extra integrator to the folded function. With this in mind, the connections summarised in Table 14.1 are obtained. In particular,

- *Deterministic case with ZOH*: Note that there are 2 fewer sampling zeros than the relative degree of the folded function (the plant transfer function plus ZOH in this case).
- *Stochastic case with instantaneous sampling*: Again note that there are 2 fewer sampling zeros than the relative degree of the folded function (the continuous-time PSD in this case).
- *Stochastic case with averaging AAF*: Here there are 2 fewer sampling zeros than the relative degree of the folded function (the continuous-time PSD of the output prior to the sampler after the averaging AAF).

Finally note that the deterministic sampling zeros are the same as the stochastic sampling zeros. The former appear in the difference equation model, whereas the latter appear in the discrete-time PSD. Of course, when the discrete-time PSD is spectrally factored to obtain a discrete-time model, the zeros are split into those inside and those outside the unit disc.

Table 14.1 Relative degree of deterministic plants and PSDs and the corresponding relative degree of discrete-time systems and PSDs. ZOH(s) denotes a zero order hold and $F(s)$ denotes an averaging AAF

	Relative degree of $G(s)$	Relative degree of the folded function	Number of sampling zeros
Deterministic case u_k ZOH $G(s)$ $\psi(t)$ Δ ψ_k	r	ZOH$(s)G(s)$ $r+1$	$r-1$
Stochastic case $\dot{v}(t)$ $G(s)$ $\psi(t)$ Δ ψ_k	r	$G(s)G(-s)$ $2r$	$2r-2$
Stochastic case $\dot{v}(t)$ $G(s)$ $F(s)$ $\psi(t)$ Δ ψ_k	r	$G(s)F(s)G(-s)F(-s)$ $2r+2$	$2r$

As an illustration, if a deterministic continuous-time system of relative degree 3 is considered, then with ZOH, the relative degree of the folded function will be 4 and there will be 2 asymptotic sampling zeros.

Similarly, if a continuous-time stochastic model having relative degree 2 is considered, then the continuous-time PSD will have relative degree 4. With instantaneous sampling, the discrete-time PSD will have 2 asymptotic sampling zeros since the folded function has relative degree 4. The corresponding discrete-time spectral factor will have 1 sampling zero. On the other hand, if an averaging AAD is used, the folded function will have relative degree 6 and hence the discrete-time PSD will have 4 asymptotic sampling zeros. The corresponding discrete-time spectral factor will have 2 asymptotic sampling zeros.

14.4 Asymptotic Sampling Zeros via Time-Domain Arguments

The central idea underlying both deterministic and stochastic sampling zeros is the observation that, for a fast sampling rate, all systems behave like an nth order integrator at high frequencies. Hence, the sampling zeros, which are a consequence of folding of high frequency components back to lower frequencies, can be understood by simply considering an nth order integrator. This case is examined below.

A stochastic nth order integrator is described by

$$\frac{d^n y(t)}{dt^n} = \dot{v}(t) \tag{14.29}$$

where $\dot{v}(t)$ is a scalar white noise process of intensity σ^2.

The system (14.29) can also be described by the linear stochastic differential equation

$$dX = AX\,dt + B\,dv_t \tag{14.30}$$

where:

$$X = \begin{bmatrix} x_1 \\ \vdots \\ x_{n-1} \\ x_n \end{bmatrix}, \qquad A = \left[\begin{array}{c|ccc} 0 & & & \\ \vdots & & I_{n-1} & \\ 0 & & & \\ \hline 0 & 0 & \cdots & 0 \end{array}\right], \qquad B = \begin{bmatrix} 0 \\ \vdots \\ 0 \\ 1 \end{bmatrix} \tag{14.31}$$

and where the output is $y(t) = x_1(t)$.

Lemma 14.7 *An exact model describing the evolution of the samples of the system* (14.30), (14.31) *is*

$$X^+ = \begin{bmatrix} 1 & \Delta & \cdots & \frac{\Delta^{n-1}}{(n-1)!} \\ 0 & 1 & \cdots & \frac{\Delta^{n-2}}{(n-2)!} \\ \vdots & \ddots & \ddots & \vdots \\ 0 & \cdots & 0 & 1 \end{bmatrix} X + \begin{bmatrix} I_{(1,0,\ldots,0,0)} \\ I_{(1,0,\ldots,0)} \\ \vdots \\ I_{(1)} \end{bmatrix} \tag{14.32}$$

where the following notation has been used:

$$I_{\underbrace{(1,0,\ldots,0)}_{m \text{ zeros}}} = \int_0^{\Delta} \int_0^{\tau_1} \cdots \int_0^{\tau_m} dv_{\tau_{m+1}} \, d\tau_m \ldots d\tau_1 \tag{14.33}$$

Proof Consider the stochastic differential equation corresponding to the last state x_n. The *exact* solution to this equation is

$$dx_n = dv_t \quad \Rightarrow \quad x_n(\Delta) = x_n(0) + \int_0^{\Delta} dv_{\tau_n} \tag{14.34}$$

If this expression is now used in the equation corresponding to x_{n-1}, it follows that:

$$\begin{aligned} dx_{n-1} = x_n \, dt \quad \Rightarrow \quad x_{n-1}(\Delta) &= x_{n-1}(0) + \int_0^{\Delta} x_n(\tau_{n-1}) \, d\tau_{n-1} \\ &= x_{n-1}(0) + x_n(0)\Delta + \int_0^{\Delta} \int_0^{\tau_{n-1}} dv_{\tau_n} \, d\tau_{n-1} \end{aligned} \tag{14.35}$$

Proceeding in this way for the other state components of the stochastic differential equation (14.30)–(14.31) leads to the result (14.32), (14.33). □

Note that the same exact discrete-time model can be obtained using the earlier results in Chap. 12. Specifically, Lemma 12.2 on p. 151 can be used. Substituting the matrices A and B given in (14.31) into Eq. (12.21) leads to

$$X^+ = \underbrace{\begin{bmatrix} 1 & \Delta & \cdots & \frac{\Delta^{n-1}}{(n-1)!} \\ 0 & 1 & \cdots & \frac{\Delta^{n-2}}{(n-2)!} \\ \vdots & \ddots & \ddots & \vdots \\ 0 & \cdots & 0 & 1 \end{bmatrix}}_{A_q} X + \underbrace{\int_0^\Delta e^{A(\Delta-\tau)} B\, dv_\tau}_{\widetilde{V}} \tag{14.36}$$

where $A_q = e^{A\Delta}$, and the elements of $\widetilde{V}$ can be obtained from Lemma 12.2, i.e.,

$$\widetilde{V} = \begin{bmatrix} \tilde{v}_1 \ \tilde{v}_2 \cdots \tilde{v}_n \end{bmatrix}^T \quad \Rightarrow \tag{14.37}$$

$$\tilde{v}_\ell = \int_0^\Delta \frac{(\Delta-\tau)^{(n-\ell)}}{(n-\ell)!}\, dv_\tau; \quad \ell = 1, \ldots, n \tag{14.38}$$

The vector $\widetilde{V}$ is a discrete-time white noise sequence satisfying

$$E\{\widetilde{V}\} = 0 \tag{14.39}$$

$$\begin{aligned} E\{\widetilde{V}\widetilde{V}^T\} &= \int_0^\Delta e^{A\eta} B B^T e^{A^T\eta}\, d\eta = \int_0^\Delta \begin{bmatrix} \frac{\eta^{n-1}}{(n-1)!} \\ \vdots \\ 1 \end{bmatrix} \begin{bmatrix} \frac{\eta^{n-1}}{(n-1)!} \\ \vdots \\ 1 \end{bmatrix}^T d\eta \\ &= \int_0^\Delta \left[\frac{\eta^{n-i}}{(n-i)!} \frac{\eta^{n-j}}{(n-j)!} \right]_{i,j=1,\ldots,n} d\eta \\ &= \left[\frac{\Delta^{2n-i-j+1}}{(n-i)!(n-j)!(2n-i-j+1)} \right]_{i,j=1,\ldots,n} \end{aligned} \tag{14.40}$$

where $[m_{ij}]_{i,j=1,\ldots,n}$ represents an $n \times n$ matrix whose entries are m_{ij}.

For example, if the case of a second order stochastic integrator is considered, then

$$E\left\{ \begin{bmatrix} \tilde{v}_1^2 & \tilde{v}_1\tilde{v}_2 \\ \tilde{v}_2\tilde{v}_1 & \tilde{v}_2^2 \end{bmatrix} \right\} = \begin{bmatrix} \frac{\Delta^3}{3} & \frac{\Delta^2}{2} \\ \frac{\Delta^2}{2} & \Delta \end{bmatrix} \tag{14.41}$$

In the above development, two alternative ways to define the noise vector $\widetilde{V}$ have been described. First, its elements were defined in terms of multiple stochastic integrals in (14.32) and (14.33). Second, using the (alternate) approach in Lemma 12.2, (14.36)–(14.38) are obtained. Both expressions are obviously equivalent. This is formally established below.

Lemma 14.8 *Consider a standard Wiener process v_t and a time period Δ. Then the following* reduction rule *for multiple integrals holds*:

$$I_{(1,\underbrace{0,\ldots,0}_{m \text{ zeros}})} = \int_0^\Delta \int_0^{\tau_m} \cdots \int_0^{\tau_1} dv_\tau \, d\tau_1 \ldots d\tau_m = \int_0^\Delta \frac{(\Delta - \tau)^m}{m!} dv_\tau; \quad m \geq 1 \tag{14.42}$$

Proof Relation (14.42) can be proven using induction. For $m = 0$, we have, by definition,

$$I_{(1)} = \int_0^\Delta dv_\tau \tag{14.43}$$

For $m = 1$:

$$I_{(1,0)} = \int_0^\Delta \int_0^{\tau_1} dv_\tau \, d\tau_1 = \int_0^\Delta \int_\tau^\Delta d\tau_1 \, dv_\tau = \int_0^\Delta (\Delta - \tau) \, dv_\tau \tag{14.44}$$

where the order of integration in the double integral has been changed.

Finally, assume that the result holds for $m = k$, and prove it for $m = k + 1$:

$$\begin{aligned} I(1, \underbrace{0,\ldots,0}_{k+1 \text{ zeros}}) &= \int_0^\Delta \left(\int_0^{\tau_{k+1}} \cdots \int_0^{\tau_1} dv_\tau \, d\tau_1 \ldots d\tau_k \right) d\tau_{k+1} \\ &= \int_0^\Delta \left(\int_0^{\tau_{k+1}} \frac{(\tau_{k+1} - \tau)^k}{k!} dv_\tau \right) d\tau_{k+1} \end{aligned} \tag{14.45}$$

where the result (14.42) has been used for $m = k$, setting $\Delta = \tau_{k+1}$. The result is obtained by changing the order of integration, i.e.,

$$\begin{aligned} I_{(1, \underbrace{0,\ldots,0}_{k+1 \text{ zeros}})} &= \int_0^\Delta \int_0^{\tau_{k+1}} \frac{(\tau_{k+1} - \tau)^k}{k!} dv_\tau \, d\tau_{k+1} \\ &= \int_0^\Delta \int_\tau^\Delta \frac{(\tau_{k+1} - \tau)^k}{k!} d\tau_{k+1} \, dv_\tau \\ &= \int_0^\Delta \frac{(\Delta - \tau)^{k+1}}{(k+1)!} dv_\tau \end{aligned} \tag{14.46}$$

□

Corollary 14.9 *As a consequence of Lemma* 14.8, *the following relation between* (14.33) *and* (14.38) *holds*:

$$\tilde{v}_\ell = I_{(1, \underbrace{0,\ldots,0}_{n-\ell \text{ zeros}})} \tag{14.47}$$

Combining Lemma 14.7 and Eq. (14.40) leads to the following.

Lemma 14.10 *For an nth order integrator, the sampled output PSD is given by*

$$\phi_{yy} = \frac{N(z)}{D(z)}\sigma_d^2 \tag{14.48}$$

where $D(z) = (z-1)^n(z^{-1}-1)^n$ *and* $N(z) = z^{1-n}B_{(2n-1)}(z)$ *and where* $B_{(2n-1)}(z)$ *is the Euler-Frobenius polynomial. Also,*

$$\sigma_d^2 = \frac{\Delta^{2n}}{(2m-1)!} \tag{14.49}$$

Proof From Lemma 14.7 and Eq. (14.40), the exact sampled-data model is

$$\begin{bmatrix} q-1 & -\Delta & \dots & \frac{-\Delta^{n-1}}{(n-1)!} \\ 0 & q-1 & \dots & \frac{-\Delta^{n-2}}{(n-2)!} \\ \vdots & \vdots & \ddots & \vdots \\ 0 & 0 & \dots & q-1 \end{bmatrix} \widehat{X} = \begin{bmatrix} \tilde{v}_1 \\ \vdots \\ \vdots \\ \tilde{v}_n \end{bmatrix} \tag{14.50}$$

$$y = \begin{bmatrix} 1 & 0 & \dots & 0 \end{bmatrix} \widehat{X} \tag{14.51}$$

where $\widetilde{V} = [\tilde{v}_1 \ \dots \ \tilde{v}_n]^T$ has covariance as in (16.20).

The output satisfies

$$y = \begin{bmatrix} \frac{1}{(q-1)^n} & 0 & \dots & 0 \end{bmatrix} \times \begin{bmatrix} (q-1)^{n-1} & \Delta(q-1)^{n-2}B_1 & \dots & \frac{\Delta^{n-1}}{(n-1)!}B_{n-1} \\ 0 & (q-1)^{n-1} & \dots & \frac{\Delta^{n-2}(q-1)}{(n-2)!}B_{n-2} \\ \vdots & \vdots & \ddots & \vdots \\ 0 & 0 & \dots & (q-1)^{n-1} \end{bmatrix} \begin{bmatrix} \tilde{v}_1 \\ \vdots \\ \vdots \\ \tilde{v}_n \end{bmatrix} \tag{14.52}$$

where $B_1, B_2, \dots$ are the Euler-Frobenius polynomials. Hence,

$$y = \frac{1}{(q-1)^n}\left\{(q-1)^{n-1}\tilde{v}_1 + \Delta B_1(q-1)^{n-2}\tilde{v}_2 + \dots + \frac{\Delta^{n-1}}{(n-1)!}B_{n-1}\tilde{v}_n\right\} \tag{14.53}$$

The result then follows by evaluating the discrete-time PSD and using spectral factorisation. □

14.5 Summary

The key points covered in this chapter are the following:

- Sampling zeros arise in the case of continuous-time stochastic power spectral densities having relative degree greater than 2.
- Stochastic sampling zeros can be conveniently characterised in the discrete-time power spectral density.
- For the case of instantaneous sampling, the discrete-time power spectral density has $2(n-m)-1$ extra zeros which asymptotically (as $\Delta \to 0$) converge to the zeros of $z^{1-r}B_{2(n-m)-1}(z)$, where $B(z)$ denotes the usual Euler–Frobenius polynomials.
- For the case of sampling after an averaging anti-aliasing filter, there are $2(n-m)$ extra zeros which asymptotically (as $\Delta \to 0$) converge to the zeros of $B_{2(n-m)}(z)$.
- The stochastic sampling zeros can also be characterised by considering the sampling of an nth order integrator continuous-time model.

Further Reading

Asymptotic sampling zeros for linear stochastic systems were first described in

Wahlberg B (1988) Limit results for sampled systems. Int J Control 48(3):1267–1283

A classic text on Kalman filtering and spectral factorization is

Anderson BDO, Moore J (1979) Optimal filtering. Prentice Hall, Englewood Cliffs

Chapter 15
Generalised Sampling Filters

In Chap. 6, it was shown that the input hold device has a significant impact on the zeros of the resultant sampled-data model. It was also demonstrated that the input hold could be designed to assign the deterministic asymptotic sampling zeros to any desired location. A similar result holds for stochastic systems: the zeros of the sampled output power spectral density (and, thus, of the corresponding sampled-data model) can be assigned by choosing a generalised anti-aliasing filter prior to instantaneous sampling.

In this chapter, the impact of the choice of the anti-aliasing filter on the resultant stochastic sampled-data model will be explored. Section 15.2 presents a filter design procedure such that the *sampling zeros* of the sampled output power spectral density are asymptotically assigned to the origin, where they no longer play any role.

15.1 Sampled-Data Models when Generalised Sampling Filters Are Deployed

Assume a sampling scheme as shown in Fig. 15.1, where a generalised anti-aliasing filter (AAF) is used prior to sampling the system output. The AAF is chosen as a generalised sampling filter (GSF), characterised by its impulse response, $h_g(t)$. Similar to Assumption 6.1 for the generalised hold case in Sect. 6.1, a class of sampling functions having support on the interval $[0, \Delta)$ will be considered.

The output sequence is obtained by sampling instantaneously the output of the filter:

$$\bar{y}_k = \bar{y}(k\Delta) = \int_{k\Delta-\Delta}^{k\Delta} y(\tau) h_g(k\Delta - \tau)\, d\tau \tag{15.1}$$

J.I. Yuz, G.C. Goodwin, *Sampled-Data Models for Linear and Nonlinear Systems*,
Communications and Control Engineering, DOI 10.1007/978-1-4471-5562-1_15,

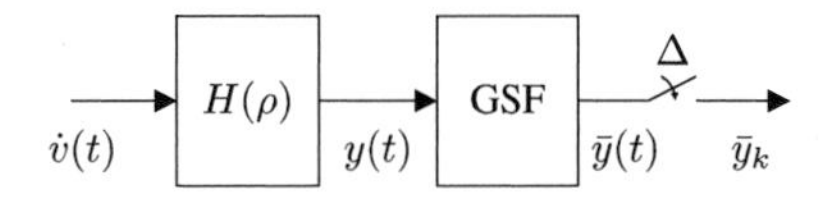

Fig. 15.1 Sampling scheme using a generalised filter

Remark 15.1 The definition of a GSF as in (15.1) is a generalisation of the *integrating filter* (11.23). For the latter special case, the following particular impulse response holds:

$$h_g(t) = \begin{cases} 1/\Delta; & 0 \le t < \Delta \\ 0; & t > \Delta \end{cases} \tag{15.2}$$

Also note that instantaneous sampling can be included in this framework by considering $h_g(t) = \delta(t)$ in (15.1).

The following result gives a discrete-time description of the sampling scheme in Fig. 15.1.

Lemma 15.2 *Consider the sampling scheme in Fig.* 15.1, *where the continuous-time system $H(\rho)$ can be expressed in state-space form as in* (12.15)–(12.16), *and the GSF has impulse response $h_g(t)$. Then the corresponding discrete-time model is given by*:

$$\delta x_k = A_\delta x_k + v_k \tag{15.3}$$

$$\bar{y}_{k+1} = C_g x_k + w_k \tag{15.4}$$

where $\delta = \frac{q-1}{\Delta}$ denotes the delta operator.

The matrices in (15.3)–(15.4) are given by:

$$A_\delta = \frac{e^{A\Delta} - I}{\Delta}, \qquad C_g = \int_0^\Delta h_g(\tau) C e^{A(\Delta-\tau)}\, d\tau \tag{15.5}$$

and v_k and w_k are white noise sequences such that:

$$E\left\{ \begin{bmatrix} v_k \\ w_k \end{bmatrix} \begin{bmatrix} v_\ell \\ w_\ell \end{bmatrix}^T \right\} = \begin{bmatrix} Q_\delta & S_\delta \\ S_\delta^T & R_\delta \end{bmatrix} \frac{\delta_K[k-\ell]}{\Delta} \tag{15.6}$$

where δ_K represents the Kronecker delta function, and where:

$$\begin{bmatrix} Q_\delta & S_\delta \\ S_\delta^T & R_\delta \end{bmatrix} \triangleq \frac{\sigma_v^2}{\Delta} \int_0^\Delta M_g(\sigma) M_g(\sigma)^T\, d\sigma \tag{15.7}$$

$$M_g(\sigma) \triangleq \begin{bmatrix} e^{A\sigma} B \\ \Delta \int_0^{\sigma} h_g(\xi) C e^{A(\sigma-\xi)} B d\xi \end{bmatrix} \tag{15.8}$$

Proof The proof follows the same lines as the proof of Lemma 12.2. First note that (15.3) can readily be obtained as for Eq. (12.17) in Lemma 12.2. Equation (15.4) can be obtained by noting that, on the interval $[k\Delta, k\Delta + \Delta)$, the system output can be expressed as

$$y(t) = Ce^{A(t-k\Delta)} x(k\Delta) + C \int_{k\Delta}^{t} e^{A(t-\eta)} B \dot{v}(\eta)\, d\eta \tag{15.9}$$

Thus, the samples of the filter output can be written as

$$\begin{aligned} \bar{y}_{k+1} &= \int_{k\Delta}^{k\Delta+\Delta} h_g(k\Delta + \Delta - \tau) y(\tau)\, d\tau \\ &= \left[\int_{k\Delta}^{k\Delta+\Delta} h_g(k\Delta + \Delta - \tau) Ce^{A(\tau-k\Delta)}\, d\tau \right] x_k + w_k \end{aligned} \tag{15.10}$$

Changing variables in the integral inside the brackets, the C_g as given in (15.5) is obtained. The noise term w_k is given by

$$w_k = \int_{k\Delta}^{k\Delta+\Delta} h_g(k\Delta + \Delta - \tau) C \int_{k\Delta}^{\tau} e^{A(\tau-\eta)} B \dot{v}(\eta)\, d\eta\, d\tau \tag{15.11}$$

Fubini's theorem (see the references at the end of the chapter) is then used to justify a change in the order of integration in the last integral to obtain:

$$\begin{bmatrix} v_k \\ w_k \end{bmatrix} = \frac{1}{\Delta} \int_{k\Delta}^{k\Delta+\Delta} \begin{bmatrix} e^{A(k\Delta+\Delta-\eta)} B \\ \Delta \int_{\eta}^{k\Delta+\Delta} h_g(k\Delta + \Delta - \tau) Ce^{A(\tau-\eta)} B\, d\tau \end{bmatrix} \dot{v}(\eta)\, d\eta \tag{15.12}$$

Equations (15.6)–(15.8) are obtained from this last expression by proceeding as in the proof of Lemma 12.2. Indeed, the matrix Q_δ in (15.7) is a scaled version of the matrix Q_q in (12.20). This occurs because the AAF does not influence the system dynamics other than via the output. □

Remark 15.3 The discrete-time model (15.3)–(15.4) can equivalently be rewritten, using the shift operator q, as:

$$qx_k = x_{k+1} = A_q x_k + \tilde{v}_k \tag{15.13}$$

$$\bar{y}_{k+1} = C_g x_k + w_k \tag{15.14}$$

where $A_q = I + \Delta A_\delta = e^{A\Delta}$ and the input noise sequence is $\tilde{v}_k = \Delta v_k$. As a consequence,

$$E\left\{ \begin{bmatrix} \tilde{v}_k \\ w_k \end{bmatrix} \begin{bmatrix} \tilde{v}_\ell \\ w_\ell \end{bmatrix}^T \right\} = \begin{bmatrix} Q_q & S_q \\ S_q^T & R_q \end{bmatrix} \delta_K[k - \ell] \tag{15.15}$$

where:

$$Q_q = \Delta Q_\delta; \qquad S_q = S_\delta; \qquad R_q = \frac{1}{\Delta} R_\delta \tag{15.16}$$

Even though Lemma 15.2 provides a sampled-data model for the system illustrated in Fig. 15.1, the resultant discrete-time description depends on two noise sequences as inputs. Using the same ideas as in Chap. 14, the quantity $\bar{y}_k$ can be expressed as the output of a system with a single white noise input; i.e., the discrete-time model can be expressed in *innovations form*.

Lemma 15.4 *The state-space model* (15.13)–(15.14) *is equivalent to the following* innovations model *in the sense that their outputs share the same second order properties*:

$$z_{k+1} = A_q z_k + K_q e_k \tag{15.17}$$

$$\bar{y}_{k+1} = C_g z_k + e_k \tag{15.18}$$

where e_k is a white noise sequence with covariance matrix

$$E\{e_k^2\} = R_q + C_g P C_g^T \tag{15.19}$$

The Kalman gain K_q in (15.17) is given by

$$K_q = (A_q P C_g^T + S_q)(R_q + C_g P C_g^T)^{-1} \tag{15.20}$$

and P is the state covariance matrix given by the discrete-time algebraic Riccati equation:

$$A_q P A_q^T - P - K_q(R_q + C_g P C_g^T)K_q^T + Q_q = 0 \tag{15.21}$$

Using the innovations form in Lemma 15.4, the sequence of output samples $\bar{y}_k$ is described by the model

$$\bar{y}_{k+1} = H_q(q) e_k \tag{15.22}$$

where

$$H_q(z) = C_g(zI_n - A_q)^{-1} K_q + 1 \tag{15.23}$$

and where e_k is a discrete-time white noise with variance (15.19).

Remark 15.5 Equation (15.23) clearly shows that the discrete-time poles depend only on A_q and, hence, only on the continuous-time system matrix A and the sampling period Δ. However, the zeros of the discrete-time model are seen to depend on C_q and K_q, and, in turn, on the GSF impulse response $h_g(t)$. This latter observation will be used in the sequel to design $h_g(t)$ to assign the zeros to specific locations.

A difficulty that we must first address is that the Kalman gain, K_q, depends upon the solution of the algebraic Riccati equation (15.21). Thus, the way that $h_g(t)$ appears in this matrix equation makes this approach difficult for design purposes. Hence, a more direct method is explored below based on spectral factorisation ideas.

Lemma 15.6 *Given the discrete-time model* (15.13)–(15.15), *the discrete-time output power spectral density* (*PSD*) *is given by*:

$$\Phi_{\bar{y}}^{q}(z) = \Delta \begin{bmatrix} C_g(zI_n - A_q)^{-1} & 1 \end{bmatrix} \begin{bmatrix} Q_q & S_q \\ S_q^T & R_q \end{bmatrix} \begin{bmatrix} (z^{-1}I_n - A_q^T)^{-1}C_g^T \\ 1 \end{bmatrix} \tag{15.24}$$

Proof The above result follows from the model (15.13)–(15.14). The output PSD of this model can be obtained in the same way as in the proof of Lemma 13.4, for the integrating filter case. □

Remark 15.7 The result in Lemma 15.6 is closely related to Lemma 15.4, noting that the output PSD of the innovations model (15.17)–(15.18) is given by

$$\Phi_{\bar{y}}^{q}(z) = H_q(z)H_q(z^{-1})\Phi_e^d \tag{15.25}$$

where the spectral factor $H_q(z)$ is given by (15.23) and Φ_e^d is the (constant) spectral density of the innovations sequence.

The previous remark shows that a stochastic sampled-data model, with a scalar noise source as input, can be obtained by spectral factorisation of the PSD (15.24). In the next section, this approach will be adopted to assign the stochastic sampling zeros of $\Phi_{\bar{y}}^{d}(z)$ (and, thus, of the spectral factor $H_q(z)$) by choosing an appropriate GSF. To achieve this result, the function $h_g(t)$ will be expressed in a special form as a linear combination of more elemental functions. This simplifies the expressions for C_g, S_q, and R_q, in Lemma 15.2.

15.2 Generalised Filters to Assign the Asymptotic Sampling Zeros

In this section, the problem of designing a GSF such that the sampling zeros of the discrete-time PSD (15.24) converge, as the sampling period goes to zero, to specific locations in the complex plane is considered. In particular, the analysis focuses on assigning the stochastic sampling zeros to the origin. This is equivalent to obtaining a discrete-time output PSD with *no* stochastic sampling zeros.

Assumption 15.8 Given a system of relative degree r, consider a GSF such that the corresponding impulse response is parameterised by:

$$h_g(t) = \begin{cases} \frac{1}{\Delta}(h_0 + \sum_{\ell=1}^{r} h_\ell \phi_\ell(t)); & t \in [0, \Delta) \\ 0; & t \notin [0, \Delta) \end{cases} \tag{15.26}$$

where the weighting coefficients $h_0, \dots, h_r \in \mathbb{R}$. The basis functions $\phi_\ell(t)$ in (15.26) (to be specified later) are required to satisfy the following condition:

$$\int_0^{\Delta} \phi_\ell(t)\, dt = 0 \tag{15.27}$$

Note that the scaling factor $1/\Delta$ has been introduced in (15.26) to resemble the *averaging* idea of the integrating filter (11.23). In fact, the averaging filter corresponds to the choice $h_0 = 1$ and $h_\ell = 0$, for $\ell = 1, \dots, r$. Condition (15.27) simplifies some of the calculations required to obtain the output PSD (15.24).

Remark 15.9 Note that Assumption 15.8 guarantees that, once the functions $\phi_\ell(t)$ in (15.26) have been chosen, then the $r + 1$ coefficients $h_0, \dots, h_r$ provide enough *degrees of freedom* to assign the r sampling zeros and the noise variance, if required.

The design procedure presented below is based on the key argument discussed earlier in Chap. 14, namely, for fast sampling rates, any system of relative degree r *behaves* at high frequencies as if it were an rth order integrator.

First consider the case of first and second order integrators. It is shown how different GSFs can be designed to assign the corresponding asymptotic stochastic sampling zeros. It is then shown that similar asymptotic sampling zeros are obtained when using the resultant GSF on more general systems, having relative degree 1 and 2, respectively.

The examples that follow are aimed at illustrating the general principle, i.e., for a stochastic system of relative degree r, a GSF can be designed based on the r-th order integrator case.

15.2.1 First Order Systems

Consider the first order integrator $H(\rho) = \rho^{-1}$. The matrices of the corresponding state-space representation (12.15)–(12.16) are, in this case, the scalars $A = 0$, and $B = C = 1$. In the sampled-data model (15.13), the matrix A_q is then equal to 1 for *any* sampling period Δ.

Example 15.10 (Integrating AAF) The impulse response, in this case, is defined in (15.2). For this choice it follows that:

$$C_g = 1 \quad \text{and} \quad \begin{bmatrix} Q_q & S_q \\ S_q^T & R_q \end{bmatrix} = \begin{bmatrix} \Delta & \frac{\Delta}{2} \\ \frac{\Delta}{2} & \frac{\Delta}{3} \end{bmatrix} \tag{15.28}$$

Substituting into (15.24), we obtain the asymptotic result in Theorem 14.4, namely,

$$\Phi_{\bar{y}}^q(z) = \frac{\Delta^2}{3!} \frac{(z + 4 + z^{-1})}{(z-1)(z^{-1}-1)} \tag{15.29}$$

A sampled-data model can be readily obtained by spectral factorisation, as in (15.25). This leads to

$$H_q(z) = \frac{\Delta}{3-\sqrt{3}} \frac{(z + 2 - \sqrt{3})}{(z-1)} \tag{15.30}$$

Example 15.11 (Piecewise Constant GSF) Choose the filter to have the impulse response corresponding to a generalised hold function as considered in Sect. 6.2. Here, $h_g(t)$ is parameterised in a slightly different way:

$$h_g(t) = \begin{cases} \frac{1}{\Delta}(h_0 + h_1); & 0 \le t < \frac{\Delta}{2} \\ \frac{1}{\Delta}(h_0 - h_1); & \frac{\Delta}{2} \le t < \Delta \\ 0; & t \notin [0, \Delta) \end{cases} \tag{15.31}$$

where $h_0, h_1 \in \mathbb{R}$. For this GSF choice, it follows that:

$$C_g = h_0 \quad \text{and} \quad \begin{bmatrix} Q_q & S_q \\ S_q^T & R_q \end{bmatrix} = \begin{bmatrix} \Delta & \frac{\Delta}{2}(h_0 + \frac{1}{4}h_1) \\ \frac{\Delta}{2}(h_0 + \frac{1}{4}h_1) & \frac{\Delta}{3}(h_0^2 + \frac{3}{4}h_0 h_1 + \frac{1}{4}h_1^2) \end{bmatrix} \tag{15.32}$$

which, on substituting into (15.24), gives

$$\Phi_{\bar{y}}^q(z) = h_0^2 \frac{\Delta^2}{3!} \frac{(z + 4 + z^{-1})}{(z-1)(z^{-1}-1)} + \frac{h_1^2}{2} \frac{\Delta^2}{3!} \frac{(-z + 2 - z^{-1})}{(z-1)(z^{-1}-1)} \tag{15.33}$$

If $h_0 = 1$ and $h_1 = \sqrt{2}$, then a sampled PSD is obtained with *no* zeros, or, equivalently, a stable spectral factor with zeros at the origin:

$$\Phi_{\bar{y}}^q(z) = \frac{\Delta^2}{(z-1)(z^{-1}-1)} \quad \Rightarrow \quad H_q(z) = \frac{\Delta z}{(z-1)} \tag{15.34}$$

Example 15.12 (Sinusoidal GSF) Another simple GSF impulse response which satisfies Assumption 15.8 is given by:

$$h_g(t) = \begin{cases} \frac{1}{\Delta}(h_0 + h_1 \pi \sin(\frac{2\pi}{\Delta} t)); & 0 \leq t < \Delta \\ 0; & t \notin [0, \Delta) \end{cases} \tag{15.35}$$

where $h_0, h_1 \in \mathbb{R}$, and the constant π is introduced as a scaling factor. For this choice, it follows that:

$$C_g = h_0 \quad \text{and} \quad \begin{bmatrix} Q_q & S_q \\ S_q^T & R_q \end{bmatrix} = \begin{bmatrix} \Delta & \frac{\Delta}{2}(h_0 + h_1) \\ \frac{\Delta}{2}(h_0 + h_1) & \frac{\Delta}{3}(h_0^2 + \frac{3}{2} h_0 h_1 + \frac{9}{8} h_1^2) \end{bmatrix} \tag{15.36}$$

Upon substituting into (15.24),

$$\Phi_{\bar{y}}^q(z) = h_0^2 \frac{\Delta^2}{3!} \frac{(z + 4 + z^{-1})}{(z-1)(z^{-1}-1)} - \frac{9h_1^2}{4} \frac{\Delta^2}{3!} \frac{(z - 2 + z^{-1})}{(z-1)(z^{-1}-1)} \tag{15.37}$$

If, for example, $h_0 = 1$ and $h_1 = 2/3$, then a sampled PSD (and a stable spectral factor) is obtained as in (15.34). As required, this discrete-time model has sampling zeros at the origin.

The GSFs obtained in Examples 15.11 and 15.12 were designed to assign the stochastic sampling zeros (of a first order integrator) to the origin. However, this GSF can also be used, for fast sampling rates, on *any* system of relative degree 1 to obtain sampling zeros *near* the origin. This principle is illustrated in the following example.

Example 15.13 Consider the continuous-time system

$$H(\rho) = \frac{1}{\rho + 2} \tag{15.38}$$

Fix the sampling period to be $\Delta = 0.1$, which corresponds to a sampling frequency about one decade above the model bandwidth. Use the same piecewise GSF obtained in Example 15.11. This leads to the following stable spectral factor of the output PSD:

$$H_q(z) = \frac{0.287(z - 2.489 \times 10^{-4})}{(z - e^{-0.2})} \tag{15.39}$$

Similarly, using the *sinusoidal* GSF described in Example 15.12 leads to

$$H_q(z) = \frac{0.287(z - 6.590 \times 10^{-5})}{(z - e^{-0.2})} \tag{15.40}$$

Note that (as expected) for both cases, the stochastic sampling zero is very close to the origin.

15.2.2 Second Order Systems

Next consider the GSF design problem for the second order integrator. This is a prelude to dealing with general systems of relative degree 2. The expressions that allow one to obtain the sampled-data model and, thus, to identify the stochastic sampling zeros, are more involved in this case than for the first order integrator. However, the design procedure previously outlined can be readily adapted as shown below.

Consider the second order integrator $H(\rho) = 1/\rho^2$. The state-space representation (12.15)–(12.16) then contains the following matrices:

$$A = \begin{bmatrix} 0 & 1 \\ 0 & 0 \end{bmatrix}, \qquad B = \begin{bmatrix} 0 \\ 1 \end{bmatrix}, \qquad C = \begin{bmatrix} 1 & 0 \end{bmatrix} \tag{15.41}$$

For any sampling period Δ, the discrete-time system matrix is given by

$$A_q = \begin{bmatrix} 1 & \Delta \\ 0 & 1 \end{bmatrix} \tag{15.42}$$

Example 15.14 (Integrating Filter) This filter is defined in (15.2). In this case, it follows that:

$$C_g = \begin{bmatrix} 1 & \frac{\Delta}{2} \end{bmatrix}, \qquad \begin{bmatrix} Q_q & S_q \\ S_q^T & R_q \end{bmatrix} = \left[\begin{array}{cc|c} \frac{\Delta^3}{3} & \frac{\Delta^2}{2} & \frac{\Delta^3}{8} \\ \frac{\Delta^2}{2} & \Delta & \frac{\Delta^2}{6} \\ \hline \frac{\Delta^3}{8} & \frac{\Delta^2}{6} & \frac{\Delta^3}{20} \end{array}\right] \tag{15.43}$$

which, upon substitution into (15.24), gives

$$\Phi_{\bar{y}}^q(z) = \Phi_{\bar{y}}^{\mathrm{IF}}(z) = \frac{\Delta^4}{5!} \frac{(z^2 + 26z + 66 + 26z^{-1} + z^{-2})}{(z-1)^2(z^{-1}-1)^2} \tag{15.44}$$

The obtained discrete-time PSD is, again, consistent with the asymptotic result in Theorem 14.4 on p. 172.

Example 15.15 (Piecewise Constant GSF) Consider a GSF defined by the impulse response:

$$h_g(t) = \begin{cases} \frac{1}{\Delta}(h_0 + h_1 + h_2); & 0 \le t < \frac{\Delta}{4} \\ \frac{1}{\Delta}(h_0 + h_1 - h_2); & \frac{\Delta}{4} \le t < \frac{\Delta}{2} \\ \frac{1}{\Delta}(h_0 - h_1 + h_2); & \frac{\Delta}{2} \le t < \frac{3\Delta}{4} \\ \frac{1}{\Delta}(h_0 - h_1 - h_2); & \frac{3\Delta}{4} \le t < \Delta \\ 0; & t \notin [0, \Delta) \end{cases} \tag{15.45}$$

where $h_0, h_1, h_2 \in \mathbb{R}$. For this choice, it follows that

$$C_g = \begin{bmatrix} h_0 & \frac{\Delta}{2}(h_0 + \frac{1}{2}h_1 + \frac{1}{4}h_2) \end{bmatrix} \tag{15.46}$$

Computing the noise PSD (15.15) and substituting in (15.24) leads to

$$\Phi_{\bar{y}}^{q}(z) = h_0^2 \Phi_{\bar{y}}^{\mathrm{IF}}(z) + h_1^2 \Phi_{\bar{y}}^{1}(z) + h_2^2 \Phi_{\bar{y}}^{2}(z) + h_1 h_2 \Phi_{\bar{y}}^{3}(z) \tag{15.47}$$

where $\Phi_{\bar{y}}^{\mathrm{IF}}(z)$ is the PSD (15.44) obtained in Example 15.14, and $\Phi_{\bar{y}}^{\ell}(z)$ $(\ell = 1, 2, 3)$ are other particular power spectral densities that do not depend on the GSF parameters.

To assign the spectrum zeros to the origin, it is simply necessary to solve for the weighting parameters h_ℓ in (15.47). Either of the following choices:

$$h_0 = 1, \qquad h_1 = \mp 9.891, \qquad h_2 = \pm 23.782, \quad \text{or} \tag{15.48}$$

$$h_0 = 1, \qquad h_1 = \mp 4.691, \qquad h_2 = \pm 5.382 \tag{15.49}$$

leads to a sampled PSD with no zeros, i.e.,

$$\Phi_{\bar{y}}^{q}(z) = \frac{\Delta^4}{5!} \frac{K}{(z-1)^2(z^{-1}-1)^2} \tag{15.50}$$

Example 15.16 (Sinusoidal GSF) Restrict the GSF impulse response to the form:

$$h_g(t) = \begin{cases} \frac{1}{\Delta}(h_0 + h_1 \pi \sin(\frac{2\pi}{\Delta} t) + h_2 \pi \sin(\frac{4\pi}{\Delta} t)); & t \in [0, \Delta) \\ 0; & t \notin [0, \Delta) \end{cases} \tag{15.51}$$

where $h_0, h_1, h_2 \in \mathbb{R}$. For this choice, it follows that

$$C_g = \begin{bmatrix} h_0 & \frac{\Delta}{2}(h_0 + h_1 + \frac{1}{2} h_2) \end{bmatrix} \tag{15.52}$$

Computing the noise PSD (15.15) and substituting into (15.24) gives

$$\Phi_{\bar{y}}^{q}(z) = h_0^2 \Phi_{\bar{y}}^{\mathrm{IF}}(z) + h_1^2 \Phi_{\bar{y}}^{1}(z) + h_2^2 \Phi_{\bar{y}}^{2}(z) + h_1 h_2 \Phi_{\bar{y}}^{3}(z) \tag{15.53}$$

where $\Phi_{\bar{y}}^{\mathrm{IF}}(z)$ is the PSD (15.44) obtained in Example 15.14, and $\Phi_{\bar{y}}^{\ell}(z)$ $(\ell = 1, 2, 3)$ are other power spectral densities that do not depend on the GSF parameters h_ℓ. Solving for these parameters to assign the zeros to the origin, it follows that either of the following choices:

$$h_0 = 1, \qquad h_1 = \pm 3.902, \qquad h_2 = \mp 9.804 \quad \text{or} \tag{15.54}$$

$$h_0 = 1, \qquad h_1 = \pm 1.902, \qquad h_2 = \mp 1.804 \tag{15.55}$$

lead to a sampled PSD with no zeros, i.e.,

$$\Phi_{\bar{y}}^{q}(z) = \frac{\Delta^4}{5!} \frac{K}{(z-1)^2(z^{-1}-1)^2} \tag{15.56}$$

The GSFs described above assign the stochastic sampling zeros of general linear models to be *close* to the origin, when using *fast* sampling rates. The following example illustrates the use of the GSFs obtained in Examples 15.15 and 15.16 for a general system of relative degree 2.

Example 15.17 Consider the following second order stochastic system:

$$H(\rho) = \frac{2}{(\rho+2)(\rho+1)} \tag{15.57}$$

First use the same piecewise GSF obtained in Example 15.15. In particular, in (15.45), choose $h_0 = 1$, $h_1 = 4.691$, and $h_2 = -5.382$. For a sampling period $\Delta = 0.1$, the following stable spectral factor is obtained:

$$H_q(z) = \frac{5.269 \times 10^{-2}(z - z_1)(z - z_1^*)}{(z - e^{-0.1})(z - e^{-0.2})} \tag{15.58}$$

where $z_1 = -0.014 + j0.081$, and where * denotes complex conjugation.

Also use the *sinusoidal* GSF obtained in Example 15.16. The sampling period is fixed to $\Delta = 0.01$. In (15.51), choose $h_0 = 1$, $h_1 = 1.902$, and $h_2 = -1.804$. The sampled-data model is then given by

$$H_q(z) = \frac{0.22 \times 10^{-10}(z - z_1)(z - z_2)}{(z - e^{-0.01})(z - e^{-0.02})} \tag{15.59}$$

where $z_1 = -1.0435 \cdot 10^{-3}$ and $z_1 = -1.0439 \cdot 10^{-3}$.

Note that, as in Example 15.13, both GSFs assign the sampling zeros very close to the origin, as expected.

15.3 Robustness Issues

This section revisits the robustness issues discussed in Chap. 7 in the context of deterministic systems. Not surprisingly, these issues carry over to the stochastic case. In particular, the results on asymptotic sampling zeros depend upon the fact that the system behaves as an rth order integrator in the frequency domain just above the Nyquist frequency. High frequency unmodelled dynamics will thus impact the location of the stochastic sampling zeros.

The following example is used here to illustrate the application of these ideas to the stochastic case.

Example 15.18 (Stochastic Systems with GSFs) Consider the presence of an unmodelled fast pole in the continuous-time stochastic system defined in (15.38). Thus, consider the following true system:

$$H(\rho) = \frac{1}{(\rho+2)(\frac{1}{\omega_u}\rho+1)} \tag{15.60}$$

The goal is to design a GSF to assign the sampling zeros to the origin in the shift domain.

The piecewise constant GSF obtained in Example 15.11 (based on a nominal system of relative degree 1) will be used.

Assume an unmodelled fast pole located at $\omega_u = 200$ rad/s. The following two cases are analysed:

1. $\Delta = 0.1$ s: This corresponds to a sampling frequency $\omega_s \approx 60$ rad/s. In this case, the unmodelled pole lies well beyond the sampling frequency, so no significant effect can be anticipated on the sampled-data model or on the ability to shift the sampling zero via the GSF. Indeed, the design described in Example 15.11 leads to the following spectral factor for the true system:

$$H_q(z) = \frac{1.1 \times 10^{-3}(z + 1.2 \times 10^{-2})(z - 5.3 \times 10^{-3})}{(z - e^{-0.2})(z - e^{-20})} \tag{15.61}$$

   In this case, a sampling frequency has been chosen *inside the bandwidth* where the assumption on the relative degree is justified.
   The sampling zeros and the fast discrete-time pole are both located close to the origin. Thus, the system can be roughly approximated by (15.39), as expected.
2. $\Delta = 0.01$ s: Next, increase the sampling frequency up to $\omega_s \approx 600$ rad/s. The unmodelled pole, in this case, should ideally be considered in the GSF design. However, assuming that the presence of the high frequency pole is unknown and using the same GSF as above leads to the following spectral factor for the true system:

$$H_q(z) = \frac{3.8 \times 10^{-5}(z + 2.1 \times 10^{-1})(z - 3.3 \times 10^{-2})}{(z - e^{-0.02})(z - e^{-2})} \tag{15.62}$$

   It can be seen, in this case, that the slowest sampling zero is far from the origin. The reason for this outcome is understandable, as the relative degree assumption on the nominal model is no longer valid at this sampling rate.

The previous example confirms the heuristic notion that the *system relative degree* and associated design procedures should be considered in terms of a *bandwidth of validity* for the nominal model of the continuous-time system.

15.4 Summary

The key points covered in this chapter are:

- Generalised sampling filters, which play a role for stochastic systems that is dual to the role played by the input hold for deterministic systems.
- Development of stochastic linear models for systems having generalised sampling filters.

- Techniques aimed at designing generalised sampling filters to assign stochastic sampling zeros.
- Robustness issues associated with the use of generalised sampling filters.

Further Reading

Further discussion on the use of generalised sampling filters in the stochastic case can be found in

Yuz JI, Goodwin GC (2005) Generalized filters and stochastic sampling zeros. In: Joint CDC-ECC'05, Seville, Spain, December 2005

Additional background related to the duality between stochastic filtering and deterministic control can be found in

Goodwin GC, Mayne DQ, Feuer A (1995) Duality of hybrid optimal regulator and hybrid optimal filter. Int J Control 61(6):1465–1471

Chapter 16
Approximate Sampled-Data Models for Linear Stochastic Systems

This chapter shows how approximate sampled-data models for stochastic linear systems can be developed. Of course, in the linear case, exact sampled-data models can be obtained as in Chaps. 12 and 13. However, results on approximate linear stochastic sampled-data models are developed here, as a prelude to the nonlinear case treated in the next chapter.

Three types of approximations are developed:

1. Frequency-domain approximation: These are achieved by mapping the continuous-time poles and zeros to the discrete-time domain and by then adding the asymptotic sampling zeros of the spectrum.
2. A time-domain approximation based on up-sampling.
3. An alternative time-domain approximation achieved by performing successively higher order integration on a model in the normal form.

A preliminary result is also presented on how to obtain stochastic sampling zeros based on the last of the above three models. It turns out that this final step requires extra considerations which are not present in the deterministic case discussed earlier in the book.

16.1 Adding the Sampling Zeros in the Frequency Domain

Section 13.5 of Chap. 13 showed that, to achieve small relative errors in the frequency domain, it is inadequate to simply map the poles and zeros of the continuous-time power spectral density (PSD) across to the discrete-time domain. Indeed, it is necessary to account (in some way) for the asymptotic sampling zeros. One obvious strategy, arising from the ideas of Chap. 14, is to simply append the asymptotic sampling zeros to an approximate model based on Euler integration (called the Euler–Maruyama scheme in the stochastic case). This scheme and other closely related ideas are discussed below.

J.I. Yuz, G.C. Goodwin, *Sampled-Data Models for Linear and Nonlinear Systems*, Communications and Control Engineering, DOI 10.1007/978-1-4471-5562-1_16,

Consider a linear continuous-time stochastic system having output PSD given by

$$\Phi^c(s) = K_0 \frac{\prod_{i=1}^{m}(s - z_i)}{\prod_{i=1}^{n}(s - p_i)}, \qquad r = n - m \tag{16.1}$$

Some candidate approximations for the associated discrete-time PSD are derived from Lemma 14.3 on p. 172. These are given below.

Approximate Model 1:

$$\overline{\Phi_1^d}\left(e^{j\omega\Delta}\right) = \frac{\Delta^r}{(r-1)!} \frac{e^{j\omega\Delta} B_{r-1}(e^{j\omega\Delta})}{(e^{j\omega\Delta} - 1)^r} \tag{16.2}$$

This model is exact when $\Phi^c(s) = 1/s^r$ (see Eq. (14.22)).

Approximate Model 2:

$$\overline{\Phi_2^d}\left(e^{j\omega\Delta}\right) = \Phi^c(j\omega) \frac{e^{j\omega\Delta} B_{r-1}(e^{j\omega\Delta})}{(r-1)!}, \qquad \omega \in \left[0, \frac{\pi}{\Delta}\right] \tag{16.3}$$

This model appends the asymptotic sampling zeros to the continuous-time PSD.

Approximate Model 3:

$$\overline{\Phi_3^d}\left(e^{j\omega\Delta}\right) = \Phi^c\left(\frac{e^{j\omega\Delta} - 1}{\Delta}\right) \frac{e^{j\omega\Delta} B_{r-1}(e^{j\omega\Delta})}{(r-1)!} \tag{16.4}$$

This model appends the asymptotic sampling zeros to a version of the continuous-time model where the poles and zeros are mapped via $s = \frac{z-1}{\Delta}$; $z = e^{j\omega}$.

Approximate Model 4:

$$\overline{\Phi_5^d}\left(e^{j\omega\Delta}\right) = K_5 \frac{\prod_{i=1}^{m}(e^{j\omega\Delta} - e^{z_i\Delta})}{\prod_{i=1}^{n}(e^{j\omega\Delta} - e^{p_i\Delta})} \frac{e^{j\omega\Delta} p_{r-1}(e^{j\omega\Delta})}{(r-1)!} \tag{16.5}$$

where K_5 is such that $\overline{\Phi_5^d}(1) = \Phi^c(0)$. This model maps the continuous-time poles to the discrete domain using an exponential. Then the asymptotic sampling zeros are added.

A simple example is used to compare the various models described above.

Example 16.1 Let the continuous-time PSD, $\Phi^c(s)$, be given by:

$$\Phi^c(s) = \frac{1}{(s+a)^2(-s+a)^2}, \qquad s = j\omega \tag{16.6}$$

Choose $a = 10$ and $\Delta = 0.01$. The following models are compared:

- *CT:* The continuous-time PSD.
- *DT:* The true discrete-time PSD.
- *Approximate Models 1 through 4:* as described above.

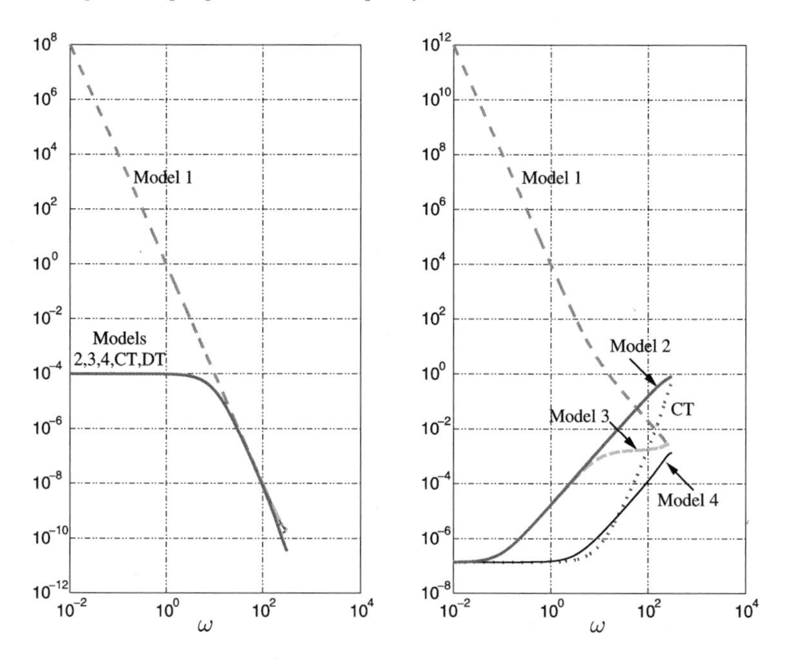

Fig. 16.1 *Figure on the left*: Magnitude of the PSDs. *Figure on the right*: Relative errors

Figure 16.1, left plot, shows the various PSDs. Note that it is virtually impossible to distinguish Models 2, 3, 4, CT, and DT on this scale. The only model which shows any discernible difference is Model 1, which is clearly only valid at high frequencies. These observations are consistent with Lemma 13.5. More informative results are shown in Fig. 16.1, right plot. This shows the relative error with respect to the true discrete-time PSD, i.e.,

$$\rho_k\left(e^{j\omega\Delta}\right) = \left|\frac{\Phi^d(e^{j\omega\Delta}) - \overline{\Phi_k^d}(e^{j\omega\Delta})}{\Phi^d(e^{j\omega\Delta})}\right| \tag{16.7}$$

Note that Model 1 yields small relative errors at high frequencies but not at other frequencies. Model 2 (which corresponds to the continuous-time spectrum modified by the sampling zeros) yields a small relative error only over a limited frequency range. The only models which give small relative errors over the complete frequency range are Model 3 (which maps the poles and zeros in the incremental domain using $(1 + \zeta\Delta)$) and Model 4 (which maps the poles and zeros in the incremental domain using $e^{\zeta\Delta}$). Moreover, the performance of these models improves as Δ is decreased. Finally, in terms of the maximal relative error over all frequencies, Model 4 performs best.

The core conclusion, illustrated by the above example, is that, to obtain an adequate discrete-time PSD (in terms of relative errors) over the complete frequency range $(0, \pi/\Delta)$, one needs to modify the continuous-time PSD in two ways:

- Map the poles and zeros, either using $e^{\zeta\Delta}$ (Model 5) or using $1+\zeta\Delta$ (Model 3); and
- add appropriate stochastic sampling zeros.

If both of these steps are taken, then an arbitrarily good model (in the sense of relative errors) is obtained as $\Delta \to 0$.

Unfortunately, the above frequency-domain ideas cannot be readily extended to the nonlinear case. Hence, time-domain approximations are next explored. These have counterparts in the nonlinear case as treated in Chap. 18.

16.2 Approximate Sampled-Data Model Based on Up-Sampling

The ideas presented here are the stochastic equivalents of the deterministic ideas presented in Sect. 9.1 of Chap. 9.

Say the ultimate goal is to use period Δ. However, assume that Δ is too large for sufficiently accurate results to be obtained from the simple Euler–Maruyama integration scheme. A possible remedy is to use an up-sampled model of period $\Delta' = \frac{\Delta}{m}$, $m > 1$.

To illustrate the ideas, consider the continuous-time stochastic linear system of Chap. 12, i.e.,

$$dx = Ax\,dt + d\omega \tag{16.8}$$

$$dy = Cx\,dt + dv \tag{16.9}$$

where

$$E\left\{\begin{bmatrix} d\omega \\ dv \end{bmatrix}\begin{bmatrix} d\omega \\ dv \end{bmatrix}^T\right\} = \begin{bmatrix} Q & 0 \\ 0 & R \end{bmatrix} dt$$

An averaging anti-aliasing filter is used.

Consider an up-sampling period $\Delta' = \frac{\Delta}{m}$. The following Euler–Maruyama approximate model is obtained at period Δ'. The model can be iterated m times to obtain a solution applicable at period Δ:

$$dx^+_{km+\ell+1} = Ax_{km+l}\Delta' + \omega'_{km+\ell}; \quad \ell = 0, \dots, m-1 \tag{16.10}$$

where $\{\omega'_{km+\ell}\}$ is a discrete-time white noise sequence having covariance $Q\Delta'$.

Recall that the averaging anti-aliasing filter operates by integrating the output over the period Δ before taking samples. This integration operation can be described by adding further states to the model. Thus, the following additional states are introduced:

$$d\bar{x}^+_{km+\ell+1} = C\bar{x}_{km+l}\Delta' + v'_{km+\ell}; \quad \ell = 0, \dots, m-1 \tag{16.11}$$

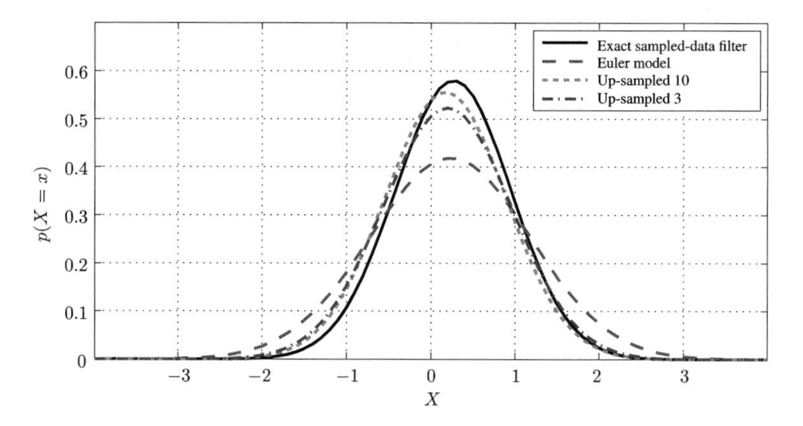

Fig. 16.2 Probability density function at the instant $t = 9$

where v_k' has covariance $R\Delta'$. Finally, note that samples are taken every m iterations of the up-sampled model. Thus

$$y_{(k+1)m+1} = y_{km+l} + [\bar{x}_{(k+1)m} - \bar{x}_{km+1}] + C\bar{x}_{(k+1)m}\Delta' + v'_{(k+1)m} \tag{16.12}$$

The use of this up-sampling idea is illustrated in the following simple example.

Example 16.2 Consider the optimal linear filtering pattern for a stochastic linear sampled-data system. Of course, the exact sampled-data model can be readily obtained, but the goal here is to illustrate the use of up-sampling.

Thus, consider the following simple continuous-time linear model:

$$dx(t) = -x(t)\,dt + d\omega(t) \tag{16.13}$$

$$dy(t) = x(t)\,dt + dv(t) \tag{16.14}$$

where $d\omega$ and dv have incremental covariance $E\{d\omega\, d\omega^T\} = 1dt$ and $E\{dv\, dv^T\} = 10dt$. Use sample period $\Delta = 1$ and an up-sample factor of m.

Note that the sampling period Δ is large relative to the system dynamics. Hence, it is not expected that an Euler–Maruyama-based approximate model (running at period Δ) will give accurate results.

Figure 16.2 shows the posterior probability density of the state estimates at $t = 9$, i.e., the probability density for $x(t)$ given output samples up to t, obtained by

1. The exact sampled-data model at period Δ,
2. The Euler–Maruyama model with period Δ,
3. An *up-sampled* Euler–Maruyama model with ratio $m = 3$, i.e., $\Delta' = \frac{1}{3}$, and
4. An *up-sampled* Euler–Maruyama model with ratio $m = 10$, i.e., $\Delta' = 0.1$.

As anticipated, the up-sampled filter converges to the true filter as the up-sampling ratio increases.

The scheme described above gives a satisfactory solution to the problem of obtaining an accurate sampled-data model (at least in the sense of providing accurate posterior densities in linear filtering). However, it does not provide insight into stochastic sampling zeros. Hence, in the next section, the ideas in Chap. 8 are utilised to build an approximate sampled-data model based on successively higher order integration.

16.3 Approximate Stochastic Sampled-Data Models Based on Successive Integration

For simplicity of exposition, we will treat the case where there are no intrinsic zeros in the system. The normal form is considered. For the stochastic case, the normal form structure is:

$$dx_1 = x_2\, dt \tag{16.15}$$

$$\vdots \tag{16.16}$$

$$dx_n = (a_1 x_1 + \cdots + a_n x_n)\, dt + dv \tag{16.17}$$

where dv is a scalar Wiener process with incremental covariance $\sigma^2\, dt$.

Successively higher integration beginning with the nth state leads to an approximate stochastic model as described below.

Lemma 16.3 *Consider the model* (16.15) *through* (16.17). *Then, beginning with Euler–Maruyama integration for the nth state and successively integrating up the chain of integrators leads to the following approximate sampled-data model*:

$$\widehat{X}^+ = \left\{ \begin{bmatrix} 1 & \Delta & \cdots & \frac{\Delta^{n-1}}{(n-1)!} \\ 0 & 1 & \cdots & \frac{\Delta^{n-2}}{(n-2)!} \\ \cdots & \ddots & \cdots & \cdots \\ 0 & 0 & \cdots & 1 \end{bmatrix} + \begin{bmatrix} \frac{\Delta^n}{n!} \\ \frac{\Delta^{n-1}}{(n-1)!} \\ \vdots \\ \Delta \end{bmatrix} \begin{bmatrix} a_1 & a_2 & \dots & a_n \end{bmatrix} \right\} \widehat{X} + \widetilde{V} \tag{16.18}$$

where $\widetilde{V}$ *is a vector white noise process having the following properties*:

$$E\{\widetilde{V}\} = 0 \tag{16.19}$$

$$E\{\widetilde{V}\widetilde{V}^T\} = \left[\frac{\Delta^{2n-i-j+1}}{(n-i)!(n-j)!(2n-i-j+1)} \right]_{i,j=1,\dots n} \tag{16.20}$$

Proof Consider the stochastic differential equation corresponding to the last state x_n. The *exact* solution is

$$x_n(\tau) = x_n(0) + \int_0^\tau (a_1 x_1 + \cdots + a_n x_n)\,dt + dv \tag{16.21}$$

The Euler–Maruyama approximate solution is then seen to be

$$\hat{x}_n(\tau) = \hat{x}_n(0) + \big[a_1\hat{x}_1(0) + \cdots + a_n\hat{x}_n(0)\big]\tau + \int_0^\tau dv \tag{16.22}$$

This expression is next used in the equation corresponding to x_{n-1}, leading to

$$\hat{x}_{n-1}(\tau) = \hat{x}_{n-1}(0) + \hat{x}_n(0)\tau + \big[a_1\hat{x}_1(0) + \cdots + a_n\hat{x}_n(0)\big]\frac{\tau^2}{2} \tag{16.23}$$

$$+ \int_0^\tau \int_0^{\tau_{n-1}} dv_{\tau_n}\,d\tau_{n-1} \tag{16.24}$$

Proceeding in this way for the other state components of the stochastic differential equation (14.30)–(14.31) leads to the result of (14.32), (14.33). □

The approximate model described in Lemma 14.10 is called the stochastic truncated Taylor series (STTS) model. The accuracy properties of the STTS class of models will be described in the next chapter in the, more general, nonlinear case.

16.4 Stochastic Sampling Zeros Revisited

The approximate sampled-data model described in Lemma 16.3 uses successive integration. For the deterministic case, it was shown in Chap. 8 that the equivalent model immediately captured the (deterministic) asymptotic sampling zeros. However, in the stochastic case, the model described in Lemma 16.3 is driven by a *vector* noise input $\widetilde{V}$, and thus there is no unique discrete noise input with which the asymptotic sampling zeros can be associated. To resolve this problem extra steps are needed. These steps include (i) using an extra level of approximation of the same order of accuracy, and (ii) collapsing the resultant driving vector noise process into a scalar driving noise process. These steps do indeed reveal the asymptotic stochastic sampling zeros as shown below. Moreover, the ideas extend, in a relatively straightforward fashion, to the nonlinear stochastic case, as shown in the next chapter. For clarity of exposition, the ideas are illustrated by a linear system of relative degree 2.

Example 16.4 Consider the second order version of (16.15) through (16.17). The approximate sampled-data model obtained by successive integration (as in Lemma 16.3) is given by

$$\begin{bmatrix} \hat{x}_1 \\ \hat{x}_2 \end{bmatrix}^+ = \begin{bmatrix} 1 + a_1\frac{\Delta^2}{2} & \Delta + a_2\frac{\Delta^2}{2} \\ a_1\Delta & 1 + a_2\Delta \end{bmatrix} \begin{bmatrix} \hat{x}_1 \\ \hat{x}_2 \end{bmatrix} + \begin{bmatrix} \tilde{v}_1 \\ \tilde{v}_2 \end{bmatrix} \tag{16.25}$$

where

$$E\left\{\begin{pmatrix}\tilde{v}_1\\ \tilde{v}_2\end{pmatrix}\begin{pmatrix}\tilde{v}_1 & \tilde{v}_2\end{pmatrix}\right\} = \begin{bmatrix}\frac{\Delta^3}{3} & \frac{\Delta^2}{2}\\ \frac{\Delta^2}{2} & \Delta\end{bmatrix} \tag{16.26}$$

Five steps are now used to obtain the asymptotic sampling zero dynamics: (i) eliminate some small terms from the STTS model, (ii) focus on the output, (iii) eliminate additional small terms which have smaller variance than the principal terms, (iv) use the Kalman filter to develop an innovations representation which has a scalar white noise driving source, and (v) consider the steady state. In the sequel, it is implicitly assumed that x_1 and x_2 have finite variance.

1. Eliminate some small terms in the STTS model.
   The first equation in (16.25) is

$$(q-1)\hat{x}_1 = \Delta\hat{x}_2 + \frac{\Delta^2}{2}[a_1\hat{x}_1 + a_2\hat{x}_2] + \tilde{v}_1; \quad \mathcal{O}^{LS}(\Delta^2). \tag{16.27}$$

   (Here the notation $\mathcal{O}^{LS}(\Delta^2)$ is used to imply that the variance of the error, i.e., the variance of $[(q-1)\hat{x}_1 - (q-1)x]$, is $k\Delta^4$.)
   Note that the second term on the right-hand side has variance of order Δ^4 and hence this term can be eliminated whilst retaining the same order of accuracy, namely, $\mathcal{O}^{LS}(\Delta^2)$. This leads to an alternative approximate model:

$$(q-1)\bar{x}_1 = \Delta\bar{x}_2 + \tilde{v}_1; \quad \mathcal{O}^{LS}(\Delta^2) \tag{16.28}$$

$$(q-1)\bar{x}_2 = \Delta[a_1\bar{x}_1 + a_2\bar{x}_2] + \tilde{v}_2; \quad \mathcal{O}^{LS}(\Delta) \tag{16.29}$$

   Then note that

$$E\left\{\left|(q-1)\bar{x}_1 - (q-1)\hat{x}_1\right|^2\right\} = k\Delta^4 \tag{16.30}$$

   and hence using the triangle inequality, it follows that

$$E\left\{\left|(q-1)\bar{x}_1 - (q-1)x_1\right|^2\right\} = k\Delta^4 \tag{16.31}$$

   It is thus clear that the order of error, $\mathcal{O}^{LS}(\Delta^2)$ has been retained.
2. Focus on the output.
   Recall that x_1 is the output.
   Operating on (16.28) by $(q-1)$ yields

$$(q-1)^2\bar{x}_1 = \Delta(q-1)\bar{x}_2 + (q-1)\tilde{v}_1; \quad \mathcal{O}^{LS}(\Delta^2) \tag{16.32}$$

   Note that the order of accuracy follows, because differencing two quantities whose difference is $\mathcal{O}(\Delta^4)$ yields another quantity whose difference is $\mathcal{O}(\Delta^4)$.
   Substituting (16.29) into (16.32) yields

$$(q-1)^2\bar{x}_1 = \Delta^2[a_1\bar{x}_1 + a_2\bar{x}_2] + \Delta\tilde{v}_2 + (q-1)\tilde{v}_1; \quad \mathcal{O}^{LS}(\Delta^2) \tag{16.33}$$

Use (16.28) to replace $\bar{x}_2$ in (16.33):

$$(q-1)^2\bar{x}_1 = \Delta^2\left\{a_1\bar{x}_1 + a_2\left(\frac{q-1}{\Delta}\bar{x}_1 - \frac{\tilde{v}_1}{\Delta}\right)\right\} + \Delta\tilde{v}_2 + (q-1)\tilde{v}_1; \quad \mathcal{O}^{LS}(\Delta^2) \tag{16.34}$$

3. Eliminate additional small terms.
   The term $\frac{\Delta^2\tilde{v}_1 a_2}{\Delta}$ in (16.34) has variance of order Δ^5.
   Using the same argument as in part (i), this term can be eliminated without impacting the order of accuracy. This leads to a further approximate model:

$$(q-1)^2\tilde{x}_1 = \Delta^2\left\{a_1\tilde{x}_1 + a_2\left(\frac{q-1}{\Delta}\right)\tilde{x}_1\right\} + \Delta\tilde{v}_2 + (q-1)\tilde{v}_1; \quad \mathcal{O}^{LS}(\Delta^2) \tag{16.35}$$

   Hence replacing $\bar{x}_1$ by $\bar{y}$ leads to the following approximate model:

$$(q-1)^2\bar{y} = \Delta^2\left\{a_1\bar{y} + a_2\left(\frac{q-1}{\Delta}\right)\bar{y}\right\} + \Delta\tilde{v}_2 + (q-1)\tilde{v}_1; \quad \mathcal{O}^{LS}(\Delta^2) \tag{16.36}$$

4. Use the Kalman filter to obtain a scalar noise source.
   Consider the following scalar process. (See the last two terms on the right-hand side of (16.36).)

$$\phi_{k+1} = (q-1)\tilde{v}_{1k} + \Delta\tilde{v}_{1k} \tag{16.37}$$

   with

$$E\left\{\begin{pmatrix}\tilde{v}_{1k}\\ \tilde{v}_{2k}\end{pmatrix}\begin{pmatrix}\tilde{v}_{1k} & \tilde{v}_{2k}\end{pmatrix}\right\} = \begin{bmatrix}\frac{\Delta^3}{3} & \frac{\Delta^2}{2}\\ \frac{\Delta^2}{2} & \Delta\end{bmatrix} \tag{16.38}$$

   or

$$\phi_{k+1} = \tilde{v}_{1(k+1)} - \tilde{v}_{1(k)} + \Delta\tilde{v}_{2(k)} \tag{16.39}$$

   Shifting one time step yields

$$\phi_k = \tilde{v}_{1k} - \tilde{v}_{1(k-1)} + \Delta\tilde{v}_{2(k-1)} \tag{16.40}$$

   A suitable state-space model for ϕ_k is then clearly seen to be

$$\bar{\bar{\bar{x}}}_{k+1} = -\tilde{v}_{1k} + \Delta\tilde{v}_{2k} \tag{16.41}$$

$$\phi_k = \bar{\bar{\bar{x}}}_k + \tilde{v}_{1(k)} \tag{16.42}$$

   which has the form

$$\bar{\bar{\bar{x}}}_{(k+1)} = \omega_k \tag{16.43}$$

$$\phi_k = \bar{\bar{\bar{x}}}_k + v_k \tag{16.44}$$

where

$$E\left\{\begin{pmatrix}\omega_k\\ v_k\end{pmatrix}\begin{pmatrix}\omega_k & v_k\end{pmatrix}\right\}=\begin{bmatrix}Q & S\\ S^T & R\end{bmatrix}=\begin{bmatrix}\frac{\Delta^3}{3} & \frac{\Delta^3}{6}\\ \frac{\Delta^3}{6} & \frac{\Delta^3}{3}\end{bmatrix} \tag{16.45}$$

since $\omega_k=-\tilde{v}_{1k}+\Delta\tilde{v}_{2k}$ and $v_k=\tilde{v}_{1k}$.

The corresponding innovations representation is obtained via the Kalman filter:

$$\hat{x}_{k+1}=K_k e_k \tag{16.46}$$

$$\phi_k=\hat{x}_k+e_k \tag{16.47}$$

where $\{e_k\}$ is a white noise innovations sequence and where $\{K_k\}$ is determined from the following general equations:

$$K_k=\left(A\Sigma_k C^T+S\right)\left(C\Sigma_k C^T+R\right)^{-1} \tag{16.48}$$

$$\Sigma_{k+1}=A\Sigma_k A^T+Q-K_k\left[C\Sigma_k C^T+R\right]K_k^T \tag{16.49}$$

For the problem considered here,

$$A=0 \tag{16.50}$$

$$C=1 \tag{16.51}$$

Hence (16.48), (16.49) can be written more compactly as

$$K_k=\frac{S}{(\Sigma_k+R)} \tag{16.52}$$

$$\Sigma_{k+1}=Q-\frac{S^2}{(\Sigma_k+R)} \tag{16.53}$$

Equations (16.46), (16.47) are thus a time-varying model, driven by the scalar white noise sequence e_k. This yields an exact model for the process ϕ_k. Moreover, e_k can be determined from ϕ_k by reordering (16.46), (16.47) to yield

$$\hat{x}_{k+1}=K_k[\phi_k-\hat{x}_k] \tag{16.54}$$

$$e_k=\phi_k-\hat{x}_k \tag{16.55}$$

Hence, e_k is determined (path-wise) from $\tilde{v}_{1k}$ and $\tilde{v}_{2k}$, which, in turn, can be determined (path-wise) from the continuous-time noise process dv.

5. Consider the steady state.
It follows from standard Kalman filtering theory that the covariance, Σ_k, converges exponentially fast to the unique positive definite solution of the following algebraic Riccati equation:

$$\Sigma=Q-\frac{S^2}{\Sigma+R} \tag{16.56}$$

(Compare with (16.53).)
The positive definite solution to (16.56) is

$$\Sigma_{ss} = (\sqrt{3})\frac{\Delta^3}{6} \tag{16.57}$$

The corresponding Kalman gain is then determined from (16.52), yielding

$$K_{ss} = \frac{1}{2+\sqrt{3}} = 2 - \sqrt{3} \tag{16.58}$$

Substituting (16.58) into (16.46), (16.47) yields the following steady-state input-output model:

$$\phi_{k+1} = [q + K_{ss}]e_k \tag{16.59}$$

$$= N_2(q)e_k \tag{16.60}$$

$N_2(q)$ is the sampling stochastic zero dynamics that arises in the case of the second order integrator and where $\{e_k\}$ is a scalar noise process.
The steady-state variance of e_k is

$$\sigma_d^2 = C\Sigma_{ss}C^T + R \tag{16.61}$$

$$= \frac{\Delta^3}{6}[2+\sqrt{3}] \tag{16.62}$$

Bringing all of the above ideas together, a steady-state model for the output y is obtained from (16.36), (16.37), (16.60) as:

$$(q-1)^2 y = \Delta^2\left\{a_1 y + a_2\left(\frac{q-1}{\Delta}\right)y\right\} + N_2(q)e_k; \quad \mathcal{O}^{LS}\big(\Delta^2\big) \tag{16.63}$$

where the noise process $\{e_k\}$ can be determined path-wise from $\{\tilde{v}_{1k}\}$ and $\{\tilde{v}_{2k}\}$ using the steady-state filter:

$$e_k = \frac{1}{N_2(q)}\big\{(q-1)\tilde{v}_{1k} + \Delta\tilde{v}_{2k}\big\} \tag{16.64}$$

Finally, the model (16.63) can be written more intuitively using the delta operator:

$$\delta^2 y = a_1 y + a_2\delta y + N_2'(\delta)e_k; \quad \mathcal{O}^{LS}\big(\Delta^2\big) \tag{16.65}$$

where $\{e_k\}$ is a white noise sequence having variance of order $\frac{1}{\Delta}$ and where $N_2'(\delta)$ denotes the stochastic sampling zero dynamics expressed in delta form.
Note the intuitive connection between (16.65) and the underlying continuous-time system.

Finally, by combining all of the ideas, an approximate model has been obtained whose output PSD satisfies

$$\phi_{yy}(z) = \frac{\bar{\sigma}^2(z-\beta)(z^{-1}-\beta)}{(z-\alpha_1)(z-\alpha_2)(z^{-1}-\alpha_1)(z^{-1}-\alpha_2)} \tag{16.66}$$

where

$$\begin{aligned}(z-\beta)(z^{-1}-\beta) &= z^{-1}(z^2+4z+1) = (z-2+\sqrt{3})(z^{-1}-2+\sqrt{3}) \\ &= B_3(z)z^{-1}\end{aligned} \tag{16.67}$$

The above development has shown (via the second order example) that, in the linear case, the asymptotic stochastic sampling zeros are revealed by the following five-step procedure beginning with the STTS model:

1. Eliminate small terms,
2. Focus on the output,
3. Use an extra step of approximation to eliminate terms which appear in the noise model and which depend upon the underlying system dynamics but whose variance is two orders of magnitude (as a function of Δ) smaller than other terms,
4. Replace the resultant vector noise process by a scalar noise process using the innovations form of the Kalman filter, and
5. Consider the steady state.

The above result is 'strong' in the sense that it describes an approximate model having path-wise convergence errors of order Δ^r (for a system of relative degree r) in the output. Of course, this result depends upon mapping the continuous-time noise process $\{dv\}$ to the discrete-time noise $\bar{v}_k$. If the specific noise process $\{e_k\}$ is replaced by another Gaussian white noise process having the same second order properties, then an approximate model is obtained which describes the statistical properties of the output of the original system (up to the order of accuracy in the path-wise convergence results).

Remark 16.5 The reader may wonder why it was necessary to follow such a tortuous path to obtain the approximate model (16.65). After all, isn't the model (16.65) simply the same as the ASZ model in which the poles are mapped via an Euler approximation and then the stochastic sampling zeros are added? This is true. However, there are two motivations for following the above course:

1. The time-domain path-wise order of accuracy has been revealed by the procedure.
2. Exactly the same steps can be followed in the nonlinear case, as is shown in the next chapter.

16.5 Summary

The key points covered in the current chapter are as follows:

- Accurate approximate sampled-data models can be obtained in the frequency domain by mapping intrinsic poles and zeros using $s = \frac{(z-1)}{\Delta}$ and adding asymptotic stochastic sampling zeros.
- For general linear stochastic systems, expressed in the nominal form, successive integration (beginning with Euler–Maruyama for the nth state) leads to an approximate sampled-data model having a vector driving noise process. This process has the same covariance as found for the nth order integrator.
- The stochastic sampling zeros are revealed by applying an additional five-step procedure.
- The resultant model is in input-output form and has a strong local error of order Δ^r in the output where r is the continuous-time relative degree.

Further Reading

For a discussion on innovations representations, see

Anderson BDO, Moore JB (1979) Optimal filtering. Dover, New York

Up-sampling for stochastic sampled-data models has been discussed in

Cea MG, Goodwin GC, Mueller C (2011) A novel technique based on up-sampling for addressing modeling issues in sampled data nonlinear filtering. In: 18th IFAC world congress, Milan, Italy

Sampled-data models based on stochastic integration were first discussed in

Yuz JI, Goodwin GC (2006) Sampled-data models for stochastic nonlinear systems. In: 14th IFAC symposium on system identification, Newcastle, Australia

Chapter 17
Stochastic Nonlinear Systems

The next step in the book will be to extend the ideas of Chap. 16 to the nonlinear case. However, before taking this step, it is necessary to first make a short diversion to describe aspects of nonlinear stochastic models. This is the appropriate extension of the ideas presented in Sect. 12.1 of Chap. 12. In the nonlinear case, several important technical issues arise. These are briefly detailed in the sequel. Readers who are discovering these ideas for the first time may find it helpful to review the material given in the references at the end of this chapter.

17.1 Background on Stochastic Differential Equations

Several concepts and results from stochastic calculus and stochastic differential equations are reviewed in this section. These topics are dealt with in a general framework, including linear systems as a particular case.

Consider a class of stochastic nonlinear continuous-time systems expressed formally as a set of stochastic differential equations:

$$\frac{dx(t)}{dt} = a(t,x) + b(t,x)\dot{v}(t) \tag{17.1}$$

$$y(t) = c(t,x) \tag{17.2}$$

where the input $\dot{v}(t)$ is a *continuous-time white noise* (CTWN) process having constant spectral density $\sigma_v^2 = 1$. The functions $a(\cdot)$ and $b(\cdot)$ are assumed analytical, i.e., $\mathcal{C}^\infty$. This latter assumption can sometimes be relaxed to *smooth enough* functions, ensuring that the required derivatives are well defined.

Note that the model structure in (17.1) is similar to the deterministic description in (9.5); i.e., the system equation is affine in the input signal.

J.I. Yuz, G.C. Goodwin, *Sampled-Data Models for Linear and Nonlinear Systems*, Communications and Control Engineering, DOI 10.1007/978-1-4471-5562-1_17,

The model (17.1)–(17.2) depends on the CTWN process $\dot{v}(t)$. However, as previously discussed in Chap. 11, white noise processes in continuous time do not exist in any meaningful sense (see Remark 11.2 on p. 141).

For a proper mathematical treatment, Eq. (17.1) should thus be understood as a stochastic differential equation (SDE):

$$dx_t = a(t, x_t)\,dt + b(t, x_t)\,dv_t \tag{17.3}$$

where the driving input to the system is the *increment* of $v_t = v(t)$, a Wiener process of unitary incremental variance. Equation (17.3) is short-hand notation for the following integral form:

$$x_t = x_0 + \int_0^t a(\tau, x_\tau)\,d\tau + \int_0^t b(\tau, x_\tau)\,dv_\tau \tag{17.4}$$

The above equation includes three terms on the right-hand side which are, respectively, an initial condition x_0 (possibly, random), a slowly varying continuous component called the *drift* term, and a rapidly varying continuous random component called the *diffusion* term.

Remark 17.1 The last integral involved in expression (17.4) cannot be interpreted in the *usual* Riemann–Stieltjes sense. In the literature, two constructions of this integral are usually considered, leading to different calculi.

- The *Ito integral* construction:

$$\int f(t)\,dv_t = \lim \sum_\ell f(t_\ell)[v_{t_{\ell+1}} - v_{t_\ell}] \tag{17.5}$$

- The *Stratonovich integral* construction:

$$\int f(t) \circ dv_t = \lim \sum_\ell f\left(\frac{t_{\ell+1} - t_\ell}{2}\right)[v_{t_{\ell+1}} - v_{t_\ell}] \tag{17.6}$$

Each of these definitions presents some advantages and disadvantages. For example, by using the Stratonovich construction (17.6), the usual *chain rule* for transformations can be applied, whereas, when using (17.5), the *Ito rule* is needed (see Sect. 17.2). On the other hand, the Ito integral (17.5) has the specific feature of *not looking into the future*. In fact, the Ito integral is a *martingale*, whereas the Stratonovich integral is not, and, conversely, every martingale can be represented as an Ito integral.

In the sequel, the analysis is restricted to the Ito construction (17.5), but equivalent results can be obtained by using the Stratonovich definition (17.6). Equation (17.4) is referred to as an Ito integral, and x_t as an Ito process described either by the integral equation (17.4) or by the SDE in (17.3).

17.2 The Ito Rule

The Ito construction of a stochastic integral in (17.5) implies an important departure point from the *usual* calculus. Specifically, the usual *chain rule* for transformations must be modified. The key point that leads to this result is given by the properties of the Wiener process $v(t)$ (see Chap. 11). In particular, the incremental variance of a Wiener process can be obtained as:

$$E\{(v(t) - v(s))^2\} = |t - s|; \quad \forall t \neq s \tag{17.7}$$

$$\Rightarrow \quad E\{dv^2\} = E\{(v(t + dt) - v(t))^2\} = dt \tag{17.8}$$

Lemma 17.2 (Ito Rule for Scalar Processes) *Consider a scalar Ito process x_t as in* (17.3), *and a transformation of this process*:

$$y = g(t, x) \tag{17.9}$$

where $g(t, x) \in C^2$, i.e., g has at least second order continuous derivatives. Then $y = y_t$ is also an Ito process, and

$$dy = \frac{\partial g}{\partial t}(t, x)\,dt + \frac{\partial g}{\partial x}(t, x)\,dx + \frac{1}{2}\frac{\partial^2 g}{\partial x^2}(t, x)(dx)^2 \tag{17.10}$$

The differential in the last term, $(dx)^2 = (dx(t))(dx(t))$, is computed according to:

$$dt \cdot dt = dt \cdot dv = dv \cdot dt = 0 \tag{17.11}$$

$$dv \cdot dv = dt \tag{17.12}$$

Note that the Ito rule arises from (17.7), where it can be seen that the variance of the increments dv is of order dt. As a consequence, the last term in (17.10) has to be considered.

Lemma 17.2 presents the derivative rule for transformations of a scalar process $x(t)$. The next result considers the general case for a vector process X_t.

Lemma 17.3 (Multidimensional Ito Rule) *Consider a vector Ito process X_t defined by the set of SDEs*

$$dX = A(t, X)\,dt + B(t, X)\,dV \tag{17.13}$$

where:

$$X = \begin{bmatrix} x_1(t) \\ \vdots \\ x_n(t) \end{bmatrix}, \qquad A(t, X) = \begin{bmatrix} a_1(t, X) \\ \vdots \\ a_n(t, X) \end{bmatrix}$$
$$B(t, X) = \begin{bmatrix} b_{11}(t, X) & \cdots & b_{1m}(t, X) \\ \vdots & \ddots & \vdots \\ b_{n1}(t, X) & \cdots & b_{nm}(t, X) \end{bmatrix} \tag{17.14}$$

and dV are increments of a multidimensional Wiener (vector) process:

$$V = V_t = \begin{bmatrix} v_1(t) & \cdots & v_m(t) \end{bmatrix}^T \tag{17.15}$$

Consider the transformation defined by the following $\mathcal{C}^2$ vector map:

$$Y = G(t, X) = \begin{bmatrix} g_1(t, X) & \cdots & g_p(t, X) \end{bmatrix}^T \tag{17.16}$$

Then $Y = Y_t$ is again an Ito process, given (component-wise) by

$$dY_k = \frac{\partial g_k}{\partial t}(t, X)\,dt + \sum_{\ell=1}^{n} \left(\frac{\partial g_k}{\partial x_\ell}(t, X)\,dx_\ell \right) + \frac{1}{2} \sum_{\ell,m=1}^{n} \frac{\partial^2 g_k}{\partial x_\ell \partial x_m}(t, X)(dx_\ell)(dx_m) \tag{17.17}$$

for all $k = 1, \ldots, p$; and where the terms in the last sum are computed according to:

$$dv_\ell \cdot dt = dt \cdot dv_m = dt \cdot dt = 0 \tag{17.18}$$

$$dv_\ell \cdot dv_m = \delta_K[\ell - m]\, dt \tag{17.19}$$

Remark 17.4 As a special case, a *linear* stochastic system, as considered in Chap. 12, can be expressed as the (vector) SDE

$$dX_t = AX_t\, dt + B\, dv_t \tag{17.20}$$

where the matrices $A \in \mathbb{R}^{n\times n}$ and $B \in \mathbb{R}^n$.

Note that, in this case, the driving input comprises increments of a single scalar Wiener process $v_t = v(t)$. The solution to this SDE can be obtained by applying the result in Lemma 17.3 for the transformation:

$$Y = e^{-At}X \quad \Rightarrow \quad d\big(e^{-At}X\big) = (-A)e^{-At}X\, dt + e^{-At}\, dX \tag{17.21}$$

In this case, all the second order derivatives in (17.17) vanish. Thus, reordering terms in (17.21) and multiplying by the *integrating factor* e^{-At}, it can be seen that:

$$e^{-At}\, dX_t - Ae^{-At}X_t\, dt = e^{-At}B\, dv_t \tag{17.22}$$

$$d\big(e^{-At}X_t\big) = e^{-At}B\, dv_t \tag{17.23}$$

$$e^{-At}X_t = X_0 + \int_0^t e^{-A\tau}B\, dv_\tau \tag{17.24}$$

Then it follows that the solution is obtained:

$$X_t = e^{At}X_0 + \int_0^t e^{A(t-\tau)}B\, dv_\tau \tag{17.25}$$

The reader will observe that this solution corresponds, exactly, to the state transition equation as derived in the deterministic linear case (see, for example, (3.24) on p. 25) where the deterministic input is replaced by the CTWN process, i.e., $u(\tau)\, d\tau = \dot{v}(\tau)\, d\tau = dv_\tau$.

17.3 Ito–Taylor Expansions

This section reviews stochastic Taylor series expansions. These expansions generalise the deterministic Taylor formula as well as the Ito stochastic rule. They lead to higher order approximations to functions of stochastic processes and, thus, will prove useful in the context of numerical solutions of SDEs in the next chapter.

The usual Taylor formula is first reviewed in the context of obtaining a deterministic sampled-data model in Chap. 9. However, here the method is re-expressed in integral form. Thus, consider the following nonlinear differential equation and its *implicit* solution in integral form:

$$\frac{dx}{dt} = a(x) \quad \Longleftrightarrow \quad x(t) = x(0) + \int_0^t a(x)\, d\tau \tag{17.26}$$

Now consider a general (continuously differentiable) function $f(x)$. Then, by using the *usual* chain rule, it follows that

$$\frac{df(x)}{dt} = a(x)\frac{\partial f}{\partial x} \quad \Longleftrightarrow \quad f(x) = f(x_0) + \int_0^t Lf(x)\,d\tau \tag{17.27}$$

where the notation $L = a(x)\frac{\partial}{\partial x}$ and $x_0 = x(0)$ has been used.

Note that the integral relation in the last equation is valid, in particular, for $f = a$. Thus, it can be used to substitute $a(x)$ into the integral on the right-hand side of (17.26), i.e.,

$$\begin{aligned} x(t) &= x_0 + \int_0^t \left(a(x_0) + \int_0^{\tau_1} La(x)\,d\tau_2 \right) d\tau_1 \\ &= x_0 + a(x_0)t + R_2 \end{aligned} \tag{17.28}$$

where the *residual* term is

$$R_2 = \int_0^t \int_0^{\tau_1} La(x)\,d\tau_2\,d\tau_1 \tag{17.29}$$

Using again the relation in (17.27), with $f = a$, to replace $a(x)$ in R_2 leads to:

$$x(t) = x_0 + a(x_0)t + (La)(x_0)\frac{t^2}{2} + R_3 \tag{17.30}$$

$$R_3 = \int_0^t \int_0^{\tau_1} \int_0^{\tau_2} L^2 a(x)\,d\tau_3\,d\tau_2\,d\tau_1 \tag{17.31}$$

This line of argument leads to the following general result.

Lemma 17.5 *Given a function $f \in C^{r+1}$ ($r + 1$ times continuously differentiable), this function can be expanded using the* Taylor formula in integral form *as follows:*

$$f\big(x(t)\big) = f(x_0) + \sum_{\ell=1}^{r} \big(L^\ell f\big)(x_0)\frac{t^\ell}{\ell!} + R_{r+1} \tag{17.32}$$

where the residual term is given by

$$R_{r+1} = \int_0^t \int_0^{\tau_1} \cdots \int_0^{\tau_r} L^{r+1} f(x)\,d\tau_{r+1} \ldots d\tau_2\,d\tau_1 \tag{17.33}$$

Note that (17.28) and (17.30) are particular cases of (17.32), considering $f(x) = x$ and, thus, $Lf(x) = a(x)$.

For the stochastic case, a similar line of reasoning can be followed. Thus, consider an Ito process

$$x_t = x_0 + \int_0^t a(x_\tau)\,d\tau + \int_0^t b(x_\tau)\,dv_\tau \tag{17.34}$$

and a transformation $f(x) \in \mathcal{C}^2$. Applying the Ito rule in (17.10) leads to:

$$\begin{aligned} f(x_t) &= f(x_0) + \int_0^t \left(a(x_\tau)\frac{\partial f(x_\tau)}{\partial x} + \frac{1}{2}b^2(x_\tau)\frac{\partial^2 f(x_\tau)}{\partial x^2} \right) d\tau \\ &\quad + \int_0^t b(x_\tau)\frac{\partial f(x_\tau)}{\partial x}\,dv_\tau \\ &= f(x_0) + \int_0^t L^0 f(x_\tau)\,d\tau + \int_0^t L^1 f(x_\tau)\,dv_\tau \end{aligned} \tag{17.35}$$

where the following operators have been used:

$$L^0 = a\frac{\partial}{\partial x} + \frac{1}{2}b^2\frac{\partial^2}{\partial x^2}, \qquad L^1 = b\frac{\partial}{\partial x} \tag{17.36}$$

Applying the Ito formula (17.35) (analogously to the deterministic case) to $f = a$ and $f = b$ in (17.34) leads to:

$$x_t = x_0 + a(x_0)\int_0^t d\tau + b(x_0)\int_0^t dv_\tau + R_2^s \tag{17.37}$$

$$\begin{aligned} R_2^s = &\int_0^t\int_0^{\tau_1} L^0 a(x_{\tau_2})\,d\tau_2\,d\tau_1 + \int_0^t\int_0^{\tau_1} L^1 a(x_{\tau_2})\,dv_{\tau_2}\,d\tau_1 \\ &+ \int_0^t\int_0^{\tau_1} L^0 b(x_{\tau_2})\,d\tau_2\,dv_{\tau_1} + \int_0^t\int_0^{\tau_1} L^1 b(x_{\tau_2})\,dv_{\tau_2}\,dv_{\tau_1} \end{aligned} \tag{17.38}$$

which is the stochastic analogue for the Taylor formula. Indeed, if the diffusion term is $b(x_t) \equiv 0$, then Eqs. (17.37)–(17.38) reduce to the deterministic expressions in (17.28)–(17.29).

It is possible to go further to obtain Ito–Taylor expansions where (17.35) is again used to replace $f = a$ and $f = b$ in (17.38). However, the expressions become increasingly involved, including multiple stochastic integrals. Thus a systematic notation is next introduced to manipulate the required multiple integrals and, hence, to obtain higher order Ito–Taylor expansions for general SDEs, by using *multi-indices* and *hierarchical sets*. This notation was developed by Kloeden and Platen (see the references at the end of the chapter). The following expression is obtained for the stochastic analogue of the expansion (17.30):

$$x_t = x_0 + aI_{(0)} + bI_{(1)} + \left(aa' + \frac{1}{2}b^2a''\right)I_{(0,0)}$$
$$+ \left(ab' + \frac{1}{2}b^2b''\right)I_{(0,1)} + ba'I_{(1,0)} + bb'I_{(1,1)} + R_3^s \qquad (17.39)$$

where:

$$a = a(x_0), \qquad a' = \frac{\partial a}{\partial x}(x_0), \qquad a'' = \frac{\partial^2 a}{\partial x^2}(x_0) \qquad (17.40)$$

$$b = b(x_0), \qquad b' = \frac{\partial b}{\partial x}(x_0) \qquad (17.41)$$

and:

$$I_{(0)} = \int_0^t d\tau_1, \qquad I_{(1)} = \int_0^t dv_{\tau_1} \qquad (17.42)$$

$$I_{(0,0)} = \int_0^t \int_0^{\tau_1} d\tau_2 \, d\tau_1, \qquad I_{(0,1)} = \int_0^t \int_0^{\tau_1} d\tau_2 \, dv_{\tau_1} \qquad (17.43)$$

$$I_{(1,0)} = \int_0^t \int_0^{\tau_1} dv_{\tau_2} \, d\tau_1, \qquad I_{(1,1)} = \int_0^t \int_0^{\tau_1} dv_{\tau_2} \, dv_{\tau_1} \qquad (17.44)$$

The following result is first presented. This result establishes the convergence of truncated Ito–Taylor expansions.

Lemma 17.6 *Consider an Ito process x_t as in* (17.34) *and its corresponding k-th order truncated Ito–Taylor expansion $x_k(t)$, around $t = t_0$. It follows that*

$$E\left\{|x_t - x_k(t)|^2\right\} \le C_k(t - t_0)^{k+1} \qquad (17.45)$$

where C_k is a constant that depends only on the truncation order k.

Proof The details of the proof can be found in the references at the end of the chapter. □

The previous lemma establishes that an Ito–Taylor expansion converges to the original Ito process in the mean square sense, as k goes to infinity. Under additional assumptions, the previous result can be strengthened to convergence with probability 1, uniformly on the interval $[t_o, t]$.

17.4 Numerical Solution of SDEs

The previous section presented Ito–Taylor expansions that allow higher order approximations of an Ito process defined by an SDE. Analogously to the deterministic case, Ito–Taylor expansions can also be used to derive discrete-time approximations to solve SDEs. In turn, these solutions can be thought of as approximate sampled-data models. The simplest of the approximations can be obtained by truncating the expansion in (17.37):

$$x_t = x_0 + a(x_0)t + b(x_0)\int_0^t dv_\tau \tag{17.46}$$

This is the stochastic equivalent of the Euler approximation for ordinary differential equations, and is called the *Euler–Maruyama approximation.*

Note that from this approximation a simple sampled-data model can be obtained as

$$\bar{x}\big((k+1)\Delta\big) = \bar{x}(k\Delta) + a\big(\bar{x}(k\Delta)\big)\Delta + b\big(\bar{x}(k\Delta)\big)\Delta v_k \tag{17.47}$$

where $\Delta v_k = v_{(k+1)\Delta} - v_{k\Delta}$ are increments of the Wiener process v_τ:

$$E\{\Delta v_k\} = 0, \qquad E\big\{(\Delta v_k)^2\big\} = \Delta \tag{17.48}$$

Naturally, other algorithms can be derived by considering more terms in the Ito–Taylor expansions.

17.5 Accuracy of Numerical Solutions

To analyse the quality of different algorithms, different criteria can be used, namely *strong* and *weak* convergence. The concept of strong convergence is defined below.

Definition 17.7 An approximation, $\bar{x}(k\Delta)$, obtained using a sampling period Δ, *converges strongly* to the process x_t at time $t = k\Delta$ if

$$\lim_{\Delta\to 0} E\big\{\big|\bar{x}(k\Delta) - x_{k\Delta}\big|\big\} = 0 \tag{17.49}$$

Furthermore, the approximation converges strongly *with order* $\gamma > 0$, if there exist a positive constant C and a sampling period $\Delta_0 > 0$ such that

$$E\{|\bar{x}(k\Delta) - x_{k\Delta}|\} \leq C\Delta^{\gamma} \tag{17.50}$$

for all $\Delta < \Delta_0$.

The error can also be bounded using the Lyapunov inequality

$$E\{|\bar{x}(k\Delta) - x_{k\Delta}|\} \leq \sqrt{E\{|\bar{x}(k\Delta) - x_{k\Delta}|^2\}} \tag{17.51}$$

This allows variances to be used in the evaluation of the order of strong convergence. For example, if two polynomials differ in one term, then the error between the polynomials is the square root of the variance of that term.

Strong convergence is also called *path-wise* convergence: the approximate solution is required to replicate the continuous-time system output (with a certain accuracy) when the same realisation of noise process is used as input.

Example 17.8 Consider again the Euler–Maruyama scheme introduced previously in (17.46). This scheme corresponds to the truncated Ito–Taylor expansion containing only the time and Wiener integrals of multiplicity 1. Thus, it can be interpreted as an order 1 strong Ito–Taylor approximation over a finite number of steps.

A weaker definition of convergence can be obtained by not considering each *path* of the process involved, but instead focusing on the associated statistical properties. This leads to the following notion of weak convergence.

Definition 17.9 A sampled-data approximation $\bar{x}(k\Delta)$, obtained using a sampling period Δ, is said to *converge weakly* to the continuous-time process x_t at time $t = k\Delta$, for function $g(\cdot)$ drawn from a class $\mathcal{T}$ of test functions, if

$$\lim_{\Delta \to 0} |E\{g(x_t)\} - E\{g(\bar{x}(k\Delta))\}| = 0 \tag{17.52}$$

Note that, if $\mathcal{T}$ contains all polynomials, this definition implies the convergence of all moments. Furthermore, the approximation converges weakly *with order* $\beta > 0$, if there exist a positive constant C and a sampling period $\Delta_0 > 0$ such that

$$|E\{g(x_k)\} - E\{g(\bar{x}(k\Delta))\}| \leq C\Delta^{\beta} \tag{17.53}$$

for all $\Delta < \Delta_0$.

Table 17.1 Summary of the different types of convergence based on time scale and accuracy measure

		Time scale	
		Fixed steps (local)	Fixed time (global)
Measure	Path-wise	Local-Strong	Global-Strong
	Moment-based	Local-Weak	Global-Weak

Also recall the definition of *local* and *global* convergence given in Chap. 9 where 'local' refers to a finite number of steps and 'global' refers to a solution over a fixed time interval.

Table 17.1 presents a summary of the preceding discussion showing the four types of convergence based on time scale and accuracy measure

Also, when special model structures are used (later the normal form will be deployed), the above convergence results can be stated component-wise for each element of the state vector. Thus, it can be said that a certain solution for an nth order SDE has local strong convergence of order $\Delta^{m_1}, \ldots, \Delta^{m_n}$.

17.6 Summary

The key ideas covered in this chapter are:

- Stochastic nonlinear differential equations—see (17.3), (17.4).
- Ito integrals—see (17.5).
- The Ito rule.
- Ito–Taylor expansions.
- Numerical solution of stochastic differential equations.
- Measures of accuracy.
- Strong and weak convergence.
- Local and global errors.

Further Reading

A comprehensive introduction to stochastic differential equations can be found in

Øksendal B (2003) Stochastic differential equations. An introduction with applications, 6th edn. Springer, Berlin

A comprehensive treatment of the solution of nonlinear stochastic differential equations, including fixed-time local and fixed-time global convergence errors, is available in

Kloeden PE, Platen E (1992) Numerical solution of stochastic differential equations. Springer, Berlin

One-step convergence errors are discussed in

Milstein G (1995) Numerical integration of stochastic differential equations, vol 313. Springer, Berlin

Chapter 18
Approximate Sampled-Data Models for Nonlinear Stochastic Systems

In this chapter, the ideas of Chap. 16 will be extended to the nonlinear case. One possible path would be to follow the methodology introduced in Sect. 16.1 of appending the linear stochastic sampling zeros to an Euler–Maruyama model. However, it is not immediately evident how this can be achieved in the nonlinear case. Thus, as an alternative, the time-domain approximate sampled models described in Sects. 16.2 and 16.3 will be extended to the nonlinear case.

18.1 Approximate Sampled-Data Models Based on Up-Sampling

Consider a model restricted to the form:

$$dx = f(x,u)\,dt + d\omega \tag{18.1}$$

$$dy = h(x)\,dt + dv \tag{18.2}$$

where $d\omega$, dv are independent Wiener processes with incremental covariance $Q\,dt$ and $R\,dt$, respectively.

Assume that an averaging anti-aliasing filter (AAF) is used. Also assume that an up-sampling rate of M is used in the Euler–Maruyama scheme at period $\frac{\Delta}{M} = \Delta'$.

Let $k = KM$, then the following approximate discrete-time model is obtained:

$$dx^+_{KM+\ell} = f(x_{KM+\ell}, u_{KM})\Delta' + w'_{KM+\ell}; \quad \ell = 0, \dots, M-1 \tag{18.3}$$

where, as usual, $dx^+_{KM+\ell}$ denotes $(x_{KM+\ell+1} - x_{KM+\ell})$ and where $\{w'_{KM+\ell}\}$ is a discrete-time white noise sequence having covariance $Q\Delta'$.

As in Sect. 16.2, it is also important to describe the action of the AAF in order to complete the model. Thus a state-space model for the AAF is included which is also updated at the up-sampling period Δ'. Since an averaging AAF is used, an appropriate model is

$$\bar{z}_k = \frac{1}{\Delta}\int_{(k-1)\Delta}^{k\Delta} dy = \frac{y(k\Delta) - y((k-1)\Delta)}{\Delta} \tag{18.4}$$

J.I. Yuz, G.C. Goodwin, *Sampled-Data Models for Linear and Nonlinear Systems*,
Communications and Control Engineering, DOI 10.1007/978-1-4471-5562-1_18,

Hence an additional state vector $\{\bar{x}_k\}$ is introduced which satisfies the up-sampled Euler–Maruyama form of (18.2), i.e.,

$$d\bar{x}^{+}_{KM+\ell} = h(x_{KM+\ell})\Delta' + v'_{KM+\ell} \tag{18.5}$$

Since the output is measured only every M samples, the output can be expressed as

$$\begin{aligned} y_{(K+1)M+1} &= y_{KM+1} + [\bar{x}_{(K+1)M} - \bar{x}_{KM+1}] \\ &\quad + h(\bar{x}_{(K+1)M})\Delta' + v'_{(K+1)M} \end{aligned} \tag{18.6}$$

where v'_k has covariance $R\Delta'$.

The composite model (18.3), (18.5), (18.6) gives an approximate sampled-data model with up-sampling ratio M.

The above idea is relatively straightforward and has, for example, been used in the context of nonlinear filtering—see the references at the end of the chapter. It is also a useful strategy to obtain a simulation of a system with period Δ. However, the method does not reveal key facts about sampling zeros in the nonlinear case. Thus, alternative schemes, which mirror the ideas in Sects. 16.3 through 16.4, are next considered. These methods expose key system properties, including the asymptotic stochastic sampling zeros.

18.2 Approximate Sampled-Data Models Based on Successive Integration

There exists a close connection between approximate solutions of stochastic differential equations and approximate models for stochastic differential equations. Indeed the latter is a recursive form of the former where the approximate solution from one step is used as the initial condition for the next step. Thus, the following sections build heavily on the ideas presented in Sect. 17.4 of Chap. 17.

Attention is restricted to the class of nonlinear stochastic systems whose model mimics the *normal form* for deterministic nonlinear systems. Also, for simplicity, it is assumed that the continuous-time model is free of intrinsic zeros.

Assumption 18.1 It is assumed that there exists a transformation $Z = \Phi(X)$ such that, in the new coordinates, the system can be expressed as:

$$dZ(t) = AZ\,dt + B\big(a(t, Z)\,dt + b(t, Z)\,dv\big) \tag{18.7}$$

$$y(t) = z_1 \tag{18.8}$$

where the matrices A and B are as in (14.31) and where dv is a Wiener process.

Renaming the state variables, the previous model can be expressed using the following set of stochastic differential equations (SDEs):

$$dx_1 = x_2\,dt \tag{18.9}$$

$$\vdots$$

$$dx_{n-1} = x_n\,dt \tag{18.10}$$

$$dx_n = a(x)\,dt + b(x)\,dv_t \tag{18.11}$$

where the output is $y = x_1$. The functions $a(\cdot)$ and $b(\cdot)$ are assumed to be analytic or $\mathcal{C}^\infty$.

The last state equation (18.11) can be expanded using a first Ito-Taylor expansion as

$$\hat{x}_n(\Delta) = \hat{x}_n(0) + a(\widehat{X}_0)\Delta + b(\widehat{X}_0)\tilde{v}_n \tag{18.12}$$

where $X_0 = [x_1(0), x_2(0), \ldots, x_n(0)]^T$ and $\tilde{v}_n$ is defined in (14.38).

Proceeding to the second last state SDE in (18.10) leads to:

$$\begin{aligned}\hat{x}_{n-1}(\Delta) &= \hat{x}_{n-1}(0) + \int_0^\tau \hat{x}_n(\tau)\,d\tau \\ &= \hat{x}_{n-1}(0) + \hat{x}_n(0)\Delta + a(\widehat{X}_0)\frac{\Delta^2}{2} + b(\widehat{X}_0)\tilde{v}_{n-1} \end{aligned}\tag{18.13}$$

where the fact that $I_{(1,0)} = \tilde{v}_{n-1}$ has been used. Proceeding in a similar way with the rest of the state components leads to truncated Ito-Taylor expansions up to the first state in (18.9), i.e.,

$$\hat{x}_1(\Delta) = \hat{x}_1(0) + \hat{x}_2(0)\Delta + \cdots + \hat{x}_n(0)\frac{\Delta^{n-1}}{(n-1)!} + a(\widehat{X}_0)\frac{\Delta^n}{n!} + b(\widehat{X}_0)\tilde{v}_1 \tag{18.14}$$

Note that the last equation corresponds, in fact, to the Ito-Taylor expansion for the system output $y = x_1$ of order n.

The truncated Ito-Taylor expansions obtained above for each of the state components can also be rewritten in terms of the δ-operator:

$$\frac{\widehat{X}^+ - \widehat{X}}{\Delta} = \delta\widehat{X} = A_\delta\widehat{X} + B_\delta a(\widehat{X}) + b(\widehat{X})V \tag{18.15}$$

where:

$$\widehat{X}^+ = \begin{bmatrix} \hat{x}_1(\Delta) \\ \hat{x}_2(\Delta) \\ \vdots \\ \hat{x}_n(\Delta) \end{bmatrix}, \quad A_\delta = \begin{bmatrix} 0 & 1 & \cdots & \frac{\Delta^{n-2}}{(n-1)!} \\ 0 & 0 & \ddots & \vdots \\ \vdots & & \ddots & 1 \\ 0 & \cdots & 0 & 0 \end{bmatrix}$$
$$B_\delta = \begin{bmatrix} \frac{\Delta^{n-1}}{n!} \\ \frac{\Delta^{n-2}}{(n-1)!} \\ \vdots \\ 1 \end{bmatrix}, \quad V = \frac{1}{\Delta}\tilde{V} \tag{18.16}$$

The approximate model (18.15) is called the stochastic truncated Taylor series (STTS) model.

The following result characterises the accuracy of the STTS model.

Theorem 18.2 *Consider the continuous-time stochastic system* (18.9)–(18.11), *with output* $y = x_1$, *and the corresponding approximate sampled-data model* (18.15), (18.16):

$$\delta\widehat{X}(k\Delta) = A_\delta\widehat{X}(k\Delta) + B_\delta a\big(\widehat{X}(k\Delta)\big) + b\big(\widehat{X}(k\Delta)\big)V_k \tag{18.17}$$

where A_δ *and* B_δ *are defined as in* (18.16), *and* $V_k = \frac{1}{\Delta}\tilde{V}_k$ *corresponds to* (14.37)–(14.38) *where the integration interval is changed to* $[k\Delta, k\Delta + \Delta)$. *(By stationarity of* v_t, *the vectors* V_k *and* V *have the same covariance.)*

The states of this model have local-strong error of order $\{\Delta^n, \Delta^{n-1}, \ldots, \Delta\}$.

Proof First, note that, in integral form, the exact solution for the last state and the associated Ito-Taylor approximation (Euler–Maruyama) are, respectively,

$$x_n(\Delta) = x_n(0) + \int_0^\Delta a(X_\tau)\,d\tau + \int_0^\Delta b(X_\tau)\,dv_\tau$$
$$\hat{x}_n(\Delta) = \hat{x}_n(0) + a(X_0)\int_0^\Delta d\tau + b(X_0)\int_0^\Delta dv_\tau$$

Therefore, considering zero initial errors, define

$$\begin{aligned} e_n(\Delta) &= x_n(\Delta) - \hat{x}_n(\Delta) \\ &= \int_0^{\Delta} \big[a(X_\tau) - a(X_0)\big]\,d\tau + \int_0^{\Delta} \big[b(X_\tau) - b(X_0)\big]\,dv_\tau \end{aligned}$$

It is known (see the references at the end of the chapter) that the Euler–Maruyama approximation has the following local-strong truncation error:

$$E\big\{|e_n(\Delta)|\big\} \in \mathcal{O}(\Delta)$$

It is then straightforward to show that the truncation errors for the remaining state equations can be characterised by

$$e_i = \begin{cases} \int_0^{\Delta}\int_0^{\tau_i} \cdots \int_0^{\tau_{n-2}} e_n(\tau_{n-1})\,d\tau_{n-1} \ldots d\tau_i, & i = 1, \ldots, n-2, \\ \int_0^{\Delta} e_n(\tau_{n-1})d\tau_{n-1}, & i = n-1 \end{cases} \tag{18.18}$$

By using the Lyapunov inequality, it follows that

$$E\big\{\big|e_i(\Delta)\big|\big\} \le E\big\{\big|e_i(\Delta)\big|^2\big\}^{1/2}$$

And therefore, by repeatedly applying the Hölder inequality,

$$\begin{aligned} &E\big\{\big|e_i(\Delta)\big|^2\big\} \\ &\quad = E\left\{\left|\int_0^{\Delta}\int_0^{\tau_i} \cdots \int_0^{\tau_{n-2}} e_n(\tau_{n-1})\,d\tau_{n-1} \ldots d\tau_i\right|^2\right\} \\ &\quad \le \Delta \cdot \int_0^{\Delta} E\left\{\left|\int_0^{\tau_i} \cdots \int_0^{\tau_{n-2}} e_n(\tau_{n-1})\,d\tau_{n-1} \ldots d\tau_{i-1}\right|^2\right\} d\tau_i \\ &\quad \le \Delta \cdot \int_0^{\Delta} \tau_i \cdot \int_0^{\tau_i} E\left\{\left|\int_0^{\tau_{i-1}} \cdots \int_0^{\tau_{n-2}} e_n(\tau_{n-1})\,d\tau_{n-1} \ldots d\tau_{i-2}\right|^2\right\} d\tau_{i-1}\,d\tau_i \\ &\quad \le \Delta \cdot \int_0^{\Delta} \tau_i \cdot \int_0^{\tau_i} \ldots \tau_{n-2} \cdot \int_0^{\tau_{n-2}} E\big\{\big|e_n(\tau_{n-1})\big|^2\big\}\,d\tau_{n-1} \ldots d\tau_{i-1}\,d\tau_i \end{aligned}$$

Finally, it follows that

$$\begin{aligned} E\big\{\big|e_i(\Delta)\big|^2\big\} &\le \Delta \cdot \int_0^{\Delta} \tau_i \cdot \int_0^{\tau_i} \ldots \tau_{n-2} \cdot \int_0^{\tau_{n-2}} C \cdot \tau_{n-1}^2\,d\tau_{n-1} \ldots d\tau_{i-1}\,d\tau_i \\ &\le \Delta \cdot \int_0^{\Delta} C \cdot \tau_i^{2+2\cdot(n-2-i+1)}\,d\tau_i \\ &\le C \cdot \Delta^{2+2\cdot(n-2-i+1)+2} \\ &\le C \cdot \Delta^{2\cdot(n-i+1)} \end{aligned}$$

The result is then obtained by taking the square root. □

Remark 18.3 From (14.40) it can be seen that the covariance structure of V_k is given by:

$$
\begin{aligned}
E\{V_k V_\ell^T\} &= \frac{1}{\Delta^2} E\{\tilde{V}_k \tilde{V}_\ell^T\} \\
&= \left[\frac{\Delta^{2n-i-j}}{(n-i)!(n-j)!(2n-i-j+1)}\right]_{i,j=1,\dots,n} \frac{\delta_K[k-\ell]}{\Delta}
\end{aligned}
\tag{18.19}
$$

The above covariance approaches the continuous-time covariance of $B\dot{v}(t)$ (see Eq. (18.7)) as the sampling period goes to zero. This result parallels the linear case discussed in Chap. 16.

Remark 18.4 It can be seen that the approximate sampled-data model presented in (18.17) closely resembles the discrete-time model obtained for the n-th order stochastic integrator in Chap. 16. In particular, the vector $\tilde{V}_k$ plays a key role as the driving input in both cases. If $a(X) \equiv 0$ and $b(X) \equiv 1$ are substituted into (18.17), then the model reduces to the *exact* sampled-data model for an n-th order continuous-time integrator as obtained in (14.36). Thus, for this very special case, the order of local-strong convergence of the sampled-data model is actually $\gamma = \infty$.

18.3 Sampling Zero Dynamics for Stochastic Nonlinear Systems

Given the similarity between the model obtained in Sect. 18.2 and the result for the nth order integrator, it would be tempting to assume that the sampled zero dynamics are immediately revealed. However, this is not the case, because it is not immediately obvious how to reduce the vector process V to a scalar input. To do this, additional levels of approximation must be used. Indeed, these steps have already been described in Chap. 16 for the linear case.

Recall the STTS model in (18.15), (18.16), which can be written explicitly as:

$$
\delta\hat{x}_1 = \hat{x}_2 + \frac{\Delta}{2}\hat{x}_3 + \cdots \frac{\Delta^{n-2}}{(n-1)!}\hat{x}_n + \frac{\Delta^{n-1}}{n!}a(\hat{x}) + b(\hat{x})v_1 \tag{18.20}
$$

$$
\delta\hat{x}_2 = \hat{x}_3 + \frac{\Delta}{2}\hat{x}_4 + \cdots \frac{\Delta^{n-2}}{(n-1)!}a(\hat{x}) + b(\hat{x})v_2 \tag{18.21}
$$

$$
\vdots
$$

$$
\delta\hat{x}_n = a(\hat{x}) + b(\hat{x})v_n \tag{18.22}
$$

where

$$E\{v_i v_j\} = \frac{1}{\Delta^2}\left[\frac{\Delta^{2n-i-j+1}}{(n-i)!(n-j)!(2n-i-j+1)}\right].$$

It is more convenient to express this model in the following equivalent form:

$$\delta\hat{x}_1 = \hat{x}_2 + \frac{\Delta}{2}\hat{x}_3 + \cdots \frac{\Delta^{n-2}}{(n-1)!}\hat{x}_n + \frac{\Delta^{n-1}}{n!}\bar{a}(\hat{x}_1,\dots,\hat{x}_n) + \Delta^{n-1}\bar{b}(\hat{x}_1,\dots,\hat{x}_n)\bar{v}_1 \tag{18.23}$$

$$\delta\hat{x}_2 = \hat{x}_3 + \frac{\Delta}{2}\hat{x}_4 + \cdots \frac{\Delta^{n-3}}{(n-2)!}\hat{x}_n + \frac{\Delta^{n-2}}{(n-1)!}\bar{a}(\hat{x}_1,\dots,\hat{x}_n) + \Delta^{n-2}\bar{b}(\hat{x}_1,\dots,\hat{x}_n)\bar{v}_2 \tag{18.24}$$

$$\vdots$$

$$\delta\hat{x}_n = \bar{a}(\hat{x}_1,\dots,\hat{x}_n) + \bar{b}(\hat{x}_1,\dots,\hat{x}_n)\bar{v}_n \tag{18.25}$$

where

$$\bar{a}(\hat{x}_1,\dots,\hat{x}_n) = a(\hat{x})$$
$$\bar{b}(\hat{x}_1,\dots,\hat{x}_n) = b(\hat{x})$$

and where

$$E\left[\begin{pmatrix}\bar{v}_1\\ \vdots\\ \bar{v}_2\end{pmatrix}\begin{pmatrix}\bar{v}_1 & \dots & \bar{v}_2\end{pmatrix}\right] = \frac{1}{\Delta}M \tag{18.26}$$

where

$$M_{ij} = \frac{1}{(n-i)!(n-j)!(2n-i-j+1)}.$$

Recall that sampling zeros are an input–output property. Hence, attention is focused on the output variable $y = x$.

Theorem 18.5 *Assume that $\bar{a}$ and $\bar{b}$ are Lipschitz continuous functions and that $x_1,\dots,x_n$ all have bounded variance. Then a steady-state model for the output having the same local-strong truncation error as that of* (18.17) *is*

$$\delta^n\bar{y} = \bar{a}\big(\bar{y},\delta\bar{y},\dots,\delta^{n-1}\bar{y}\big) + \bar{b}\big(\bar{y},\delta\bar{y},\dots,\delta^{n-1}\bar{y}\big)N_n(\delta)e \tag{18.27}$$

where $\bar{y}$ is an approximation to y, e is a scalar white noise process, and $N_n(\delta)$ is the linear zero dynamics associated with a stochastic system of relative degree n.

Proof The proof mirrors the steps for the linear case described in Sect. 16.4. The proof contains five key steps as follows:

(i) Eliminate small terms in the state-space model,
(ii) Focus on the output,
(iii) Introduce extra approximations without changing the order of the error,
(iv) Use the Kalman filter to obtain a single driving noise, and
(v) Consider the steady state.

These steps are performed below. (For clarity of exposition, the case $n = 3$ is used to develop the ideas.)

(i) Eliminate some small terms.
Note that the terms $\frac{\Delta^{n-1}}{n!}a(\hat{x}), \dots, \Delta a(\hat{x})$ in (18.23)–(18.26) have smaller variance than all other terms and hence can be eliminated without effecting the accuracy of the state-space model.
(Recall that mean square error measurements can be translated to variance measurements when the difference between two polynomials is only one term, as is the case here.)
Eliminating those terms leads to:

$$\delta\bar{x}_1 = \bar{x}_2 + \frac{\Delta}{2}\bar{x}_3 + \Delta^2(\bar{b}_1\bar{v}_1); \quad \mathcal{O}^{LS}(\Delta^2) \tag{18.28}$$

$$\delta\bar{x}_2 = \bar{x}_3 + \Delta(\bar{b}_2\bar{v}_2); \quad \mathcal{O}^{LS}(\Delta) \tag{18.29}$$

$$\delta\bar{x}_3 = \bar{a} + \bar{b}\bar{v}_3; \quad \mathcal{O}^{LS}(1) \tag{18.30}$$

(ii) Focus on the output.
Operating on Eq. (18.28) by δ leads to

$$\delta^2\bar{x}_1 = \delta\bar{x}_2 + \frac{\Delta}{2}\delta\bar{x}_3 + \Delta^2\delta(\bar{b}\bar{v}_1); \quad \mathcal{O}^{LS}(\Delta) \tag{18.31}$$

Operating by δ again yields

$$\delta^3\bar{x}_1 = \delta^2\bar{x}_2 + \frac{\Delta}{2}\delta^2\bar{x}_3 + \Delta^2\delta^2(\bar{b}\bar{v}_1); \quad \mathcal{O}^{LS}(\Delta^0) \tag{18.32}$$

Also, from (18.29),

$$\delta^2\bar{x}_2 = \delta\bar{x}_3 + \Delta\delta(\bar{b}\bar{v}_2); \quad \mathcal{O}^{LS}(\Delta^0) \tag{18.33}$$

and finally from (18.30),

$$\delta^2\bar{x}_3 = \delta\bar{a} + \delta(\bar{b}\bar{v}_3); \quad \mathcal{O}^{LS}(\Delta^{-1}) \tag{18.34}$$

Substituting (18.33) and (18.34) into (18.32), and using (18.30) leads to

$$\delta^3\bar{x}_1 = \big(\delta\bar{x}_3 + \Delta\delta(\bar{b}\bar{v}_2)\big) + \frac{\Delta}{2}\big(\delta\bar{a} + \delta(\bar{b}\bar{v}_3)\big) + \Delta^2\delta^2(\bar{b}\bar{v}_1) \tag{18.35}$$

$$= (\bar{a} + \bar{b}\bar{v}_3) + \Delta\delta(\bar{b}\bar{v}_2) + \frac{\Delta}{2}\delta\bar{a} + \frac{\Delta}{2}\delta(\bar{b}\bar{v}_3) + \Delta^2\delta^2(\bar{b}\bar{v}_1) \tag{18.36}$$

$$= \bar{a} + \frac{\Delta}{2}\delta\bar{a} + \Delta^2\delta^2(\bar{b}\bar{v}_1) + \Delta\delta(\bar{b}\bar{v}_2) + \frac{\Delta}{2}\delta(\bar{b}\bar{v}_3) + \bar{b}\bar{v}_3 \tag{18.37}$$

which has an error of the order of Δ^0.

(iii) Introduce extra approximations.
First consider the terms $\Delta(\delta\bar{a})$.
Note that

$$\Delta\delta\bar{a} = \bar{a}^+ - \bar{a} \tag{18.38}$$

$$\simeq \frac{\partial\bar{a}}{\partial\bar{x}}\big[\bar{x}^+ - \bar{x}\big] \tag{18.39}$$

However, $(\bar{x}^+ - \bar{x})$ is at least of the order of Δ.
Hence, these terms can be eliminated without changing the order of the error, which was already the order of Δ.
Next, consider the term $\Delta\delta(\bar{b}\bar{v}_*)$, where '*' denotes 1, 2, or 3. It follows that:

$$\Delta\delta(\bar{b}\bar{v}_*) = \bar{b}^+\bar{v}_*^+ - \bar{b}\bar{v} \tag{18.40}$$

$$= \bar{b}^+\big(\bar{v}_*^+ - \bar{v}_*\big) + \bar{v}_*\big(\bar{b}^+ - \bar{b}\big) \tag{18.41}$$

$$= \bar{b}^+(\Delta\delta\bar{v}_*) + \bar{v}_*\big(\bar{b}^+ - \bar{b}\big) \tag{18.42}$$

Note that $(\bar{b}^+ - \bar{b})$ can be approximated as follows:

$$\bar{b}^+ - \bar{b} \simeq \frac{\partial\bar{b}}{\partial\bar{x}}\big(\bar{x}^+ - \bar{x}\big) \tag{18.43}$$

which again has an extra Δ in every term. Thus, the last term in (18.42) can be eliminated and the $\bar{b}^+$ in the first term can be replaced by $\bar{b}$ without changing the order of the error.
Hence, $\Delta\delta(\bar{b}\bar{v}_*)$ can be replaced by $\bar{b}(\Delta\delta\bar{v}_*)$ without changing the order of the error. The same argument applies to $\Delta^2\delta^2(\bar{b}\bar{v})$.
Finally, it is noted that:

$$\bar{a}(\bar{x}_1, \bar{x}_2, \bar{x}_3) = \bar{a}\big(\bar{x}_1, \delta\bar{x}_1, \delta^2\bar{x}_1\big) + \frac{\partial\bar{a}}{\partial x_2}(\bar{x}_2 - \delta\bar{x}_1) + \frac{\partial\bar{a}}{\partial x_3}(\bar{x}_3 - \delta\bar{x}_2); \quad \mathcal{O}^{LS}(\Delta) \tag{18.44}$$

However, the terms $(\bar{x}_2 - \delta\bar{x}_1)$ and $(\bar{x}_3 - \delta\bar{x}_2)$ are at least of order Δ^1 and hence can be eliminated without changing the order of error in (18.36).
A similar argument applies to $\bar{b}(\bar{x}_1, \bar{x}_2, \bar{x}_3)$.

Thus, finally $\delta^3\bar{x}_1$ can be expressed as:

$$\delta^3\bar{x}_1 = \bar{a}(\bar{x}_1, \delta\bar{x}_1, \delta^2\bar{x}_1) + \bar{b}(\bar{x}_1, \delta\bar{x}_1, \delta^2\bar{x}_1)\left\{\Delta^2\delta^2\bar{v}_1 + \Delta\delta\bar{v}_2 + \frac{1}{2}\Delta\delta\bar{v}_3 + \bar{v}_3\right\};$$
$$\mathcal{O}^{LS}(\Delta^0) \tag{18.45}$$

Note that this has the same accuracy as the model (18.23)–(18.26).
Finally, note that $\bar{x}_1 = y$; $\mathcal{O}^{LS}(\Delta^3)$, $\delta\bar{x}_1 = \delta y$; $\mathcal{O}^{LS}(\Delta^2)$, $\delta^2 x_1 = \delta^2 y$; $\mathcal{O}^{LS}(\Delta^0)$ and $\delta^3\bar{x}_1 = \delta^3 y$; $\mathcal{O}^{LS}(\Delta^0)$. Hence, the following approximate model for y is obtained:

$$\delta^3\bar{y} = \bar{a}(\bar{y}, \delta\bar{y}, \delta^2\bar{y}) + \bar{b}(\bar{y}, \delta\bar{y}, \delta^2\bar{y})\left\{\Delta^2\delta^2\bar{v}_1 + \Delta\delta\bar{v}_2 + \frac{1}{2}\Delta\delta\bar{v}_3 + \bar{v}_3\right\};$$
$$\mathcal{O}^{LS}(\Delta^0) \tag{18.46}$$

where $\bar{y}$ is the approximate solution for the true output y.

(iv) Use the Kalman filter.
Consider the final term in curly brackets on the right-hand side of (18.46).
Define

$$\bar{\bar{v}} = \Delta^2\delta^2 v_1 + \Delta\delta\bar{v}_2 + \frac{1}{2}\Delta\delta\bar{v}_3 + \bar{v}_3 \tag{18.47}$$

Using the Kalman filter (as in Chap. 16), the coloured noise process $\bar{\bar{v}}$ can be described via a time-varying model driven by a single white noise source (the innovations process). Thus,

$$\bar{\bar{v}}_t = N_{2t}(t)\omega_t \tag{18.48}$$

where N_{2t} is the time-varying innovations model and $\{\omega_t\}$ is a white noise process of time-varying variance.
Substituting into (18.46) leads to

$$\delta^3\bar{y} = \bar{a}(\bar{y}, \delta\bar{y}, \delta^2\bar{y}) + \bar{b}(\bar{y}, \delta\bar{y}, \delta^2\bar{y})\{N_{3t}(\delta)\omega_t\} \tag{18.49}$$

Also note that the process $\{\omega_t\}$ can be obtained from $\{\bar{\bar{v}}_t\}$ by inverting (18.48). Thus, $\{\omega_t\}$ is defined path-wise.

(v) Consider the steady state.
The Kalman filter and the associated innovations variance both converge exponentially fast. Moreover, the innovations model converges to the minimum phase spectral function of the noise process defined in (18.48). Hence, in steady state, the model (18.46) can be re-expressed as

$$\delta^3\bar{y} = \bar{a}(\bar{y}, \delta\bar{y}, \delta^2\bar{y}) + \bar{b}(\bar{y}, \delta\bar{y}, \delta^2\bar{y})\{N_3(\delta)\omega\} \tag{18.50}$$

where $N_3(\delta)$ is the stochastic sampling zero dynamics for a system of relative degree 3 as defined in Chap. 14. □

Remark 18.6 A key conclusion from Theorem 18.5 is that the same (asymptotic) stochastic zero dynamics apply as in the linear stochastic case.

18.4 Summary

The key ideas discussed in the current chapter are:

- Use of up-sampling to obtain an approximate model.
- Development of the stochastic truncated Taylor series (STTS) approximate sampled-data models for continuous-time systems of relative degree r (expressed in the normal form) by successively higher order integration beginning with Euler–Maruyama for the rth state.
- Development of an input-output sampled-data model for stochastic nonlinear systems which explicitly contains the approximate stochastic sampling zero dynamics for the given relative degree.
- The remarkable observation (as in the deterministic nonlinear case) that the extra asymptotic stochastic sampling zero dynamics are identical to those for linear stochastic systems having the same relative degree.

Further Reading

The use of up-sampling in the context of nonlinear filtering is discussed in

Cea M, Goodwin GC, Müller C (2011) A novel technique based on up-sampling for addressing modeling issues in sampled data nonlinear filtering. In: 18th IFAC world congress, Milan, Italy

Conditions for existence of diffeomorphisms that transform stochastic linear systems to different canonical forms can be found in

Pan Z (2002) Canonical forms for stochastic nonlinear systems. Automatica 38(7):1163–1170

The extension of one-step convergence errors to fixed-step convergence errors and the proof of Theorem 18.5 first appeared in

Carrasco DS (2014) PhD thesis, University of Newcastle, Australia (in preparation)

Chapter 19
Applications of Approximate Stochastic Sampled-Data Models

This chapter explores several applications of approximate sampled-data models. It is shown that an understanding of the role of sampling zero dynamics can be crucial in obtaining accurate results in some applications.

19.1 When Are Stochastic Sampling Zeros Important?

The reader is referred to the discussion in Sect. 10.1 of Chap. 10 regarding a similar question for the deterministic case. It was shown in Chap. 10 that the answer depended on the 'design bandwidth', i.e., on whether or not the algorithm under consideration is sensitive to the model accuracy in the region near the sampling frequency. As shown below, a similar conclusion holds in the stochastic case.

19.2 Models for System Identification

As a specific example, the issues associated with the use of sampled-data models in continuous-time identification from sampled data are examined. Sampled-data models are deployed to estimate the parameters of an underlying continuous-time system. In this context, one may hope that, if samples are taken *quickly enough*, then the difference between discrete-time and continuous-time processing would become vanishingly small. Alas, this is not always true, as shown below.

To set the scene, let θ represent the *true* parameter vector of the continuous-time model

$$\mathcal{M}:\quad y(t) = G(\rho,\theta)u(t) + H(\rho,\theta)\dot{v}(t) \tag{19.1}$$

where $G(\rho,\theta), H(\rho,\theta)$ are transfer functions in the continuous-time differential operator $\rho, u(t)$ is a piecewise constant continuous-time input, and $\dot{v}(t)$ is a continuous-time white noise process.

J.I. Yuz, G.C. Goodwin, *Sampled-Data Models for Linear and Nonlinear Systems*, Communications and Control Engineering, DOI 10.1007/978-1-4471-5562-1_19,

Say a set of sampled data $\{u_k = u(k\Delta), y_k = y(k\Delta)\}$ is given, and the goal is to identify a sampled-data model

$$\mathcal{M}_\Delta: \quad y_k = G_\delta(\delta, \hat{\theta})u_k + H_\delta(\delta, \hat{\theta})v_k \tag{19.2}$$

where $\hat{\theta}$ is a vector with the parameters to be estimated. The goal is to design a system identification algorithm such that $\hat{\theta}$ converges to the corresponding continuous-time parameters, as Δ goes to zero, i.e.,

$$\hat{\theta} \xrightarrow{\Delta \to 0} \theta \tag{19.3}$$

As shown in earlier chapters, there is an inherent *loss of information* when using discrete-time model representations. In the time domain, the use of sampled data implies that information about the intersample behaviour of the system is lost. In the frequency domain, this fact translates to the aliasing effect, where high frequency components are folded back to low frequencies in such a way that is not possible to distinguish among them.

To fill the gap between systems evolving in continuous time and their sampled-data representations, extra assumptions on the continuous-time model and signals are needed. This leads to the issue of sampling zeros. Ignoring these sampling zeros in naive models can lead to erroneous results.

19.3 Effect of Sampling Zeros in Stochastic System Identification

A particular case where the *naive* use of sampled-data models can lead to erroneous results is in the identification of continuous-time stochastic systems where the noise model has relative degree greater than zero. In this case, it was shown in Chaps. 14 through 18 that the sampled-data model will have *stochastic sampling zero dynamics*. These sampling zero dynamics play a crucial role in obtaining unbiased parameter estimates in the identification of such systems from sampled data when the full bandwidth is used for identification purposes.

To illustrate, consider an auto-regressive system having model

$$E(\rho)y(t) = \dot{v}(t) \tag{19.4}$$

where $\dot{v}(t)$ represents a continuous-time white noise (CTWN) process, and $E(\rho)$ is a polynomial in the differential operator $\rho = \frac{d}{dt}$, i.e.,

$$E(\rho) = \rho^n + a_{n-1}\rho^{n-1} + \cdots + a_0 \tag{19.5}$$

If a simple derivative replacement model is used, then this is equivalent to replacing the derivatives in the continuous-time model by divided differences. This leads to

$$\left(\delta^2 + \bar{a}_1\delta + \bar{a}_0\right)y_k = \omega_k \tag{19.6}$$

where $\{\omega_k\}$ is a discrete-time white noise (DTWN) sequence.

The parameters can then, in principle, be estimated using ordinary least squares with cost function of the form

$$J_{LS} = \sum_{k=1}^{N} \left[\left(\delta^2 + \bar{a}_1 \delta + \bar{a}_0 \right) y(k\Delta) \right]^2 \tag{19.7}$$

The idea is illustrated by a simple example.

Example 19.1 Consider the continuous-time system defined by the *nominal model*

$$E(\rho) y(t) = \dot{v}(t) \tag{19.8}$$

where $\dot{v}(t)$ is a CTWN process with (constant) spectral density equal to 1, and

$$E(\rho) = \rho^2 + 3\rho + 2 \tag{19.9}$$

For simulation purposes, a sampling frequency $\omega_s = 250$ rad/s was used. Note that this frequency is two decades above the fastest system pole, located at $s = -2$. A set of $N_{\text{sim}} = 250$ Monte Carlo simulations were studied, using $N = 10000$ data points in each run.

Test 1: Ordinary least squares was used as in (19.7). The continuous-time parameters were extracted from the delta form. The following (mean) parameter estimates were obtained:

$$\begin{bmatrix} \hat{a}_1 \\ \hat{a}_0 \end{bmatrix} = \begin{bmatrix} 1.9834 \\ 1.9238 \end{bmatrix} \tag{19.10}$$

Note that these estimates are far from the continuous-time parameters. In particular, it can be seen that the estimate $\hat{a}_1 = 1.9834$ is clearly biased with respect to the continuous-time value $a_1 = 3$.

Of course the above procedure has ignored the sampling zeros. Indeed the exact discrete-time model describing the continuous-time system (19.4) takes the following generic form:

$$E(\delta) y(k\Delta) = F(\delta) w_k \tag{19.11}$$

where w_k is a DTWN process, and E and F are polynomials in the operator δ.

As shown in Chap. 14, the polynomial E in Eq. (19.11) is *well behaved* in the sense that it converges naturally to its continuous-time counterpart. This relationship is most readily portrayed using the delta form, where

$$E_\delta(\delta) = \delta^n + \bar{a}_{n-1} \delta^{n-1} + \cdots + \bar{a}_0 \tag{19.12}$$

As the sampling period Δ goes to zero,

$$\lim_{\Delta \to 0} \bar{a}_i = a_i; \quad i = n-1, \ldots, 0 \tag{19.13}$$

However, as also seen in Chap. 14, the polynomial F contains the stochastic sampling zeros, with no continuous-time counterpart. Thus, to obtain the correct estimates—say via the prediction error method—one needs to minimise a filtered least squares cost function:

$$J_{\mathrm{PEM}} = \sum_{k=1}^{N} \left[\frac{E(\delta) y(k\Delta)}{F(\delta)} \right]^2 \tag{19.14}$$

We note the key role played by the sampling zeros in this expression. A simplification can be applied, when using high sampling rates, by replacing the polynomial $F(\delta)$ by its asymptotic expression.

Example 19.2 (Example 19.1 Continued)

Test 2: Least squares estimation of the parameters was performed, but with prefiltering of the data by the asymptotic sampling zero polynomial; i.e., the following sequence of filtered output samples was utilised:

$$y_F(k\Delta) = \frac{1}{1 + (2 - \sqrt{3})q^{-1}} y(k\Delta) \tag{19.15}$$

Again, the continuous-time parameter estimates were extracted from the delta model. The following estimates were obtained for the coefficients of the polynomial (19.9):

$$\begin{bmatrix} \hat{a}_1 \\ \hat{a}_0 \end{bmatrix} = \begin{bmatrix} 2.9297 \\ 1.9520 \end{bmatrix} \tag{19.16}$$

These are almost exactly the continuous-time parameters. The small residual bias in this case can be explained by the use of the asymptotic sampling zero in (19.15), while the sampling period Δ is finite.

19.4 Restricted Bandwidth Estimation

The reader is reminded of the comments made in Sect. 10.1 of Chap. 10 regarding the use of schemes where model fidelity in the vicinity of the Nyquist frequency is important. Clearly, this idea underpins the results obtained above. However, it also suggests an alternative remedy to the problem. Rather than performing estimation using the full data bandwidth, where sampling zeros play a crucial role, it seems heuristically reasonable that it may simply be necessary to restrict the data bandwidth and ignore sampling zeros. This idea is explored below.

The following result on maximum likelihood estimation using a restricted bandwidth in the frequency domain will be utilised.

Lemma 19.3 *Assume that a given set of input-output data* $\{u_k = u(k\Delta), y_k = y(k\Delta)\}$, $k = 0 \dots N_d$, *is generated by the* exact *discrete-time model*

$$y_k = G_q(q, \theta)u_k + H_q(q, \theta)v_k \tag{19.17}$$

where v_k *is a Gaussian DTWN sequence,* $v_k \sim N(0, \sigma_w^2)$.

The data is transformed to the frequency domain yielding the discrete-time Fourier transforms U_ℓ *and* Y_ℓ *of the input and output sequences, respectively.*

The maximum likelihood estimate of θ, when considering frequency components up to $\omega = \omega_{\max} \le \frac{\omega_s}{2}$, is given by

$$\hat{\theta}_{ML} = \arg\min_\theta L(\theta) \tag{19.18}$$

where $L(\theta)$ is the negative logarithm of the likelihood function of the data given θ, i.e.,

$$\begin{aligned} L(\theta) &= -\log p(Y_0, \dots, Y_{n_{\max}} | \theta) \\ &= \sum_{\ell=0}^{n_{\max}} \frac{|Y_\ell - G_q(e^{j\omega_\ell \Delta}, \theta)U_\ell|^2}{\lambda_v^2 |H_q(e^{j\omega_\ell \Delta}, \theta)|^2} + \log\bigl(\pi\lambda_v^2 \bigl|H_q\bigl(e^{j\omega_\ell \Delta}, \theta\bigr)\bigr|^2\bigr) \end{aligned} \tag{19.19}$$

where $\lambda_v^2 = \Delta N_d \sigma_v^2$, and $n_{\max}$ is the index associated with $\omega_{\max}$.

Proof Equation (19.17) can be expressed in the frequency domain as

$$Y_\ell = G_q\bigl(e^{j\omega_\ell \Delta}, \theta\bigr)U_\ell + H_q\bigl(e^{j\omega_\ell \Delta}, \theta\bigr)V_\ell \tag{19.20}$$

where Y_ℓ, U_ℓ, and V_ℓ are discrete Fourier transforms, e.g.,

$$Y_\ell = Y\bigl(e^{j\omega_\ell \Delta}\bigr) = \Delta \sum_{k=0}^{N_d - 1} y_k e^{-j\omega_\ell k \Delta}; \qquad \omega_\ell = \frac{2\pi}{\Delta}\frac{\ell}{N_d} \tag{19.21}$$

Assuming that the DTWN sequence $v_k \sim N(0, \sigma_w^2)$, then V_ℓ are (asymptotically) independent and have a circular complex Gaussian distribution (see the references at the end of the chapter). Thus, the frequency-domain noise sequence V_ℓ has zero mean and variance $\lambda_v^2 = \Delta N_d \sigma_v^2$. It follows that Y_ℓ is also complex Gaussian and satisfies

$$Y_\ell \sim N\bigl(G_q\bigl(e^{j\omega_\ell \Delta}, \theta\bigr)U_\ell, \lambda_v^2 \bigl|H_q\bigl(e^{j\omega_\ell \Delta}, \theta\bigr)\bigr|^2\bigr) \tag{19.22}$$

The corresponding probability density function is given by

$$p(Y_\ell) = \frac{1}{\pi\lambda_v^2 |H_q(e^{j\omega_\ell \Delta}, \theta)|^2} \exp\left\{-\frac{|Y_\ell - G_q(e^{j\omega_\ell \Delta}, \theta)U_\ell|^2}{\lambda_v^2 |H_q(e^{j\omega_\ell \Delta}, \theta)|^2}\right\} \tag{19.23}$$

If the elements Y_ℓ are considered within a *limited bandwidth*, i.e., up to some maximum frequency $\omega_{\max}$ indexed by $n_{\max}$ with $\omega_{\max} = \omega_s \frac{n_{\max}}{N_d} \leq \frac{\omega_s}{2}$, the appropriate log-likelihood function is given by:

$$L(\theta) = -\log p(Y_0, \dots, Y_{n_{\max}}) = -\log \prod_{\ell=0}^{n_{\max}} p(Y_\ell)$$
$$= \sum_{\ell=0}^{n_{\max}} \frac{|Y_\ell - G_q(e^{j\omega_\ell \Delta}, \theta)U_\ell|^2}{\lambda_v^2 |H_q(e^{j\omega_\ell \Delta}, \theta)|^2} + \log\bigl(\pi \lambda_v^2 \bigl|H_q\bigl(e^{j\omega_\ell \Delta}, \theta\bigr)\bigr|^2\bigr) \qquad (19.24)$$

□

Remark 19.4 The logarithmic term in the log-likelihood function (19.19) plays a key role in obtaining consistent estimates of the true system. This term can be neglected *only* if:

- The noise model is assumed to be known. In this case H_q does not depend on θ and, thus, plays no role in the minimisation (19.18); or
- The frequencies ω_ℓ are equidistantly distributed over the full frequency range $[0, \frac{2\pi}{\Delta})$. This is equivalent to considering the *full bandwidth* case in (19.19), i.e., $n_{\max} = \frac{N}{2}$ (or N, because of periodicity). This yields

$$\frac{2\pi}{N_d} \sum_{\ell=0}^{N_d-1} \log\bigl|H_q\bigl(e^{j\omega_\ell \Delta}, \theta\bigr)\bigr|^2 \xrightarrow{N_d \to \infty} \int_0^{2\pi} \log\bigl|H_q\bigl(e^{j\omega}, \theta\bigr)\bigr|^2 d\omega \qquad (19.25)$$

The last integral is equal to zero for any monic, stable, and inversely stable transfer function $H_q(e^{j\omega}, \theta)$.

Remark 19.5 In the previous lemma, the discrete-time model (19.17) has been expressed in terms of the shift operator q. The results apply mutatis mutandis when the model is reparameterised using the δ-operator:

$$G_q\bigl(e^{j\omega_\ell \Delta}\bigr) = G_\delta(\gamma_\omega) = G_\delta\left(\frac{e^{j\omega_\ell \Delta} - 1}{\Delta}\right) \qquad (19.26)$$

Example 19.6 (Example 19.1 Continued) The system identification problem discussed above is repeated based on the model (19.6), but where the data is restricted to a certain *bandwidth of validity*. For example, the usual rule of thumb is to consider up to one decade above the fastest nominal system pole, in this case, 20 rad/s. The resultant (mean of the) parameter estimates obtained by the procedure described in

Lemma 19.3 are found to be:

$$\begin{bmatrix} \hat{a}_1 \\ \hat{a}_0 \end{bmatrix} = \begin{bmatrix} 3.0143 \\ 1.9701 \end{bmatrix} \tag{19.27}$$

Note that these estimates are essentially equal to the (continuous-time) true values.

Finally, the frequency-domain procedure outlined above is also robust to the presence of unmodelled fast poles.

Example 19.7 (Example 19.1 Continued) Consider the true system to be

$$(\rho^2 + a_1\rho + a_0)\left(\frac{2\rho}{\omega_u} + 1\right)y = \dot{\omega} \tag{19.28}$$

where the pole $(\frac{2\rho}{\omega_u} + 1)$ is not included in the model. Take $\omega_u = 50$ for the sake of illustration. Restrict the estimation bandwidth up to 20 rad/s. In this case, the mean parameter estimates are found to be:

$$\begin{bmatrix} \hat{a}_1 \\ \hat{a}_0 \end{bmatrix} = \begin{bmatrix} 2.9285 \\ 1.9409 \end{bmatrix} \tag{19.29}$$

Again, it can be seen that the estimates are very close to the true values of $a_1 = 3$, $a_0 = 2$. A more general situation is shown in Fig. 19.1. The figure shows the parameter estimates obtained using the limited frequency-domain estimation procedure, with the same restricted bandwidth used before, $\omega_{\max} = 20$ rad/s, for different locations of the unmodelled fast pole ω_u, as in (19.28). It can be seen that, provided the unmodelled pole is located well above the system dynamics and the bandwidth is restricted to be below the Nyquist folding frequency, accurate estimates are obtained by ordinary least squares ignoring both the unmodelled pole and the sampling zero.

19.5 Identification of Continuous-Time State-Space Models from Nonuniformly Fast-Sampled Data

This section explores the application of the expectation-maximisation (EM) algorithm to identify continuous-time state-space models from non-uniformly fast-sampled data. The following key issues are exploited in this application:

- Since the observation interval is not fixed, the anti-aliasing filter averages the output over the sample period, irrespective of its length.
- A consequence of the variable sampling period and choice of anti-aliasing filter is that the associated discrete-time model is time varying.

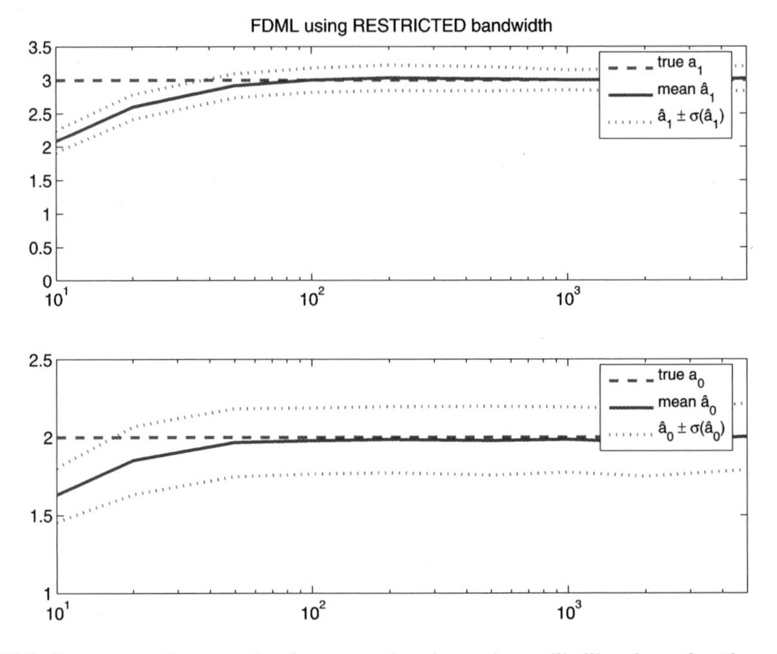

Fig. 19.1 Parameter estimates using frequency-domain maximum likelihood as a function of unmodelled pole

- However, by use of incremental models, the sampling period is made explicit. Moreover, the resulting incremental model parameters, though still time varying, converge to the fixed underlying continuous-time parameters when the maximum sampling interval is small.
- Thus, the trick is to estimate the incremental model parameters as if they were constant and then interpret them as the underlying continuous-time parameters.

19.5.1 Continuous-Time System Description

Consider a general multiple-input multiple-output linear time-invariant continuous-time system subject to stochastic disturbances. An appropriate stochastic differential equation (SDE) model is:

$$dx(t) = A_c x(t)\,dt + B_c u(t)\,dt + dw(t) \tag{19.30}$$

$$dz(t) = C_c x(t)\,dt + D_c u(t)\,dt + dv(t) \tag{19.31}$$

where $u(t) \in \mathbb{R}^{n_u}$, $x(t) \in \mathbb{R}^n$, $y(t) \in \mathbb{R}^{n_y}$ are the input, the system state, and the output signal, respectively; the system matrices are $A_c \in \mathbb{R}^{n \times n}$, $B_c \in \mathbb{R}^{n \times n_u}$, $C_c \in \mathbb{R}^{n_y \times n}$, and $D_c \in \mathbb{R}^{n_y \times n_u}$; and the incremental state disturbance $dw(t)$ and incremental measurement disturbance $dv(t)$ are stochastic processes that are independent (in time), zero mean, and have Gaussian distribution (i.e., $w(t)$ and $v(t)$ are Wiener processes) such that

$$E\left\{\begin{bmatrix} dw(t) \\ dv(t) \end{bmatrix}\begin{bmatrix} dw(s) \\ dv(s) \end{bmatrix}^T\right\} = \begin{cases} \begin{bmatrix} Q_c & 0 \\ 0 & R_c \end{bmatrix} dt; & t = s \\ 0; & t \neq s \end{cases} \tag{19.32}$$

where $Q_c \in \mathbb{R}^{n \times n}$ is a positive semi-definite matrix, and $R_c \in \mathbb{R}^{n_y \times n_y}$ is a positive definite matrix. The initial state of the system is assumed to be independent of $dw(t)$ and $dv(t)$, and Gaussian distributed having mean μ_o and covariance Σ_o.

The focus here is on obtaining an estimate of the system parameter

$$\theta^c = \{A_c, B_c, C_c, D_c, Q_c, R_c\} \tag{19.33}$$

from a finite set of input-output samples $\{u_k = u(t_k), y_k = y(t_k)\}$, where $k \in \{0, \dots, N\}$. A key departure from earlier work in this book is that the samples are non-uniformly spaced, i.e., the sampling interval depends on the discrete-time index k:

$$\Delta_k = t_{k+1} - t_k > 0; \quad \text{for all } k \in 0, \dots, N-1 \tag{19.34}$$

The continuous-time input to the system, $u(t)$, is assumed to be generated from an input sequence, u_k, by a zero order hold (ZOH) device, i.e.,

$$u(t) = u_k; \quad t_k \leq t < t_{k+1} \tag{19.35}$$

The available data is collected on the continuous-time interval $[0, T_f]$, where $T_f = t_N = \sum_{k=0}^{N-1} \Delta_k$.

It is assumed that the sampling intervals $\{\Delta_k\}$ are uniformly bounded by a positive constant $\bar{\Delta}$, i.e., there exists

$$\bar{\Delta} \triangleq \max_k \Delta_k \tag{19.36}$$

such that $0 < \Delta_k \leq \bar{\Delta}$, for all $k \in 0, \dots, N-1$.

In order to take the sampling process into account, an *exact* sampled-data model corresponding to (19.30)–(19.31) is first obtained.

19.5.2 Sampled-Data Model

As emphasised in Chaps. 11 and 12, the sampling process of the system output must be dealt with carefully. From (19.30), we see that the output has a *pure CTWN component*. Instantaneous sampling of this output would lead to a discrete-time sequence having infinite variance. Thus, as made clear earlier in the book, some kind of *pre-filtering* is needed prior to sampling. Hence, an averaging filter is included at the system output before instantaneous sampling. The filter output is given by:

$$\bar{y}(t_{k+1}) = \frac{1}{t_{k+1}-t_k}\int_{t_k}^{t_{k+1}} \frac{dz(\tau)}{d\tau}\,d\tau = \frac{1}{\Delta_k}\int_{t_k}^{t_{k+1}} dz(\tau) = \frac{z(t_{k+1})-z(t_k)}{\Delta_k} \tag{19.37}$$

An additional advantage of this sampling strategy is that, as the sampling rate is increased, the sampled output $\bar{y}$ is consistent with the continuous-time output (19.30), i.e.,

$$\bar{y}(t) = \lim_{\Delta_k\to 0} \frac{z(t_{k+1})-z(t_k)}{\Delta_k} = \frac{dz(t)}{dt} \tag{19.38}$$

The incremental discrete-time model is obtained as in Sect. 12.3, leading to

$$dx_k^+ = A_k^\delta x_k \Delta_k + B_k^\delta u_k \Delta_k + dw_k^+ \tag{19.39}$$

$$\bar{y}_{k+1}\Delta_k = dz_k^+ = C_k^\delta x_k \Delta_k + D_k^\delta u_k \Delta_k + dv_k^+ \tag{19.40}$$

where, as usual, the increments are defined as

$$df_k^+ = f_{k+1} - f_k \tag{19.41}$$

The matrices are given by:

$$A_k^\delta = \frac{e^{A_c\Delta_k} - I}{\Delta_k}, \qquad B_k^\delta = \left[\frac{1}{\Delta_k}\int_0^{\Delta_k} e^{A_c\eta}\,d\eta\right] B_c \tag{19.42}$$

$$C_k^\delta = C_c\left[\frac{1}{\Delta_k}\int_0^{\Delta_k} e^{A_c\eta}\,d\eta\right] D_k^\delta = D_c + C_c\left[\frac{1}{\Delta_k}\int_0^{\Delta_k}\int_0^{\xi} e^{A\eta}\,d\eta\,d\xi\right] B_c \tag{19.43}$$

and the covariance structure of the noise vector is given by

$$E\left\{\begin{bmatrix} dw_\ell^+ \\ dv_\ell^+ \end{bmatrix}\begin{bmatrix} dw_k^+ \\ dv_k^+ \end{bmatrix}\right\} = \begin{bmatrix} Q_k^\delta & S_k^\delta \\ (S_k^\delta)^T & R_k^\delta \end{bmatrix} \Delta_k\, \delta_K[\ell - k] \tag{19.44}$$

where

$$\begin{bmatrix} Q_k^\delta & S_k^\delta \\ (S_k^\delta)^T & R_k^\delta \end{bmatrix} = \frac{1}{\Delta_k}\int_0^{\Delta_k} \begin{bmatrix} e^{A_c\eta} & 0 \\ C_c\int_0^\eta e^{A_c\xi}\,d\xi & I \end{bmatrix} \begin{bmatrix} Q_c & 0 \\ 0 & R_c \end{bmatrix} \begin{bmatrix} e^{A_c\eta} & 0 \\ C_c\int_0^\eta e^{A_c\xi}\,d\xi & I \end{bmatrix}^T d\eta \tag{19.45}$$

Remark 19.8 As discussed in Chap. 12, when the maximum sampling period $\bar{\Delta}$ goes to zero, the incremental model (19.39)–(19.40) converges to the SDE representation. In particular,

$$\lim_{\bar{\Delta}\to 0} A_k^\delta = A_c, \qquad \lim_{\bar{\Delta}\to 0} B_k^\delta = B_c, \qquad \lim_{\bar{\Delta}\to 0} C_k^\delta = C_c, \qquad \lim_{\bar{\Delta}\to 0} D_k^\delta = D_c \tag{19.46}$$

and

$$\lim_{\bar{\Delta}\to 0}\begin{bmatrix} Q_k^\delta & S_k^\delta \\ (S_k^\delta)^T & R_k^\delta \end{bmatrix} = \begin{bmatrix} Q_c & 0 \\ 0 & R_c \end{bmatrix} \tag{19.47}$$

A consequence of this observation is that, if the (variable) sample period is made explicit as in (19.39), (19.40), then, even though non-uniform sampling is used, the resulting discrete-time model converges to the state-space model (19.30)–(19.31) as the sampling rate is increased.

Next consider the (time-varying) parameter

$$\theta_k^\delta = \left\{A_k^\delta, B_k^\delta, C_k^\delta, D_k^\delta, Q_k^\delta, R_k^\delta, S_k^\delta\right\} \tag{19.48}$$

Then, from Remark 19.8, we have that, for every k,

$$\lim_{\bar{\Delta}\to 0}\theta_k^\delta = \lim_{\Delta_k\to 0}\theta_k^\delta = \theta^c \tag{19.49}$$

where $\bar{\Delta}$ is the maximum sampling interval and θ_c is the continuous-time parameter (19.33).

Thus, the exact time-varying sampled-data model (19.39)–(19.45) can be used to identify the continuous (time-invariant) system parameters from non-uniformly fast-sampled data. Specifically, the time-varying matrices θ_k^δ are replaced by a constant parameter θ. Note, however, that the time-varying quantity, Δ_k, must be used in the model.

19.5.3 Maximum Likelihood Identification and the EM Algorithm

A commonly used strategy for parameter estimation is to maximise the likelihood function. The likelihood function is the conditional probability of the data Y given the parameter vector θ. When the measurements are Gaussian distributed, it is common to maximise the logarithm of the likelihood function:

$$\ell(\theta) = \log p(Y|\theta) \tag{19.50}$$

19.5.4 EM Algorithm

The EM algorithm is an iterative method used to maximise the log-likelihood function (19.50). In fact, if the state sequence X is chosen as the *hidden data*, the log-likelihood can be decomposed as

$$\log p(Y|\theta) = \mathcal{Q}(\theta,\hat{\theta}_i) - \mathcal{H}(\theta,\hat{\theta}_i) \tag{19.51}$$

where $\hat{\theta}_i$ is an available estimate of the parameter θ, and

$$\mathcal{Q}(\theta, \hat{\theta}_i) = E\{\log p(X, Y|\theta)|Y, \hat{\theta}_i\} \tag{19.52}$$

$$\mathcal{H}(\theta, \hat{\theta}_i) = E\{\log p(X|Y, \theta)|Y, \hat{\theta}_i\} \tag{19.53}$$

Using Jensen's inequality it is easy to show that $\mathcal{H}(\theta, \hat{\theta}_i) \leq \mathcal{H}(\hat{\theta}_i, \hat{\theta}_i)$. As a consequence, if one finds a value of θ that maximises (or increases) the function $\mathcal{Q}(\theta, \hat{\theta}_i)$, then one can generate an iterative procedure to maximise (or to increase) the log-likelihood function (19.50).

The EM algorithm can be summarised by the following steps (here $\hat{\theta}_i$ is used to denote the parameter estimate at iteration i of the algorithm):

1. Start with an initial estimate of the system parameter $\hat{\theta}_0$.
2. *E-step*: Obtain the function $\mathcal{Q}(\theta, \hat{\theta}_i)$, defined in (19.52), which is the expected value of the complete data (X, Y) given the observed data Y and an available estimate $\hat{\theta}_i$. Note that Y is chosen as the (non-uniformly sampled) time-domain data and the hidden data X is chosen as the state sequence. Hence, the E-step involves the use of a Kalman smoother to estimate the state sequence.
3. *M-step*: Maximise the function $\mathcal{Q}(\theta, \hat{\theta}_i)$ with respect to the parameter θ. This yields a new parameter estimate, i.e.,

$$\hat{\theta}_{i+1} = \arg\max_{\theta} \mathcal{Q}(\theta, \hat{\theta}_i) \tag{19.54}$$

   In the case of incremental models, the new estimate θ_i that maximises $\mathcal{Q}(\theta, \hat{\theta}_i)$ in (19.52) can be obtained explicitly (i.e., analytically).
4. Go to step 2 (the E-step), increasing the index $i \to i + 1$, and iterate until convergence (or some predefined level of accuracy) is achieved.

Full details of the EM algorithm for incremental state-space models go beyond the scope of this chapter. The interested reader is referred to the literature cited at the end of the chapter.

The key aspects of the problem here are:

1. The use of an underlying continuous-time SDE model.
2. The need for an anti-aliasing filter.
3. The fact that the time-varying sample period Δ_k is used explicitly in the model.
4. The fact that the continuous-time parameter can be directly estimated from sampled data.

The ideas are illustrated by the following example.

Example 19.9 Consider a state-space continuous-time system as in (19.30)–(19.31), where the matrices are given by:

$$A_c = \begin{bmatrix} -6.00 & -2.50 \\ 2.00 & 0 \end{bmatrix}, \qquad B_c = \begin{bmatrix} 2.00 \\ 0 \end{bmatrix} \tag{19.55}$$

$$C_c = \begin{bmatrix} 1.00 & 1.50 \end{bmatrix}, \qquad D_c = 0 \tag{19.56}$$

The spectral densities of the continuous-time process and measurement noise are respectively given by:

$$Q_c = \begin{bmatrix} 0.5 & 0 \\ 0 & 0.5 \end{bmatrix}, \qquad R_c = 0.01 \tag{19.57}$$

The corresponding transfer function from the deterministic input $u(t)$ to the output is given by

$$G(s) = \frac{2(s+3)}{(s+1)(s+5)} \tag{19.58}$$

The EM algorithm using incremental and shift operator models is compared at three different (uniform) values for the sampling periods $\Delta = 0.05$, 0.015, and 0.005.

To simulate data from the associated continuous-time system the exact sampled-data model obtained in Sect. 19.5.2 is used. The continuous-time input $u(t)$ is generated using a pseudo-random binary sequence of length 200 with standard deviation, $\sigma_u = 9$, passed through a ZOH of fixed sampling period $\Delta = 0.15$. This continuous-time input is also used for the other sampling periods, resampling it at the corresponding rate to obtain u_k. A fixed continuous-time horizon, $T_f = 30$, is used. Thus, there are $N = 600$, 2000, 6000, and 20000 data points for each case, respectively.

Notice that, even though more data is available as the sampling period Δ is reduced, the discrete measurement noise variance changes with Δ (see, for example, Eq. (19.44)). However, the discrete noise power spectral density is roughly invariant with respect to Δ. Thus, as might be expected, it is the (continuous-time) observation period that primarily affects the estimation accuracy, and not the number of samples.

Figure 19.2 shows the frequency response of the true continuous-time system, G, the exact sampled-data model, G_{disc}, and the two identified models using the shift operator model G_{qe} and incremental model G_{ie}, for the sampling period $\Delta = 0.05$. Each of these models has been estimated using the corresponding version of the EM algorithm. For each case, the estimated model is the *average* of the models obtained for each of 50 realisations of the noise processes. In Fig. 19.2 the difference between the magnitude of the exact sampled-data model and those of the estimated models is hardly noticeable. This shows that, for this sampling rate, both implementations of the EM algorithm provide the same accuracy. This is not surprising, because shift operator and incremental models are equivalent representations. The equivalence is only affected by finite numerical precision in the computation of the estimates, which is not an important issue for 'low' sampling rates. However, as the sampling rate is increased (in order to identify the continuous-time parameters), a clear difference appears between the estimated models.

The results are shown in Fig. 19.2 for $\Delta = 0.05$. In particular, the deterministic part of the estimated model, obtained using the EM formulation based on the incremental form of the sampled-data model, is given by:

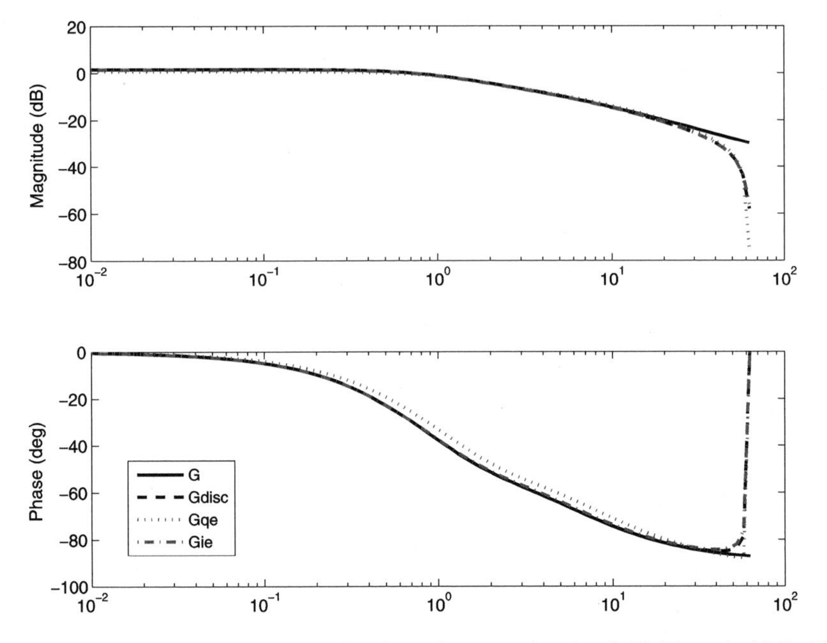

Fig. 19.2 Comparison between real and estimated systems for $\Delta = 0.05$ (Example 19.9). The *thick black line* corresponds to the true continuous-time system G. The *dashed black line* corresponds to the exact discrete-time model G_{disc}. The *dotted line* and the *dot-dashed line* correspond to the estimations obtained using shift operator G_{qe} and incremental models G_{ie}, respectively

$$\begin{aligned}\hat{G}_c(s) &= \hat{C}_c(sI - \hat{A}_c)^{-1}\hat{B}_c + \hat{D}_c \\ &= \frac{1.741(s+2.850)}{(s+4.506)(s+0.9702)} + 0.0483 \quad (\Delta = 0.05)\end{aligned} \tag{19.59}$$

Figure 19.3 shows the results for $\Delta = 0.015$. It can be seen that a clear difference now appears between the estimated models. In fact, the incremental version of the EM algorithm provides a good estimation of the continuous-time system:

$$\hat{G}_c(s) = \frac{1.951(s+2.77)}{(s+4.954)(s+0.9679)} + 0.0153 \quad (\Delta = 0.015) \tag{19.60}$$

On the other hand, the estimation using shift operator models fails to recover the true system. In fact, several of the models obtained for the different realisations of the noise processes are unstable.

In Fig. 19.4, similar results are shown for $\Delta = 0.005$; i.e., incremental models provide a good estimate of the system, whereas the shift operator formulation shows poor performance. In fact, the incremental form leads to the following estimates:

$$\hat{G}_c(s) = \frac{2.066(s+2.661)}{(s+5.198)(s+1.049)} + 0.0051 \quad (\Delta = 0.005) \tag{19.61}$$

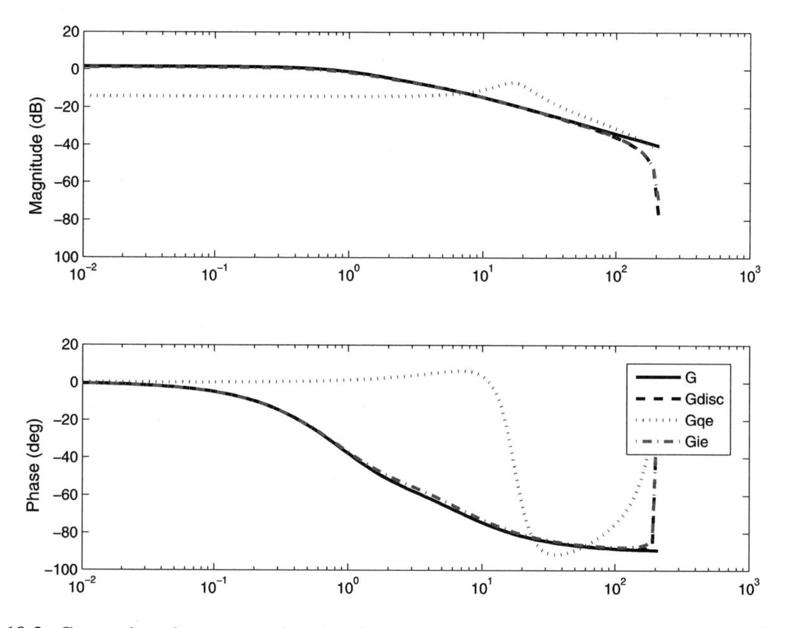

Fig. 19.3 Comparison between real and estimated system for $\Delta = 0.015$ (Example 19.9). The *thick black line* corresponds to the true continuous-time system G. The *dashed black line* corresponds to the exact discrete-time model G_{disc}. The *dotted line* and the *dot-dashed line* correspond to the estimations obtained using shift operator G_{qe} and incremental models G_{ie}, respectively

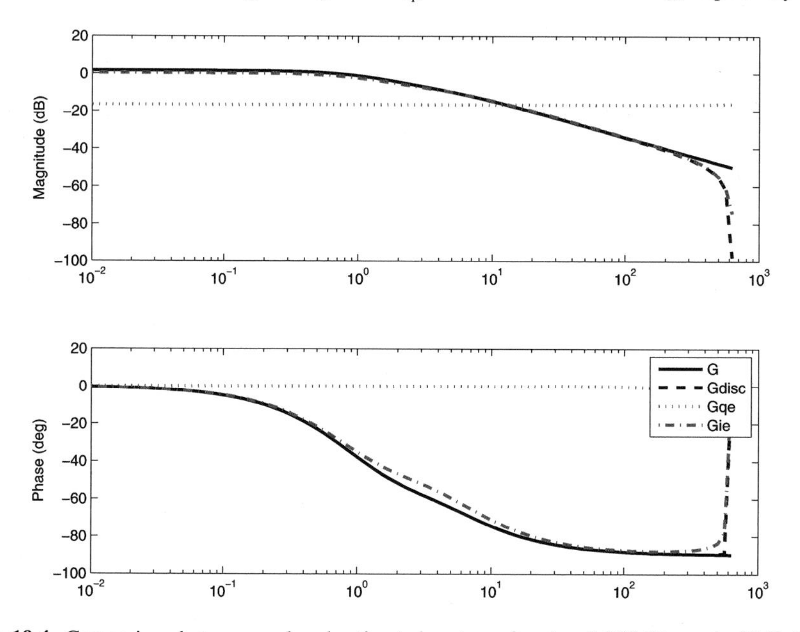

Fig. 19.4 Comparison between real and estimated systems for $\Delta = 0.005$ (Example 19.9). The *thick black line* corresponds to the true continuous-time system G. The *dashed black line* corresponds to the exact discrete-time model G_{disc}. The *dotted line* and the *dot-dashed line* correspond to the estimations obtained using shift operator G_{qe} and incremental models G_{ie}, respectively

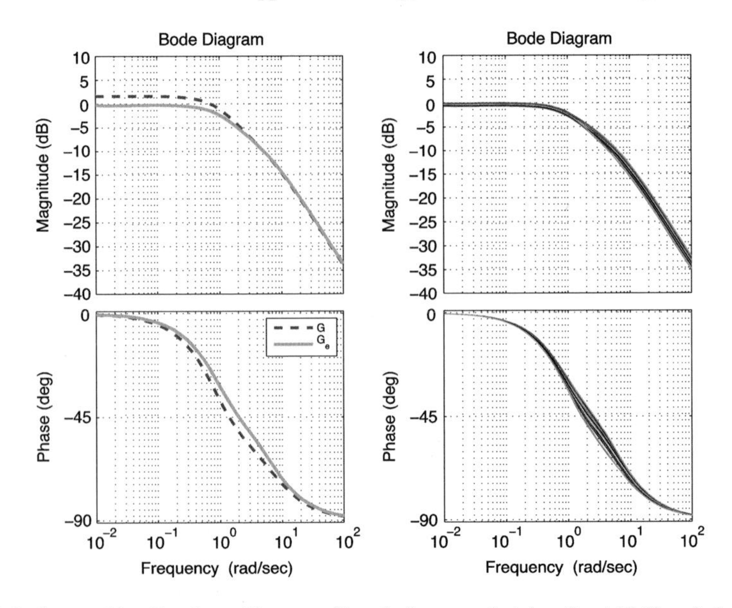

Fig. 19.5 System identification with non-uniformly fast-sampled data for ±10 % variation of the *nominal* sampling interval (Example 19.10). The *left plot* shows the real model and the average model corresponding to the 20 realisations shown in the *right plot*

Thus, the proposed incremental formulation of the EM algorithm provides a good estimate of the continuous-time system. Moreover, this implementation is clearly more numerically robust than the shift operator formulation when using fast sampling rates. In fact, the proposed incremental model formulation shows good performance up to 10 times the sampling rate at which shift operator formulation fails.

Example 19.10 Consider the same continuous-time system as in Example 19.9. However, in this case non-uniformly sampled data is used.

The sampling instants are generated as $t_k = \Delta(k + \eta_k)$, where an average sampling period $\Delta = 0.006$ is used, and η_k is a random sequence, uniformly distributed on the interval $[-r/100, r/100]$. This is a $\pm r$ % variability in the average sampling period. Note that these sampling instants are exactly known by the algorithm.

The results are shown in Fig. 19.5 for $r = 10$ (variation of ±10 %). The left plot shows the frequency response of the true system G and the (average) estimated system G_e. The plot on the right side of the figure shows the estimated systems corresponding to each of 20 realisations of the experiment. It can be seen that, in all cases, the proposed implementation of the EM algorithm gives an excellent estimate of the continuous-time system parameters even though non-uniformly sampled data has been utilised.

19.6 Summary

The key ideas discussed in the current chapter are the following:

- Some procedures, especially those which depend on model fidelity in the vicinity of the Nyquist frequency, require that asymptotic sampling zeros be accounted for.
- It is illustrated that ignoring asymptotic sampling zeros when estimating the parameters of a continuous-time stochastic system will lead to bias errors.
- A key observation is that the fidelity of the models at high frequencies plays an important role in obtaining models suitable for continuous-time system identification. In particular, it has been shown that unmodelled high frequency dynamics (including stochastic sampling zeros) in the sampled-data model can have a critical impact on the quality of the estimation process when using sampled data.
- Incremental models are used to identify the parameters of continuous-time state-space models from non-uniformly sampled data.

Further Reading

The fact that the ordinary least squares method leads to biased estimates in the identification of continuous-time AR models from sampled data was first observed in

Söderström T, Fan H, Carlsson B, Bigi S (1997) Least squares parameter estimation of continuous-time ARX models from discrete-time data. IEEE Trans Autom Control 42(5):659–673

That the cause of the difficulty was sampling zeros was first pointed out in

Larsson EK (2003) Identification of stochastic continuous-time systems. PhD thesis, Division of Systems and Control, Uppsala University, Sweden

Larsson EK (2005) Limiting sampling results for continuous-time ARMA systems. Int J Control 78(7):461–473

The resolution of the problem using asymptotic sampling zero dynamics as a pre-filter is discussed in more detail in

Yuz JI, Goodwin GC (2008) Robust identification of continuous-time systems from sampled data. In: Garnier H, Wang L (eds) Continuous-time model identification from sampled data. Springer, Berlin

Robust estimation using a limited frequency range is described in

Aguero JC et al (2012) Dual time-frequency domain system identification. Automatica 48(12):3031–3041

Yuz JI, Goodwin GC (2008) Robust identification of continuous-time systems from sampled data. In: Garnier H, Wang L (eds) Continuous-time model identification from sampled data. Springer, Berlin

Circular complex distributions are described in

Brillinger DR (1974) Fourier analysis of stationary processes. Proc IEEE 62(12):1628–1643
Brillinger DR (1981) Time series: data analysis and theory. McGraw-Hill, New York

Further information regarding frequency-domain identification can be found in

Gillberg J, Ljung L (2009) Frequency-domain identification of continuous-time ARMA models from sampled data. Automatica 45(6):1371–1378
Pintelon R, Schoukens J (2007) Frequency domain maximum likelihood estimation of linear dynamic errors-in-variables models. Automatica 43(4):621–630
Pintelon R, Schoukens J, Rolain Y (2008) Frequency-domain approach to continuous-time system identification: some practical aspects. In: Garnier H, Wang L (eds) Continuous-time model identification from sampled data. Springer, Berlin, pp 215–248

The section on identification based on non-uniformly fast-sampled data draws heavily on

Yuz JI, Alfaro J, Agüero JC, Goodwin GC (2011) Identification of continuous-time state-space models from non-uniform fast-sampled data. IET Control Theory Appl 5(7):842–855

Other papers describing alternative approaches to the identification of continuous systems from non-uniform data include

Ding F, Qiu L, Chen T (2009) Reconstruction of continuous-time systems from their non-uniformly sampled discrete-time systems. Automatica 45(2):324–332
Gillberg J, Ljung L (2010) Frequency domain identification of continuous-time output error models, part II: non-uniformly sampled data and B-spline output approximation. Automatica 46(1):11–18
Larsson E, Söderström T (2002) Identification of continuous-time AR processes from unevenly sampled data. Automatica 38:709–718

Part III
Embellishments and Extensions

Chapter 20
The Euler–Frobenius Polynomials

Contributed by Diego S. Carrasco
School of Electrical Engineering and Computer Science
The University of Newcastle, Australia

This section presents a historical account of what are generally called *Euler–Frobenius* polynomials. Definitions and properties found in the literature are summarised. The intention is to clarify the difference between *Euler*, *Eulerian*, and *Euler–Frobenius* numbers, fractions, and polynomials found in the mathematics literature.

Searching for *Euler–Frobenius* polynomials in the literature will not provide many references. The reason is that the polynomials referred to here as *Euler–Frobenius* have been rediscovered and redefined many times in history. For instance, Sobolev calls them *Euler* polynomials, Frobenius and Carlitz call them *Eulerian*, and Reimer, Dubeau et al., and Weller et al. call them *Euler–Frobenius* polynomials. Adding to the confusion is the fact that there exist other polynomials, different from the three referred to above, also called *Euler* polynomials. Furthermore, several generalisations of these polynomials are also called *Eulerian*, *Euler–Frobenius*, or *Frobenius–Euler*, dropping the *generalised* prefix.

20.1 Euler–Frobenius Polynomials

Leonhard Euler first introduced the polynomials in 1749 in his work "*Remarques sur un beau rapport entre les series des puissances tant direct que reciproques*" while describing a method of computing values of the zeta function at negative integer points. Euler introduced them in the following way:

$$\sum_{k=0}^{\infty}(k+1)^n \cdot t^k = \frac{A_n(t)}{(1-t)^{n+1}} \tag{20.1}$$

J.I. Yuz, G.C. Goodwin, *Sampled-Data Models for Linear and Nonlinear Systems*, Communications and Control Engineering, DOI 10.1007/978-1-4471-5562-1_20,

For $n = 0, \ldots, 4$, the following expressions were obtained:

$$1 + 2^0 t + 3^0 t^2 + 4^0 t^3 + \cdots = \frac{1}{1-t} \tag{20.2}$$

$$1 + 2^1 t + 3^1 t^2 + 4^1 t^3 + \cdots = \frac{1}{(1-t)^2} \tag{20.3}$$

$$1 + 2^2 t + 3^2 t^2 + 4^2 t^3 + \cdots = \frac{t+1}{(1-t)^3} \tag{20.4}$$

$$1 + 2^3 t + 3^3 t^2 + 4^3 t^3 + \cdots = \frac{t^2+4t+1}{(1-t)^4} \tag{20.5}$$

$$1 + 2^4 t + 3^4 t^2 + 4^4 t^3 + \cdots = \frac{t^3+11t^2+11t+1}{(1-t)^5} \tag{20.6}$$

Frobenius named the polynomials, $A_n(t)$, the *Eulerian polynomials*, and proved that $A_n(t)$ has only simple negative real roots. Frobenius also showed that these polynomials could be written in terms of the Stirling numbers of the second kind, $S(n,k)$, by the following expression:

$$A_n(t) = \sum_{k=1}^{n} k! \cdot S(n,k) \cdot (t-1)^{n-k} \tag{20.7}$$

The most simple way to define all of these polynomials, and distinguish between them, is in terms of their generating function.

The *Euler–Frobenius* polynomials are defined by the following exponential generating function:

$$\sum_{n=0}^{\infty} A_n(t) \cdot \frac{x^n}{n!} = \frac{t-1}{t-e^{(t-1)x}} \tag{20.8}$$

The first few polynomials are given by:

$$A_0(t) = 1 \tag{20.9}$$

$$A_1(t) = 1 \tag{20.10}$$

$$A_2(t) = t + 1 \tag{20.11}$$

$$A_3(t) = t^2 + 4t + 1 \tag{20.12}$$

$$A_4(t) = t^3 + 11t^2 + 11t + 1 \tag{20.13}$$

The polynomials $A_n(t)$ are palindromic, with simple negative real roots. Being palindromic polynomials, the following identity holds:

$$A_n(t) = t^{n-1} \cdot A_n\left(t^{-1}\right) \tag{20.14}$$

The *Euler–Frobenius* polynomials can be computed by the recurrence:

$$A_0(t) = 1 \tag{20.15}$$

$$A_n(t) = t(1-t) \cdot A'_{n-1}(t) + \big(1 + (n-1)t\big) \cdot A_{n-1}(t) \tag{20.16}$$

or they can be defined inductively by

$$A_0(t) = 1 \tag{20.17}$$

$$A_n(t) = \sum_{k=0}^{n-1} \binom{n}{k} A_k(t) \cdot (t-1)^{n-k-1} \tag{20.18}$$

or, considering the notation $A^n = A_n(t)$, by

$$\big(A + (t-1)\big)^n - tA_n(t) = \begin{cases} (1-t), & n = 0 \\ 0, & n > 0 \end{cases} \tag{20.19}$$

They can also be computed explicitly by

$$A_o(t) = 1 \tag{20.20}$$

$$A_n(t) = \frac{(1-t)^{n+1}}{t} \left(t \frac{d}{dt}\right)^{n-1} \frac{t}{(1-t)^2}, \quad n > 0 \tag{20.21}$$

Note that, to be consistent with the numbering of the polynomials, the last definition is shifted when compared to the original expression given in Sobolev (1977).

20.2 Euler–Frobenius Numbers

The polynomials $A_n(t)$ can also be written as

$$A_n(t) = \sum_{k=0}^{n} A_{n,k} \cdot t^k \tag{20.22}$$

where $A_{n,k}$ are the *Euler–Frobenius* numbers, most commonly known as *Eulerian* numbers.

Table 20.1 Eulerian numbers

	$k=0$	1	2	3	4	5
$n=0$	1					
1	1	0				
2	1	1	0			
3	1	4	1	0		
4	1	11	11	1	0	
5	1	26	66	26	1	0

These numbers are defined by (see Foata 2008 and Graham et al. 1994)

$$A_{n,k} = \sum_{i=0}^{k} \binom{n+1}{i} (k-i+1)^n (-1)^i \tag{20.23}$$

The *Eulerian* numbers can also be obtained by the following recurrence relation:

$$A_{n,k} = (k+1) \cdot A_{n-1,k} + (n-k) \cdot A_{n-1,k-1} \tag{20.24}$$

$$A_{n,0} = 1 \tag{20.25}$$

$$A_{n,k} = 0 \tag{20.26}$$

Table 20.1 illustrates the first few numbers.

20.3 Euler–Frobenius Fractions

The name *Euler–Frobenius fractions* is less common in the literature, but is kept here for clarity of exposition. They can be found under the name of *Eulerian polynomials* in the work of Carlitz and *Euler–Frobenius numbers* in the work of Kim. They were first defined by Euler in his treaty *Institutiones Calculi Differentialis, Sect. 173, Chap. 7, Part II.*

The *Euler–Frobenius fractions* $H_n(t)$ are defined by the following exponential generating function:

$$\sum_{n=0}^{\infty} H_n(t) \cdot \frac{x^n}{n!} = \frac{1-t}{e^x - t} \tag{20.27}$$

The first few fractions are given by:

$$H_0(t) = 1 \tag{20.28}$$

$$H_1(t) = \frac{1}{t-1} \tag{20.29}$$

$$H_2(t) = \frac{t+1}{(t-1)^2} \tag{20.30}$$

$$H_3(t) = \frac{t^2+4t+1}{(t-1)^3} \tag{20.31}$$

$$H_4(t) = \frac{t^3+11t^2+11t+1}{(t-1)^4} \tag{20.32}$$

The *Euler–Frobenius fractions* can be generated inductively, considering the notation $H^n = H_n(t)$, by

$$(H+1)^n - t \cdot H_n(t) = 0, \qquad H_0(t) = 1 \tag{20.33}$$

and possess the property that

$$\frac{dH_n(t)}{dt} = n \cdot H_{n-1}(t) \tag{20.34}$$

which implies that the polynomials $H_n(t)$ form an Appell sequence.

The connection between *Euler–Frobenius* polynomials and fractions is straightforward:

$$A_n(t) = (t-1)^n \cdot H_n(t) \tag{20.35}$$

Finally, since the polynomials $A_n(t)$ are palindromic, the following identity can be established:

$$H_n\left(t^{-1}\right) = (-1)^n \cdot H_n(t) \tag{20.36}$$

20.4 Combinatorial Interpretation of Eulerian Numbers

The Eulerian numbers $A_{n,k}$ have a special property relating to combinatorial analysis (see Comtet 1974; Graham et al. 1994; and Olver et al. 2010, Sect. 26.14).

A permutation of a set $\mathcal{N}$ is a bijective map of $\mathcal{N}$ onto itself. The set of all permutations of $\mathcal{N}$ is denoted by $\mathfrak{S}(\mathcal{N})$. If $\mathcal{N} = \{1, \dots, n\}$, a permutation $\sigma = \sigma_1\sigma_2\cdots\sigma_n \in \mathfrak{S}(\mathcal{N})$ can be thought of as an ordered selection or arrangement of these integers where the integer in position j is σ_j, e.g., if $\mathcal{N} = \{1,2,3\}$ then 231 is the permutation $\sigma_1 = 2$, $\sigma_2 = 3$, $\sigma_3 = 1$.

The following are a subset of properties defined for any permutation:

- An *ascent* of a permutation is a pair of adjacent elements for which $\sigma_j < \sigma_{j+1}$.

- A *descent* of a permutation is a pair of adjacent elements for which $\sigma_j > \sigma_{j+1}$.
- An *excedance* in a permutation is a position j for which $\sigma_j > j$.
- A *weak excedance* in a permutation is a position j for which $\sigma_j \geq j$.

The Eulerian number $A_{n,k}$ then represents:

- the number of permutations $\sigma \in \mathfrak{S}(\mathcal{N})$ that have k ascents (descents).
- the number of permutations $\sigma \in \mathfrak{S}(\mathcal{N})$ with exactly k excedances.
- the number of permutations $\sigma \in \mathfrak{S}(\mathcal{N})$ with exactly $k+1$ weak excedances.

The total number of permutations of a set with n elements is well known to be $n!$. In the same way, the sum of all the permutations that have up to n ascents is the total number of permutations. Therefore, the *Eulerian* polynomials and numbers satisfy the following identity:

$$A_n(1) = \sum_{k=0}^{n} A_{n,k} = n! \tag{20.37}$$

20.5 Euler and Bernoulli Polynomials

The *Euler* polynomials are defined by the following exponential generating function:

$$\sum_{k=0}^{\infty} E_n(t) \cdot \frac{x^k}{k!} = \frac{2e^{tx}}{e^x + 1} \tag{20.38}$$

where the first few polynomials are given by:

$$E_0(t) = 1 \tag{20.39}$$

$$E_1(t) = t - \frac{1}{2} \tag{20.40}$$

$$E_2(t) = t^2 - t \tag{20.41}$$

$$E_3(t) = t^3 - \frac{3}{2}t^2 + \frac{1}{4} \tag{20.42}$$

$$E_4(t) = t^4 - 2t^3 + t \tag{20.43}$$

The Euler polynomials have only one root in the interval $t \in (0, 1)$.

The *Euler* numbers E_n are defined in terms of the *Euler* polynomials $E_n(t)$ by evaluating $t = 1/2$ and normalising by 2^n, i.e.,

$$E_n = 2^n \cdot E_n\left(\frac{1}{2}\right) \tag{20.44}$$

The first few *Euler* numbers are:

$$E_0 = 1 \tag{20.45}$$

$$E_1 = 0 \tag{20.46}$$

$$E_2 = -1 \tag{20.47}$$

$$E_3 = 0 \tag{20.48}$$

$$E_4 = 5 \tag{20.49}$$

These are also the coefficients of the Taylor–Maclaurin series

$$\frac{1}{\cosh(t)} = \frac{2e^x}{e^{2x}+1} = \sum_{k=0}^{\infty} E_n \cdot \frac{x^k}{k!} \tag{20.50}$$

On the other hand, the *Bernoulli* polynomials are defined by the following exponential generating function:

$$\sum_{k=0}^{\infty} B_n(t) \cdot \frac{x^k}{k!} = \frac{x \cdot e^{tx}}{e^x - 1} \tag{20.51}$$

The first few *Bernoulli* polynomials are:

$$B_0(t) = 1 \tag{20.52}$$

$$B_1(t) = t - \frac{1}{2} \tag{20.53}$$

$$B_2(t) = t^2 - t + \frac{1}{6} \tag{20.54}$$

$$B_3(t) = t^3 - \frac{3}{2}t^2 + \frac{1}{2}t \tag{20.55}$$

$$B_4(t) = t^4 - 3t^3 + t^2 - \frac{1}{30} \tag{20.56}$$

The well-known *Bernoulli* numbers are given by

$$B_n = B_n(0) \tag{20.57}$$

Finally, it is important to note that the *Euler* and *Bernoulli* numbers satisfy

$$E_{2n+1} = B_{2n+1} = 0, \quad n = 1, 2, \ldots \tag{20.58}$$

In addition, similar to the *Euler–Frobenius* fractions, both *Euler* and *Bernoulli* polynomials form an Appell sequence, i.e., they satisfy the identities

$$\frac{dE_n(t)}{dt} = n \cdot E_{n-1}(t) \tag{20.59}$$

$$\frac{dB_n(t)}{dt} = n \cdot B_{n-1}(t) \tag{20.60}$$

There exists a vast literature on many other properties and relations that apply between *Euler* and *Bernoulli* polynomials and numbers. However, only basic definitions and properties will suffice for the purpose of this chapter. We ask the reader to see the references for further details.

20.6 Generalised Eulerian Polynomials

Carlitz introduced a generalisation of the *Euler–Frobenius* fractions that he named *Eulerian polynomials*. The generalisation is defined by

$$\sum_{k=0}^{\infty} H_n(u,t) \cdot \frac{x^k}{k!} = \frac{(1-t)e^{ux}}{e^x - t} \tag{20.61}$$

These polynomials can be computed inductively by means of the following identity:

$$H_n(u,t) = \sum_{k=0}^{n} \binom{n}{k} H_k(t) \cdot u^{n-k} \tag{20.62}$$

where $H_k(t)$ are the *Euler–Frobenius* fractions previously defined. The first few polynomials are given by:

$$H_0(u,t) = 1 \tag{20.63}$$

$$H_1(u,t) = u + \frac{1}{t-1} \cdot u^0 \tag{20.64}$$

$$H_2(u,t) = u^2 + \frac{1}{t-1} \cdot 2u + \frac{t+1}{(t-1)^2} \cdot u^0 \tag{20.65}$$

$$H_3(u,t) = u^3 + \frac{1}{t-1} \cdot 3u^2 + \frac{t+1}{(t-1)^2} \cdot 3u + \frac{t^2+4t+1}{(t-1)^3} \cdot u^0 \tag{20.66}$$

It is then straightforward to see that

$$H_n(0,t) = H_n(t) \tag{20.67}$$

$$(t-1)^n \cdot H_n(0,t) = A_n(t) \tag{20.68}$$

On the other hand, by simply comparing their respective generating functions, the polynomials defined by Carlitz have the following property:

$$H_n(u,-1) = E_n(u) \tag{20.69}$$

where $E_n(u)$ are the Euler polynomials previously defined. The *Eulerian* polynomials also satisfy

$$\frac{\partial H_n(u,t)}{\partial u} = n \cdot H_{n-1}(u,t) \tag{20.70}$$

Carlitz proved several multiplication and summation properties of the *Eulerian* polynomials, e.g., the following summation identity holds:

$$\sum_{k=0}^{n} \binom{n}{k} H_k(u,t) \cdot H_{n-k}(v,-t) = H_n\left(u+v,t^2\right) \tag{20.71}$$

In addition, the following multiplication formula holds provided $t \neq 1$, $z \neq 1$, $tz \neq 1$:

$$\begin{aligned} H_m(u,t)H_n(u,z) &= H_{m+n}(u,tz) \\ &\quad + \frac{t(1-z)}{1-tz} \sum_{r=1}^{m} \binom{m}{r} H_r(t) H_{m+n-r}(u,tz) \\ &\quad + \frac{z(1-t)}{1-tz} \sum_{s=1}^{n} \binom{n}{s} H_s(z) H_{m+n-s}(u,tz) \end{aligned} \tag{20.72}$$

On the other hand, when $tz = 1$, $t \neq 1$, the following relationship holds:

$$H_m(u,t) \cdot H_n\left(u,t^{-1}\right) = (t-1) \cdot \sum_{r=1}^{m} \binom{m}{r} H_r(t) \frac{B_{m+n-r+1}(t)}{m+n-r+1}$$

$$+\left(t^{-1}-1\right)\cdot\sum_{s=1}^{n}\binom{n}{s}H_s\left(t^{-1}\right)\frac{B_{m+n-s+1}(t)}{m+n-s+1}$$

$$+(-1)^{n+1}\frac{m!n!}{(m+n+1)!}(1-t)H_{m+n+1}(t) \qquad (20.73)$$

Several other identities are provided in Carlitz (1963).

20.7 Summary

This chapter has given a thorough presentation of the *Euler–Frobenius* polynomials. The definitions and properties presented here should allow the reader to easily identify the different types of numbers, polynomials, and fractions present in the literature.

Further Reading

Material on generating functions can be found in

Wilf HS (2005) Generatingfunctionology, 3rd edn. CRC Press, Boca Raton

Material on combinatorial analysis and Euler, Bernoulli, and Eulerian polynomials and numbers can be found in

Comtet L (1974) Advanced combinatorics: the art of finite and infinite expansions. Reidel, Dordrecht

Graham RL, Knuth DE, Patashnik O (1994) Concrete mathematics, 2nd edn. Addison-Wesley, Reading

Olver FWJ, Lozier DW, Boisvert RF, Clark CW (eds) (2010) NIST handbook of mathematical functions. Cambridge University Press, New York. Print companion to http://dlmf.nist.gov/

Riordan J (1958) Introduction to combinatorial analysis. Wiley, New York

Euler's work and a glimpse into his life can be found in

Euler L (1755) Institutiones calculi differentialis cum eius usu in analysi finitorum ac doctrina serierum. Available online: http://eulerarchive.maa.org/pages/E212.html

Euler L (1768) Remarques sur un beau rapport entre les series des puissances tant directes que reciproques. Mem Acad Sci Berl 17:83–106. Available online: http://eulerarchive.maa.org/pages/E352.html

Gautschi W (2008) Leonhard Euler: his life, the man, and his works. SIAM Rev 50(1):3–33

Extensive material on *Eulerian/Euler–Frobenius* polynomials can be found in

Carlitz L (1959) Eulerian numbers and polynomials. Math Mag 32:247–260
Carlitz L (1963) The product of two Eulerian polynomials. Math Mag 36:37–41
Dubeau F, Savoie J (1995) On the roots of orthogonal polynomials and Euler–Frobenius polynomials. J Math Anal Appl 196(1):84–98
Foata D (2008) Eulerian polynomials: from Euler's time to the present. Invited address at the 10th Annual Ulam Colloquium, University of Florida, February 18, 2008
Frobenius FG (1910) Über die Bernoullischen Zahlen und die Eulerschen Polynome. Sitzungsber Kon Preuss Akad Wiss Berl, Phys Math Kl 1910:809–847
Janson S (2013) Euler–Frobenius numbers and rounding. arXiv:1305.3512
He TX (2012) Eulerian polynomials and B-splines. J Comput Appl Math 236(15):3763–3773
Kim T (2012) Identities involving Frobenius–Euler polynomials arising from non-linear differential equations. J Number Theory 132(12):2854–2865
Kim DS, Kim T (2012) Some new identities of Frobenius–Euler numbers and polynomials. J Inequal Appl 2012:307. doi:10.1186/1029-242X-2012-307
Luschny P. Eulerian polynomials. Available online: http://www.luschny.de/math/euler/EulerianPolynomials.html
Sobolev SL (1977) On the roots of Euler polynomials. Dokl Akad Nauk SSSR 235:935–938 [Sov Math Dokl 18:935–938]
Reimer M (1985) The main roots of the Euler–Frobenius polynomials. J Approx Theory 45:358–362
Weller SR, Moran W, Ninness B, Pollington AD (2001) Sampling zeros and the Euler–Frobenius polynomials. IEEE Trans Autom Control 46(2):340–343
Weisstein EW. Euler polynomial. From MathWorld—a Wolfram web resource. http://mathworld.wolfram.com/EulerPolynomial.html

Chapter 21
Models for Intersample Response

This book has principally focused on the problem of describing the *at-sample* response of a sampled-data system. However, in some applications, it is also important to be able to model and understand the response that occurs between samples. To do this, it is actually first necessary to be able to model the at-sample response. Thus, the ideas presented here build on earlier results. Two views of the intersample response will be presented: in the frequency domain and in the time domain.

21.1 Frequency Domain

For the case of linear systems, one can relate the frequency response of the discrete-time signal and the continuous-time signal prior to sampling.

To illustrate, consider the signals in Fig. 21.1.

The relationship among different signals is given in the results below.

Lemma 21.1 *The continuous-time and discrete-time signals in Fig.* 21.1 *satisfy the identity*

$$\Phi(j\omega) = \frac{1}{\Delta} H(j\omega) U_d\bigl(e^{j\omega\Delta}\bigr), \tag{21.1}$$

where $H(s) = \frac{1-e^{-s\Delta}}{s}$ *is the transfer function of a zero order hold, and* $\Phi(j\omega) = \int_{-\infty}^{\infty} \phi(t) e^{-j\omega t}\, dt$ *and* $U_d(e^{j\omega\Delta}) = \Delta \sum_{k=-\infty}^{\infty} u_k e^{-j\omega k \Delta}$ *are the continuous-time and discrete-time Fourier transforms of* $\phi(t)$ *and* u_k*, respectively.*

Proof The result follows immediately from the definition of $H(j\omega)$ and $U_d(e^{j\omega})$. □

J.I. Yuz, G.C. Goodwin, *Sampled-Data Models for Linear and Nonlinear Systems*, Communications and Control Engineering, DOI 10.1007/978-1-4471-5562-1_21,

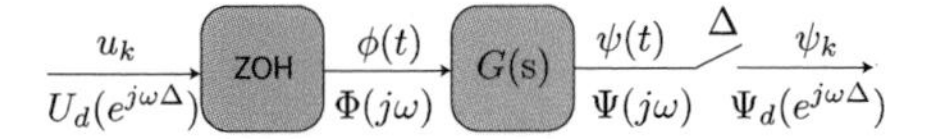

Fig. 21.1 Signals involved in a sampling process: ZOH represents a zero order hold, $F(s)$ is any linear system, and Δ is the sampling time

Note that $\Phi(j\omega)$ is the product of a non-periodic function $H(j\omega)$ and a periodic function $U_d(e^{j\omega\Delta})$.

Lemma 21.2 *The Fourier transform of the discrete-time signals in Fig.* 21.1 *satisfies*

$$\Psi_d(e^{j\omega\Delta}) = [GH]_q U_d(e^{j\omega\Delta}), \tag{21.2}$$

where $\Psi_d(e^{j\omega\Delta})$ *is the (scaled) discrete-time Fourier transform given by* $\Psi_d(e^{j\omega\Delta}) = \Delta \sum_{-\infty}^{\infty} \psi_k e^{-j\omega k \Delta}$ *and* $[GH]_q$ *denotes the folded frequency response of* GH, *that is,*

$$[GH]_q = \frac{1}{\Delta} \sum_{\ell=-\infty}^{\infty} G(j\omega + j2\pi\ell/\Delta) H(j\omega + j2\pi\ell/\Delta) \tag{21.3}$$

Proof Fact 21.2 follows from the observation that the discrete transfer function is the folded version of GH. □

Lemma 21.3 *The continuous-time and discrete-time signals in Fig.* 21.1 *satisfy*

$$\Psi_d(e^{j\omega\Delta}) = \big[G(j\omega)\Phi(j\omega)\big]_q \tag{21.4}$$

where $[G\Phi]_q$ *denotes the folded form of* $G\Phi$, *that is,*

$$[G\Phi]_q = \frac{1}{\Delta} \sum_{\ell=-\infty}^{\infty} G(j\omega + j2\pi\ell/\Delta) \Phi(j\omega + j2\pi\ell/\Delta) \tag{21.5}$$

Proof The result follows from the usual folding produced by sampling. □

The preceding results have important implications in the digital control of continuous-time processes. For example, it follows that:

$$\frac{\Psi_d(e^{j\omega})}{\Psi(j\omega)} = \frac{[GH]_q}{\frac{1}{\Delta}(GH)} = \frac{\frac{1}{\Delta}\sum_{\ell=-\infty}^{\infty} G(j(\omega + \frac{2\pi}{\Delta}\ell))H(j(\omega + \frac{2\pi}{\Delta}\ell))}{\frac{1}{\Delta}G(j\omega)H(j\omega)} \tag{21.6}$$

As described in detail elsewhere in the book, folding (as evident in the numerator of the above expression) leads to sampling zeros. For example, if the relative degree of G is even, then there is an asymptotic sampling zero at $\omega = \frac{\pi}{\Delta}$. In this case, the ratio of $\Psi(j\omega)$ (the continuous-time frequency response) and $\Psi_d(e^{j\omega\Delta})$ (the discrete-time frequency response) can become very large near the Nyquist frequency. Hence, the continuous-time and discrete-time signals will be very different.

For this reason it is generally never appropriate to have significant content in $\Psi_d(e^{j\omega\Delta})$ near the Nyquist frequency, as doing so leads to a very large (approaching ∞ as $\Delta \to 0$) component in the continuous-time response at this frequency.

Two illustrations of this idea are presented below.

Example 21.4 Consider a continuous-time plant with transfer function

$$G(s) = \frac{1}{s(s+1)} \tag{21.7}$$

The sampling period is chosen as $\Delta = 0.1$.

It is desired that the closed-loop response time be one sample period. Recall the discussion in Sect. 10.1 of Chap. 10 which suggests that one cannot ignore the sampling zeros in this case. The exact sampled-data model in shift operator form is

$$G_q(z) = 0.0048\frac{z+0.967}{(z-1)(z-0.905)} \tag{21.8}$$

When the following feedback controller is used, then the discrete-time response reaches the setpoint in one sample period:

$$C_q(z) = 208.33\frac{z-0.905}{z+0.967} \tag{21.9}$$

The sampled response of the closed loop for a step reference signal is shown in Fig. 21.2.

The corresponding continuous-time response is also shown in Fig. 21.2. Note the large intersample response. Indeed, the continuous-time response has near infinite energy at $\omega = \frac{\pi}{\Delta}$, as predicted above.

Example 21.5 (Repetitive (or Iterative Learning) Control) The basic idea of this controller is to track a periodic discrete-time reference signal. However, the design bandwidth is then comparable to the sample period. Two consequences are that (i) sampling zero dynamics cannot be ignored, and (ii) there is the strong possibility

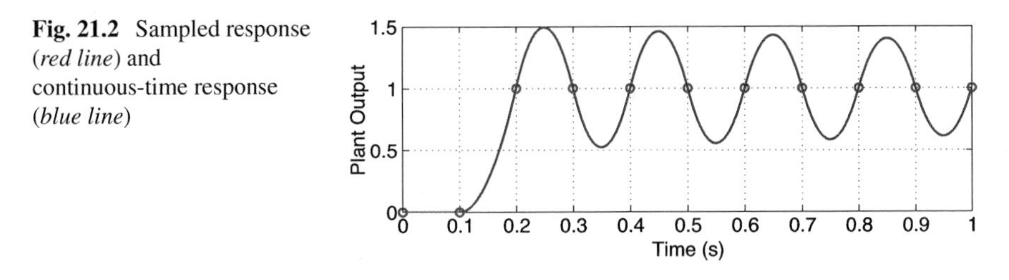

Fig. 21.2 Sampled response (*red line*) and continuous-time response (*blue line*)

of large intersample behaviour. One option to reduce the undesirable intersample behaviour is to filter the control action so that the tracking bandwidth is made small relative to the Nyquist frequency.

21.2 Time Domain

It is also possible to capture the intersample response in the time domain using lifting ideas. The key idea is to utilise a form of series-to-parallel conversion. To illustrate, say that the final goal is to utilise a sampling period of Δ. Then, it is convenient to employ an up-sampling period $\Delta' = \frac{\Delta}{m}$, where m is an integer ≥ 1. To illustrate, consider the following discrete-time deterministic system with model appropriate to period Δ':

$$x_{k+1} = A'_q x_k + B'_q u_k \tag{21.10}$$

This is a standard sampled-data model and will necessarily include sampling zero dynamics when relevant.

Then, over the period Δ,

$$\begin{bmatrix} x_{k+1} \\ x_{k+2} \\ \vdots \\ x_{k+m} \end{bmatrix} = \begin{bmatrix} 0 & \cdots & 0 & A'_q \\ & & & \\ 0 & \cdots & 0 & (A'_q)^m \end{bmatrix} \begin{bmatrix} x_{k-m+1} \\ \\ x_{k-1} \\ x_k \end{bmatrix} + \begin{bmatrix} B'_q & 0 & \cdots & 0 \\ A'_q B'_q & B'_q & \ddots & \vdots \\ & & \ddots & 0 \\ (A'_q)^{m-1} B'_q & & & B'_q \end{bmatrix} \begin{bmatrix} u_k \\ u_{k+1} \\ \vdots \\ u_{k+m-1} \end{bmatrix} \tag{21.11}$$

It is then possible to constrain the relationship between the inputs. For example, a ZOH has the property

$$u_k = u_{k+1} = \cdots = u_{k+m-1} \tag{21.12}$$

Substituting into (21.11) leads to the following state-space system:

$$x_{(k+1)m} = \left(A'_q\right)^m x_{km} + \left[B'_q + A'_q B'_q \cdots + \left(A'_q\right)^{m-1} B'_q\right] u_{km} \tag{21.13}$$

from which the other intersample values of the state are given by the output equation

$$\begin{bmatrix} x_{km} \\ x_{km+1} \\ \vdots \\ x_{km+m-1} \end{bmatrix} = \begin{bmatrix} I_n \\ A'_q \\ \vdots \\ (A'_q)^{m-1} \end{bmatrix} x_{km} + \begin{bmatrix} 0 \\ B'_q \\ \vdots \\ B'_q + A'_q B'_q + \cdots + (A'_q)^{m-2} B'_q \end{bmatrix} u_{km} \tag{21.14}$$

One can then carry out any required design ($\mathcal{H}_2, \mathcal{H}_\infty, \ldots$) in a hybrid form with sampled control and deploying a cost function that reflects the intersample response. Similar ideas can be applied to optimal filtering. Indeed, the hybrid optimal filtering problem and the hybrid regulator problem are dual to each other with the anti-aliasing filter in the hybrid filter being dual to the hold circuit in the hybrid regulator (see the references at the end of the chapter).

21.3 Summary

The key ideas in this chapter are as follows:

- There exist simple ways of describing the frequency content of the intersample response in sampled-data systems.
- Sampling zeros have a major impact on these relationships.
- In the time domain the intersample response can be modelled using lifting techniques.

Further Reading

Intersample response can be modelled using modified Z-transforms

Jury EI (1958) Sampled-data control systems. Wiley, New York

The frequency-domain ideas described in Sect. 21.1 are explained in greater detail in

Goodwin GC, Graebe SF, Salgado ME (2001a) Control system design. Prentice Hall, Upper Saddle River, Chap. 4

Further insights into repetitive control are given in

Goodwin GC, Graebe SF, Salgado ME (2001b) Control system design. Prentice Hall, Upper Saddle River, p 375

Intersample issues associated with generalised holds are described in

Feuer A, Goodwin GC (1994) Generalized sample hold functions—frequency-domain analysis of robustness, sensitivity, and intersample difficulties. IEEE Trans Autom Control 39(5):1042–1047
Feuer A, Goodwin GC (1996) Sampling in digital signal processing and control. Birkhäuser, Boston

More information on lifting can be found in

Chen T, Francis BA (1995) Optimal sampled-data control systems. Springer, London

Duality between hybrid optimal filtering and the hybrid regulator problem is described in

Goodwin GC, Mayne DQ, Feuer A (1995) Duality of hybrid optimal regulator and hybrid optimal filter. Int J Control 61(6):1465–1471

Chapter 22
Approximate Sampled-Data Models for Fractional Order Systems

This chapter extends the ideas presented in Chaps. 5 and 8 to systems having fractional order dynamics. Specifically, fractional order Euler–Frobenius polynomials are defined and used to characterise the asymptotic sampling zeros for fractional systems as the sampling period tends to zero. Note that this chapter involves extra technicalities and is thus aimed at readers who have a specific interest in this class of problems.

22.1 Historical Perspective

Fractional calculus originated in the same era as traditional calculus. However, fractional order systems (FOSs) have only recently attracted the attention of the engineering community with an increasing number of applications (see the references given at the end of the chapter).

This chapter considers the consequences of the sampling process for a FOS. Different approaches have been proposed in the literature to obtain approximate sampled models for this kind of system: e.g., (i) analytical computation of the output (which is, in general, hard to obtain explicitly), (ii) direct approaches where the fractional differentiator is replaced by a discrete-time operator and then the resultant model is truncated to a rational discrete-time approximation, and (iii) indirect approaches where the FOS is approximated by a rational continuous-time model which is then discretised (e.g., via the CRONE approach).

This chapter adopts the philosophy of earlier chapters and focuses on the specific problem of characterising the asymptotic sampling zeros for the FOS.

The following notation is used: $\lceil a \rceil$ denotes the smallest positive integer greater than a, $\lfloor a \rfloor$ denotes the largest integer smaller than a. The fractional part of a is defined as $\{a\} = a - \lfloor a \rfloor$.

The next section reviews relevant aspects of fractional calculus.

J.I. Yuz, G.C. Goodwin, *Sampled-Data Models for Linear and Nonlinear Systems*, Communications and Control Engineering, DOI 10.1007/978-1-4471-5562-1_22,

22.2 Fractional Calculus Background

The definition of the Mittag–Leffler function is first presented. For a FOS this function plays a similar role to that played by the exponential function in the analysis of differential equations of integer order.

The Mittag–Leffler function of two parameters is defined by the series expansion

$$E_{\alpha,\beta}(t) = \sum_{k=0}^{\infty} \frac{t^k}{\Gamma(\alpha k + \beta)} \tag{22.1}$$

where $\Gamma(\cdot)$ is Euler's Gamma function defined by

$$\Gamma(\nu) = \int_0^{\infty} e^u u^{\nu-1}\, du \tag{22.2}$$

The Laplace transform of the Mittag–Leffler function satisfies the following equation (see the references at the end of the chapter):

$$\int_0^{\infty} e^{-st} t^{\alpha k+\beta-1} E_{\alpha,\beta}^{(k)}\left(a t^{\alpha}\right) dt = \frac{k! s^{\alpha-\beta}}{(s^{\alpha} - a)^{k+1}} \tag{22.3}$$

where $E_{\alpha,\beta}^{(k)}(z)$ is the kth derivative of the Mittag–Leffler function.

The Riemann–Liouville approach is adopted for fractional derivatives. Within this approach, the fractional derivative of order $\nu > 0$ is defined as

$${}_a\mathcal{D}_t^{\nu} f(t) = \frac{1}{\Gamma(1 - \{\nu\})} \left(\frac{d}{dt}\right)^{\lceil \nu \rceil} \int_a^t (t-\tau)^{-\{\nu\}} f(\tau)\, d\tau \tag{22.4}$$

When $\nu \in [0, 1)$, the fractional derivative becomes

$${}_a\mathcal{D}_t^{\nu} f(z) = \frac{1}{\Gamma(1-\nu)} \frac{d}{dt} \int_a^t (t-\tau)^{-\nu} f(\tau)\, d\tau \tag{22.5}$$

The Laplace transform of the Riemann–Liouville operator applied to a function $f(t)$ is

$$\mathcal{L}\{{}_a\mathcal{D}_t^\alpha f(t)\} = s^\alpha F(s) - \sum_{k=0}^{\lceil\alpha\rceil-1} s^k \left[{}_a\mathcal{D}_t^{\alpha-k-1} f(t)\right]_{t=a} \tag{22.6}$$

where $F(s)$ is the Laplace transform of $f(t)$.

As an alternative, the fractional derivative can also be defined as a limit. This is known as the Grunwald–Letnikov differointegral operator:

$$_aD_t^\alpha f(t) = \lim_{h\to 0} \frac{1}{h^\alpha} \sum_{k=0}^{\infty} (-1)^k \binom{\alpha}{k} f(t-kh) \tag{22.7}$$

Note that, under regularity conditions on $f(t)$, both the Riemann–Liouville and Grunwald–Letnikov approaches coincide. The Grunwald–Letnikov representation will be used in the sequel to approximate fractional differential operators in the discrete domain.

22.3 Sampling Zeros for Fractional Order Systems

In the remainder of the chapter, a FOS is considered which is described by the following fractional order differential equation:

$$a_{n0}\mathcal{D}_t^{\nu_{a_n}} y(t) + \cdots + a_{00}\mathcal{D}_t^{\nu_{a_0}} y(t) = b_{m0}\mathcal{D}_t^{\nu_{b_m}} u(t) + \cdots + b_{00}\mathcal{D}_t^{\nu_{b_0}} u(t) \tag{22.8}$$

where $y(t)$, $u(t)$ are the system output and input, respectively, and where $\nu_{a_n} = \max_{i=0\cdots n}\{\nu_{a_i}\}$, $\nu_{a_n} > \max_{i=0\cdots m}\{\nu_{b_i}\}$. The system is assumed to have a commensurate order ν; i.e., all the exponents $\{\nu_{a_k}\}$ and $\{\nu_{b_k}\}$ are integer multiples of $\nu \in \mathbb{R}$. Utilising (22.6), the Laplace transform of the output of the system (22.8) can be expressed in transfer function form, i.e., $Y(s) = G(s)U(s)$, where

$$\begin{aligned} G(s) &= \frac{b_m s^{\nu_{bm}} + \cdots + b_0 s^{\nu_{b0}}}{a_n s^{\nu_{a0}} + \cdots + a_0 s^{\nu_{a0}}} \\ &= \frac{b_m (s^\nu)^{n_{bm}} + \cdots + b_0 (s^\nu)^{n_{b0}}}{a_n (s^\nu)^{n_{an}} + \cdots + a_0 (s^\nu)^{n_{a0}}} \end{aligned} \tag{22.9}$$

and where $U(s)$ is the Laplace transform of the input $u(t)$. Note that the system relative degree is $\alpha = \nu_{a_n} - \nu_{b_m} = (n_{a_n} - n_{b_m})\nu$ and the initial conditions are assumed to be zero (see references at the end of the chapter for more details on initial conditions of the FOS).

The exact sampled-data model for a FOS can be obtained by applying (3.27), as in the case of integer order systems. However, even though (3.27) applies for both integer and fractional time derivatives, a difference arises regarding how the integration path is closed. For the fractional case, (3.27) can be solved by residues closing the integral path in the right half-plane. This leads to the usual aliasing formula (3.40). On the other hand, if the integral path is closed in the complex left half-plane, then the integral (3.27) has branch points which do not lead to an analytical representation as in (3.30). This is a key departure from the integer order case.

The result presented below depends upon the observation that, at fast sampling rates, any FOS can be approximated by a *fractional order* integrator. This extends the result presented in Chap. 5 for the integer order case.

Theorem 22.1 *Let $G(s)$ be a linear system having fractional relative degree α. When the sampling period tends to zero, the discrete system behaves as a fractional order integrator of order α, i.e.,*

$$G_d(z) \xrightarrow{\Delta\to 0} \mathcal{ZOH}\left\{\frac{1}{s^\alpha}\right\} = (1 - z^{-1})\frac{\Delta^\alpha}{\Gamma(\alpha+1)}\sum_{k=1}^{\infty} k^\alpha z^{-k} \tag{22.10}$$

Proof The proof follows by applying a change of variables $\eta = s\Delta$, i.e.,

$$G_q(z) = \frac{(z-1)}{z}\int_{\gamma\Delta - j\infty}^{\gamma\Delta + j\infty} \frac{e^\eta}{z - e^\eta} G\left(\frac{\eta}{\Delta}\right)\frac{d\eta}{\eta} \tag{22.11}$$

where, using (22.9),

$$G\left(\frac{\eta}{\Delta}\right) = \frac{b_m}{a_n}\left(\frac{\Delta}{\eta}\right)^\alpha \left(\frac{1 + \cdots + \frac{b_0}{b_m}(\frac{\Delta}{\eta})^{(n_{b_m} - n_{b_0})\nu}}{1 + \cdots + \frac{a_0}{a_n}(\frac{\Delta}{\eta})^{(n_{a_n} - n_{a_0})\nu}}\right) \tag{22.12}$$

Hence, letting the sampling period tend to zero, it follows that

$$\lim_{\Delta\to 0} \Delta^{-\alpha} G_q(z) = \frac{(z-1)}{z}\int_{\gamma\Delta - j\infty}^{\gamma\Delta + j\infty} \frac{e^\eta}{z - e^\eta}\frac{b_m}{a_n}\frac{1}{\eta^\alpha}\frac{d\eta}{\eta} \tag{22.13}$$

which corresponds to the sampled model of an α-order integrator, where $\alpha = (n_{a_n} - n_{b_m})\nu$ is the system relative degree. □

Theorem 22.1 motivates the approximation of a system of (fractional) relative degree α by an α-order integrator for fast sampling rates. Then, the Riemann–Liouville fractional derivative can be used to obtain the step response of $1/s^\alpha$ which is $t^\alpha/\Gamma(\alpha+1)$. However, a problem associated with the result in Theorem 5.1 for a FOS is that the asymptotic discrete-time system cannot be represented as a rational transfer function. To deal with this problem, an approximation to the result in Theorem 5.1 is developed. The definition of the Euler–Frobenius polynomials in (5.19)–(5.22) is first extended to the fractional order case.

Definition 22.2 The fractional Euler–Frobenius polynomials of order α ($\in \mathbb{R}^+$) are defined as

$$B_\alpha(z^{-1}) = b_0^\alpha + \cdots + b_{\lfloor\alpha\rfloor}^\alpha z^{-\lfloor\alpha\rfloor} \tag{22.14}$$

where the coefficients b_ℓ^α are given by

$$b_\ell^\alpha = (-1)^\ell \sum_{i=1}^{\ell+1} \binom{\alpha+1}{\ell+1-i} \binom{-\{\alpha\}-1}{i-1} i^{\lfloor\alpha\rfloor} \tag{22.15}$$

Remark 22.3 The fractional Euler–Frobenius polynomials for various α's are as follows:

$$B_\alpha(z^{-1}) = 1 \quad (\alpha \in [0,1)) \tag{22.16}$$

$$B_\alpha(z^{-1}) = 1 + \{\alpha\}z^{-1} \quad (\alpha \in [1,2)) \tag{22.17}$$

$$B_\alpha(z^{-1}) = 1 + (3\{\alpha\}+1)z^{-1} + \{\alpha\}^2 z^{-2} \quad (\alpha \in [2,3)) \tag{22.18}$$

$$\begin{aligned} B_\alpha(z^{-1}) = {} & 1 + (4+7\{\alpha\})z^{-1} + (1+4\{\alpha\}+6\{\alpha\}^2)z^{-2} \\ & + \{\alpha\}^3 z^{-3} \quad (\alpha \in [3,4)) \end{aligned} \tag{22.19}$$

Note that, for any fixed α, if $\alpha \to \lfloor\alpha\rfloor^+$ or $\alpha \to \lceil\alpha\rceil^-$, then the fractional Euler–Frobenius polynomials tend to the usual Euler–Frobenius polynomials (see Chap. 5) of integer order $\lfloor\alpha\rfloor^+$ and $\lceil\alpha\rceil^-$, respectively.

The following preliminary result connects the fractional Euler–Frobenius polynomials and the $\mathcal{Z}$-transform of a fractional combinatorial.

Lemma 22.4 *Consider the discrete-time sequence*

$$f[k] = \begin{cases} 0; & k = 0 \\ (-1)^k \binom{-\{\alpha\}-1}{k-1} k^{\lfloor \alpha \rfloor}; & k \geq 1 \end{cases} \tag{22.20}$$

where $\alpha > 0$. The $\mathcal{Z}$-transform of the above sequence, $F(z) = \mathcal{Z}\{f[k]\}$, is given by

$$\mathcal{Z}\{f[k]\} = \sum_{k=1}^{\infty} (-1)^k \binom{-\{\alpha\}-1}{k-1} k^{\lfloor \alpha \rfloor} z^{-k} = \frac{z^{-1} B_\alpha(z^{-1})}{(1-z^{-1})^\alpha} \tag{22.21}$$

Proof To compute the $\mathcal{Z}$-transform, it is noted that

$$\begin{aligned} \mathcal{Z}\{f[k]\} &= \sum_{k=1}^{\infty} (-1)^{k-1} \binom{-\{\alpha\}-1}{k-1} k^{\lfloor \alpha \rfloor} z^{-k} \\ &= \left(-z\frac{d}{dz}\right)^{\lfloor \alpha \rfloor} \sum_{k=1}^{\infty} (-1)^{k-1} \binom{-\{\alpha\}-1}{k-1} z^{-k} \\ &= \left(-z\frac{d}{dz}\right)^{\lfloor \alpha \rfloor} z^{-1} \sum_{k=0}^{\infty} \binom{-\{\alpha\}-1}{k} z^{-k} \\ &= \left(-z\frac{d}{dz}\right)^{\lfloor \alpha \rfloor} z^{-1} \left(1-z^{-1}\right)^{-(1+\{\alpha\})} \end{aligned} \tag{22.22}$$

where the generalised binomial theorem has been used. Proceeding by induction shows that

$$\left(-z\frac{d}{dz}\right)^{\lfloor \alpha \rfloor} z^{-1} \left(1-z^{-1}\right)^{-(1+\{\alpha\})} = \frac{z^{-1} \Lambda_\alpha(z^{-1})}{(1-z^{-1})^{1+\alpha}} = F(z) \tag{22.23}$$

where $\Lambda_\alpha(^{-1})$ is a polynomial in z^{-1} of order $\lfloor \alpha \rfloor$ defined by

$$\Lambda_\alpha\left(z^{-1}\right) = \lambda_0^\alpha + \cdots + \lambda_{\lfloor \alpha \rfloor}^\alpha z^{-\lfloor \alpha \rfloor} \tag{22.24}$$

Specifically, assume that (22.23)–(22.24) hold true for $\alpha \in [n-1, n)$, and then show that they also hold for $\alpha \in [n, n+1)$. In fact, in this case $\lfloor \alpha \rfloor = n$ and $\{\alpha\} = \alpha - n \in [0, 1)$. Then (22.23) reduces to

$$\begin{aligned}\left(-z\frac{d}{dz}\right)^n&\left[\frac{z^{-1}}{(1-z^{-1})^{1+\alpha-n}}\right]\\&=\left(-z\frac{d}{dz}\right)\underbrace{\left(-z\frac{d}{dz}\right)^{n-1}\left[\frac{z^{-1}}{(1-z^{-1})^{1+\alpha-n}}\right]}\\&=\left(-z\frac{d}{dz}\right)\left[\frac{z^{-1}(\lambda_0+\cdots+\lambda_{n-1}z^{-(n-1)})}{(1-z^{-1})^{\alpha}}\right]\\&=\frac{z^{-1}p(z^{-1})}{(1-z^{-1})^{1+\alpha}}\end{aligned}\tag{22.25}$$

where (22.23) has been applied. Recall that this result was assumed true for $\alpha \in [n-1, n)$. The result in (22.25) then follows on differentiating and multiplying by $-z$ to obtain, in the numerator, a polynomial $p(\cdot)$ in z^{-1} of order $n = \lfloor \alpha \rfloor$. Finally, notice that, for $\alpha \in [0, 1)$,

$$\left(-z\frac{d}{dz}\right)^0 z^{-1}\left(1-z^{-1}\right)^{-(1+\alpha)} = \frac{z^{-1}\Lambda_\alpha(z^{-1})}{(1-z^{-1})^{1+\alpha}}\tag{22.26}$$

where $\Lambda_\alpha(z^{-1}) = 1$. Thus (22.23)–(22.24) hold for all α.

Next, compute the coefficients of the polynomials $\Lambda_\alpha(z^{-1})$. To do this, note that $F(z) = \mathcal{Z}\{f_k\}$ in (22.23) can be interpreted as the step response of the following transfer function:

$$H(z) = \frac{z^{-1}\Lambda_\alpha(z^{-1})}{(1-z^{-1})^{\alpha}}\tag{22.27}$$

From (22.27), and for any input $U(z)$, the corresponding output is $Y(z)$ which satisfies:

$$z^{-1}\Lambda_\alpha\left(z^{-1}\right)U(z) = \left(1-z^{-1}\right)^\alpha Y\left(z^{-1}\right)\tag{22.28}$$

$$\sum_{\ell=0}^{\lfloor\alpha\rfloor}\lambda_\ell^\alpha z^{-\ell-1}U(z) = \sum_{j=0}^{\infty}(-1)^j\binom{\alpha}{j}z^{-j}Y(z)\tag{22.29}$$

For any fixed $k < \lceil \alpha \rceil$,

$$\sum_{\ell=0}^{k-1}\lambda_\ell^\alpha u[k-\ell-1] = \sum_{j=0}^{k}(-1)^j\binom{\alpha}{j}y[k-j]\tag{22.30}$$

$$\sum_{i=0}^{k-1}\lambda_i^\alpha u[k-i-1] = \sum_{i=0}^{k}(-1)^{k-i}\binom{\alpha}{k-i}y[i]\tag{22.31}$$

On the other hand, the step response of $H(z)$ is

$$f[k] = \begin{cases} 0; & k = 0 \\ (-1)^{k-1}\binom{-\{\alpha\}-1}{k-1}k^{\lfloor\alpha\rfloor}; & k \geq 1 \end{cases} \tag{22.32}$$

Then, replacing $u[k]$ in (22.31) by the unit step $\mu[k] = 1$, and $y[k] = f[k]$ given in (22.32), leads to

$$\begin{aligned} \sum_{\ell=0}^{k} \lambda_\ell^\alpha &= \sum_{i=1}^{k} (-1)^{k-i}\binom{\alpha}{k-i}(-1)^{i-1}\binom{-\{\alpha\}-1}{i-1} i^{\lfloor\alpha\rfloor} \\ &= (-1)^{k-1}\sum_{i=1}^{k}\binom{\alpha}{k-i}\binom{-\{\alpha\}-1}{i-1} i^{\lfloor\alpha\rfloor} \end{aligned} \tag{22.33}$$

Then

$$\begin{aligned} \lambda_{k-1}^\alpha &= -\sum_{i=0}^{k-1}\lambda_i^\alpha + (-1)^{k-1}\sum_{i=1}^{k}\binom{\alpha}{k-i}\binom{-\{\alpha\}-1}{i-1} i^{\lfloor\alpha\rfloor} \\ &= -(-1)^{k-2}\sum_{i=1}^{k-1}\binom{\alpha}{k-1-i}\binom{-\{\alpha\}-1}{i-1} i^{\lfloor\alpha\rfloor} \\ &\quad + (-1)^{k-1}\sum_{i=1}^{k}\binom{\alpha}{k-i}\binom{-\{\alpha\}-1}{i-1} i^{\lfloor\alpha\rfloor} \\ &= (-1)^{k-1}\left[\sum_{i=1}^{k-1}\left[\binom{\alpha}{k-1-i} + \binom{\alpha}{k-i}\right]\binom{-\{\alpha\}-1}{i-1} i^{\lfloor\alpha\rfloor}\right. \\ &\quad \left. + \binom{\alpha}{0}\binom{-\{\alpha\}-1}{k-1} k^{\lfloor\alpha\rfloor}\right] \end{aligned} \tag{22.34}$$

Thus,

$$\lambda_{k-1}^\alpha = (-1)^{k-1}\sum_{i=1}^{k}\binom{\alpha+1}{k-i}\binom{-\{\alpha\}-1}{i-1} i^{\lfloor\alpha\rfloor} \tag{22.35}$$

It can be seen that $\Lambda_\alpha(z^{-1})$ as in (22.24) are the same polynomials as defined in (22.14) and (22.15). □

The above result is next used to characterise the asymptotic sampling zeros of an α-order integrator.

Theorem 22.5 *An approximate discrete transfer function for a fractional integrator of order* $\alpha > 1$ *is given by*

$$\mathcal{ZOH}\left\{\frac{1}{s^\alpha}\right\} \approx \Delta^\alpha \frac{\Gamma(\{\alpha\}+1)}{\Gamma(\alpha+1)} \frac{\Sigma_\alpha(z^{-1})}{(1-z^{-1})^\alpha} \tag{22.36}$$

where

$$\Sigma_\alpha(z^{-1}) = z^{-1} B_\alpha(z^{-1}) + 0.5\{\alpha\}(\{\alpha\}-1)(1-z^{-1})z^{-1} B_{\alpha-1}(z^{-1}) \tag{22.37}$$

Proof The exact discrete model of a fractional integrator of order $\alpha > 1$ is given in Theorem 22.1, i.e.,

$$H(z) = (1-z^{-1}) \frac{\Delta^\alpha}{\Gamma(\alpha+1)} \sum_{k=1}^{\infty} k^\alpha z^{-k} \tag{22.38}$$

Focus on the approximation of the infinite sum $\sum_{k=1}^{\infty} k^\alpha z^{-k}$. In particular, the terms of the series are approximated for large values of k. Begin with the binomial

$$\binom{\{\alpha\}+k-1}{k-1} = \frac{\Gamma(\{\alpha\}+k)}{\Gamma(k)\Gamma(\{\alpha\}+1)} \tag{22.39}$$

Then use an asymptotic expansion (see references at the end of the chapter) of Euler's gamma function, given by

$$\Gamma(z) = \sqrt{2\pi} z^{z-1/2} e^{-z} \left[1 + \frac{1}{12z} + \frac{1}{288z^2} + \cdots\right] \tag{22.40}$$

Substituting (22.40) into (22.39) yields

$$\begin{aligned}
\frac{\Gamma(\{\alpha\}+k)}{\Gamma(k)} &= k^{\{\alpha\}} e^{-\{\alpha\}} \left(1 + \frac{\{\alpha\}}{k}\right)^{\{\alpha\}+k-1/2} \\
&\quad \times \frac{1 + \frac{1}{12(\{\alpha\}+k)} + \frac{1}{288(\{\alpha\}+k)^2} + \cdots}{1 + \frac{1}{12k} + \frac{1}{288k^2} + \cdots} \\
&= k^{\{\alpha\}} e^{-\{\alpha\}} \left(1 + \frac{\{\alpha\}}{k}\right)^{\{\alpha\}+k-1/2} \left[1 + \mathcal{O}(k^{-2})\right]
\end{aligned} \tag{22.41}$$

where

$$\begin{aligned}
\log\left(1+\frac{\{\alpha\}}{k}\right)^{\{\alpha\}+k-1/2} \\
&= (\{\alpha\}+k-1/2)\log\left(1+\frac{\{\alpha\}}{k}\right) \\
&= (\{\alpha\}+k-1/2)\sum_{\ell=1}^{\infty}(-1)^{\ell+1}\frac{\{\alpha\}^{\ell}}{k^{\ell}\ell} \\
&= (\{\alpha\}+k-1/2)\frac{\{\alpha\}}{k}-\frac{\{\alpha\}^2}{2k}+\mathcal{O}(k^{-2})
\end{aligned} \tag{22.42}$$

Neglecting terms of order k^{-2}, (22.42) can be approximated by

$$\log\left(1+\frac{\{\alpha\}}{k}\right)^{\{\alpha\}+k-1/2} \approx \{\alpha\}+\frac{\{\alpha\}(\{\alpha\}-1)}{2k} \tag{22.43}$$

Then, applying the exponential function to (22.43) leads to

$$\left(1+\frac{\{\alpha\}}{k}\right)^{\{\alpha\}+k-1/2} \approx e^{\{\alpha\}+\{\alpha\}(\{\alpha\}-1)/2k} = e^{\alpha}e^{1+\frac{\{\alpha\}(\{\alpha\}-1)}{2k}} \tag{22.44}$$

Taking the first two terms of the Taylor expansion of the exponential function yields

$$\left(1+\frac{\{\alpha\}}{k}\right)^{\{\alpha\}+k-1/2} \approx e^{\{\alpha\}}\left(1+\frac{\{\alpha\}(\{\alpha\}-1)}{2k}\right) \tag{22.45}$$

In summary, using (22.44) and (22.41) for large values of k, we get

$$\frac{\Gamma(\{\alpha\}+1)}{\Gamma(k)} \approx k^{\{\alpha\}}\left(1+\frac{\{\alpha\}(1+\{\alpha\})}{2k}\right) \tag{22.46}$$

Thus, in (22.39),

$$\begin{aligned}
\frac{\Gamma(\{\alpha\}+1)}{\Gamma(k)\Gamma(\{\alpha\}+1)} &= \binom{\{\alpha\}+k-1}{k-1} \\
&\approx \frac{k^{\{\alpha\}}}{\Gamma(\{\alpha\}+1)}\left(1+\frac{\{\alpha\}(1+\{\alpha\})}{2k}\right)
\end{aligned} \tag{22.47}$$

From the last approximation (22.47), it follows that

$$\begin{aligned}
\frac{k^{\alpha}}{\Gamma(\{\alpha\}+1)} &\approx \left(1+\frac{\{\alpha\}(\{\alpha\}-1)}{2k}\right)^{-1}\binom{\{\alpha\}+k-1}{k-1} \\
&\approx \left(1-\frac{\{\alpha\}(\{\alpha\}-1)}{2k}\right)\binom{\{\alpha\}+k-1}{k-1}
\end{aligned} \tag{22.48}$$

Performing the summation over all $k \in \mathbb{N}$, we have

$$\frac{1}{\Gamma(\{\alpha\}+1)} \sum_{k=1}^{\infty} k^{\{\alpha\}} z^{-k} \approx \sum_{k=1}^{\infty} \binom{\{\alpha\}+k-1}{k-1} z^{-k} - \frac{\{\alpha\}(\{\alpha\}-1)}{2} \sum_{k=1}^{\infty} \binom{\{\alpha\}+k-1}{k-1} \frac{z^{-k}}{k} \tag{22.49}$$

Next, using the property

$$\binom{j+\ell}{\ell} = (-1)^{\ell} \binom{-j-1}{\ell} \tag{22.50}$$

leads to

$$\begin{aligned} \frac{1}{\Gamma(\{\alpha\}+1)} \sum_{k=1}^{\infty} k^{\{\alpha\}} z^{-k} &\approx \sum_{k=1}^{\infty} \binom{-\{\alpha\}-1}{k-1} (-1)^{k-1} z^{-k} \\ &\quad - \frac{\{\alpha\}(\{\alpha\}-1)}{2} \sum_{k=1}^{\infty} \binom{-\{\alpha\}-1}{k-1} (-1)^{k-1} \frac{z^{-k}}{k} \end{aligned} \tag{22.51}$$

$$\approx \underbrace{\sum_{k=1}^{\infty} \binom{-\{\alpha\}-1}{k-1} (-1)^{k-1} z^{-k}}_{z^{-1}(1-z^{-1})^{-(1+\{\alpha\})}} + \frac{\{\alpha\}(\{\alpha\}-1)}{2} \int_{-\infty}^{z} \sum_{k=1}^{\infty} \binom{-\{\alpha\}-1}{k-1} (-1)^{k-1} \eta^{-k-1} d\eta \tag{22.52}$$

Applying the generalised binomial theorem,

$$\frac{1}{\Gamma(\{\alpha\}+1)} \sum_{k=1}^{\infty} k^{\{\alpha\}} z^{-k} \approx z^{-1}\left(1-z^{-1}\right)^{-(1+\{\alpha\})} + \frac{\{\alpha\}(\{\alpha\}-1)}{2} \int_{-\infty}^{z} \eta^{-2}\left(1-\eta^{-1}\right)^{-(1+\{\alpha\})} d\eta \tag{22.53}$$

and applying the operator $(-z\frac{d}{dz})^{\lfloor\alpha\rfloor}$ yields

$$\frac{1}{\Gamma(\{\alpha\}+1)}\sum_{k=1}^{\infty}k^{\alpha}z^{-k}\approx\left(-z\frac{d}{dz}\right)^{\lfloor\alpha\rfloor}z^{-1}\left(1-z^{-1}\right)^{-(1+\{\alpha\})}$$
$$+\frac{\{\alpha\}(\{\alpha\}-1)}{2}\left(-z\frac{d}{dz}\right)^{\lfloor\alpha\rfloor-1}z^{-1}\left(1-z^{-1}\right)^{-(1+\{\alpha\})} \tag{22.54}$$

Then, using Lemma 22.4, the factors involved in (22.54) can be expressed as

$$\left(1-z^{-1}\right)\left(-z\frac{d}{dz}\right)^{\lfloor\alpha\rfloor}z^{-1}\left(1-z^{-1}\right)^{-(1+\{\alpha\})}=\frac{z^{-1}B_{\alpha}(z^{-1})}{(1-z^{-1})^{\alpha}} \tag{22.55}$$

where Lemma 22.4 and the fractional Euler–Frobenius polynomials $B_{\alpha}(z^{-1})$ given in Definition 22.2 have been used. Finally, the proof is completed by substituting (22.55) into (22.54) and (22.38) to obtain

$$\mathcal{Z}\left\{\frac{1}{s^{\alpha}}\right\}\approx\Delta^{\alpha}\frac{\Gamma(\{\alpha\}+1)}{\Gamma(\alpha+1)}$$
$$\times\frac{z^{-1}B_{\alpha}(z^{-1})+0.5\{\alpha\}(\{\alpha\}-1)(1-z^{-1})z^{-1}B_{\alpha-1}(z^{-1})}{(1-z^{-1})^{\alpha}} \tag{22.56}$$

□

Theorem 22.5 describes the connection between the fractional Euler–Frobenius polynomials and approximate models for a fractional order integrator. It is important to note that, in the integer case, the approximate equivalence in (22.36) is actually an equality.

Theorem 22.5, when combined with the previous result in Theorem 22.1, allows one to characterise the asymptotic sampling zeros of a general linear FOS having relative degree $\alpha\in\mathbb{R}^{+}$.

In particular, following the same methodology as outlined in Chap. 8 for the integer order case, one can append the asymptotic sampling zeros to a simple derivative replacement model. Thus, for a fractional order continuous transfer function $G(s)$, one obtains

$$G_d(z)=\frac{\Gamma(\{\alpha\}+1)}{\Gamma(\alpha+1)}\Sigma_{\alpha}\left(z^{-1}\right)G\left(\frac{1-z^{-1}}{\Delta}\right) \tag{22.57}$$

22.4 Approximate Discrete-Time Models

Theorem 22.5 provides an approximate transfer function for a fractional α-integrator. In this section it is shown how one can use this result to obtain approximate sampled-data models for more general systems. In the sequel, three sampled models for a FOS are considered:

(ESD) The *exact sampled-data* model is given by (3.29). For the fractional case, this model is not a rational transfer function. Nevertheless, in the examples presented later, a high order expansion is used for the aliasing formula (3.29) to obtain an accurate approximation of the ESD model (in particular, an expansion of order 2, i.e., in (3.40) on p. 29, $\ell = -2, \dots, 2$).

(SDR) The *simple derivative replacement* model is obtained by replacing the continuous-time derivative by the Euler approximation, i.e.,

$$G_d^{\text{SDR}}(z) = G\left(\frac{1 - z^{-1}}{\Delta}\right) \tag{22.58}$$

(ASZ) The *asymptotic sampling zeros* model is obtained by appending the fractional order Euler-Fröbenious polynomial (of order of the relative degree of the continuous-time system) to the SDR model. This leads to

$$G_d^{\text{ASZ}}(z) = \frac{\Gamma(\{\alpha\} + 1)}{\Gamma(\alpha + 1)} \Sigma_\alpha\left(z^{-1}\right) G_d^{\text{SDR}}(z) \tag{22.59}$$

where $\alpha \in \mathbb{R}$ is the relative degree of $G(s)$ (not necessarily an integer number) and $\Sigma_\alpha(z^{-1})$ is given in (22.37). This ASZ model has previously been analysed for the case of systems with integer order derivatives in Chap. 8.

Two examples are presented to illustrate the use of the approximate asymptotic sampling zeros in building approximate discrete models for the FOS. The measures of the error are the same frequency-domain relative error measures as used in Sect. 8.1.1 of Chap. 8:

$$R_1^i(\omega) = \left| \frac{G_d^{\text{ESD}}(e^{j\omega\Delta}) - G_d^i(e^{j\omega\Delta})}{G_d^{\text{ESD}}(e^{j\omega\Delta})} \right| \tag{22.60}$$

$$R_2^i(\omega) = \left| \frac{G_d^{\text{ESD}}(e^{j\omega\Delta}) - G_d^i(e^{j\omega\Delta})}{G_d^i(e^{j\omega\Delta})} \right| \tag{22.61}$$

Example 22.6 Consider the FOS given by

$$G(s) = \frac{s^{1.2} + 1}{s^{4.7} + 6} \tag{22.62}$$

which has relative degree $\alpha = 3.5$. The asymptotic sampling zero dynamics are given by (22.37), i.e.,

$$\Sigma_{3.5}\left(z^{-1}\right) = 1.375z^{-1} + 8.0625z^{-2} + 3.65625z^{-3} + 0.03125z^{-4} \tag{22.63}$$

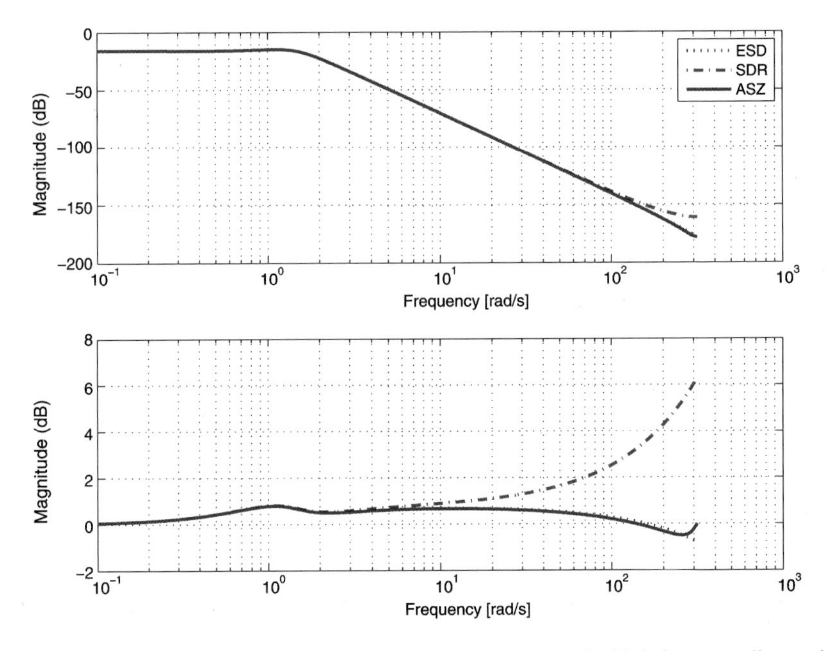

Fig. 22.1 Magnitude and phase Bode plots of the system in Example 22.6, for a sampling period $\Delta = 0.01$

Next a comparison is made between the SDR and ASZ models and the ESD model. Figure 22.1 shows the magnitude and phase Bode plots. We see that the closest approximate model to ESD is ASZ at high frequencies.

The relative errors (22.60) and (22.61) are plotted in Fig. 22.2. Figure 22.2 shows that the SDR model is actually more accurate than the ASZ model at low frequencies, and the converse is true at high frequencies. Moreover, the simulations show that the *crossing frequency* of the relative error measures does not change when the sampling rate is increased. This observation mirrors known results for the integer-derivative case.

22.5 Summary

The key points covered in this chapter are:

- A review of fractional calculus.
- Description of fractional order continuous-time models.
- Extension of the Euler–Frobenius polynomials to the fractional case.
- Development of sampling zeros for fractional order systems.
- Development and comparison of various approximate models based on appending the asymptotic sampling zeros for fractional order systems.

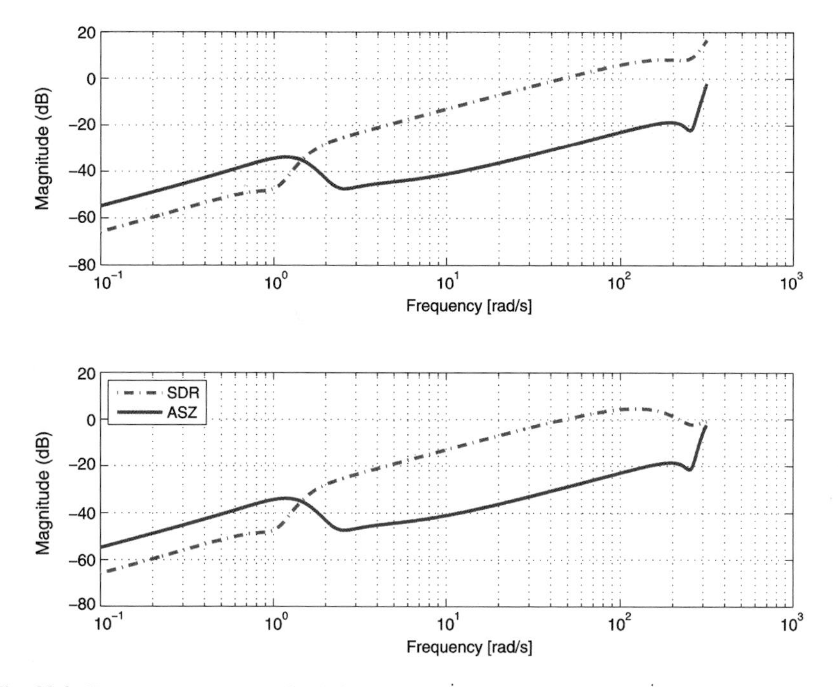

Fig. 22.2 Frequency response of relative errors $R_1^i(\omega)$ (*top plot*) and $R_2^i(\omega)$ (*bottom plot*) for $i \in \{\text{SDR, ASZ}\}$ (Example 22.6)

Further Reading

Background on fractional calculus can be found in

Kilbas A, Srivastava M, Trujillo J (2006) Theory and application of fractional differential equations. Elsevier, Amsterdam

Miller KS, Ross B (1993) An introduction to the fractional calculus and differential equations. Wiley, New York

Oldham K, Spanier J (1974) The fractional calculus. Academic Press, New York

Podlubny I (1999) Fractional differential equations. Academic Press, New York

Applications of fractional calculus in engineering can be found in

Baleanu D, Guvenc ZB, Tenreiro-Machado JA (eds) (2009) New trends in nanotechnology and fractional calculus applications. Springer, New York

Caponetto R, Dongola G, Fortuna L, Petras I (2010) Fractional order systems: modeling and control applications. World Scientific, Singapore

Monje CA, Chen Y, Vinagre BM, Xue D, Feliu V (2010) Fractional-order systems and controls: fundamentals and applications. Springer, Berlin

Sabatier J, Agrawal OP, Tenreiro-Machado JA (2007) Advances in fractional calculus: theoretical developments and applications in physics and engineering. Springer, Berlin

Other approaches for obtaining approximate sampled-data models for fractional order systems can be found in

Aoun M, Malti R, Levron F, Oustaloup A (2004) Numerical simulations of fractional systems: an overview of existing methods and improvements. In: Nonlinear dynamics, pp 117–131

Chen YQ, Moore KL (2002) Discretization schemes for fractional-order differentiators and integrators. IEEE Trans Circuits Syst I, Fundam Theory Appl 49(3):363–367

Chen YQ, Vinagre BM, Podlubny I (2004) Continued fraction expansion approaches to discretizing fractional order derivatives—an expository review. Nonlinear Dyn 38(1):155–170

Lubich C (1986) Discretized fractional calculus. SIAM J Math Anal 17(3):704–719

Maione G (2008) Continued fractions approximation of the impulse response of fractional-order dynamic systems. IET Control Theory Appl 2(7):564–572

Maione G (2011) High-speed digital realizations of fractional operators in the delta domain. IEEE Trans Autom Control 56(3):697–702

Valério D, da Costa JS (2005) Time-domain implementation of fractional order controllers. In: Control theory and applications. IEE proceedings, pp 539–552

Yucra E, Yuz JI (2013) Sampling zeros of discrete models for fractional order systems. IEEE Trans Autom Control 58(9):2383–2388. doi:10.1109/TAC.2013.2254000

Index

J.I. Yuz, G.C. Goodwin, *Sampled-Data Models for Linear and Nonlinear Systems*,
Communications and Control Engineering, DOI 10.1007/978-1-4471-5562-1,

Printed by Printforce, the Netherlands